Building Physics – Heat, Air and Moisture

Ihr persönlicher E-Book Code

ISBN 9783433034293 – Building Physics
(inkl. E-Book als PDF)

Das von Ihnen erworbene E-Book ist im ePDF-Format. Das Format ist nicht mit Amazon Endgeräten oder Apps kompatibel.

Beachten Sie: Sobald der Code verwendet wurde, sind eine Rückgabe oder der Weiterverkauf ausgeschlossen.

Das E-Book können Sie unter www.wiley-vch.de/ebooks/einlösen.
Eine ausführliche Anleitung zum Herunterladen des E-Books finden Sie unter
http://www.wiley-vch.de/publish/dt/ebooks

Building Physics – Heat, Air and Moisture

Fundamentals, Engineering Methods, Material Properties and Exercises

Hugo Hens

Fourth, revised Edition

Author:

Prof. em. Hugo S.L.C. Hens, PhD
University of Leuven
Department of Civil Engineering
Kasteelpark Arenberg
3001 Heverlee
Belgium

Cover: New Bioscience Centre of the KULeuven, Belgium, under construction

All books published by **Ernst & Sohn** are carefully produced. Nevertheless, authors, editors, and publisher do not warrant the information contained in these books, including this book, to be free of errors. Readers are advised to keep in mind that statements, data, illustrations, procedural details or other items may inadvertently be inaccurate.

Library of Congress Card No.: applied for

British Library Cataloguing-in-Publication Data
A catalogue record for this book is available from the British Library.

Bibliographic information published by the Deutsche Nationalbibliothek
The Deutsche Nationalbibliothek lists this publication in the Deutsche Nationalbibliografie; detailed bibliographic data are available on the Internet at http://dnb.d-nb.de.

© 2024 Ernst & Sohn GmbH, Rotherstraße 21, 10245 Berlin, Germany

All rights reserved (including those of translation into other languages). No part of this book may be reproduced in any form – by photoprinting, microfilm, or any other means – nor transmitted or translated into a machine language without written permission from the publishers. Registered names, trademarks, etc. used in this book, even when not specifically marked as such, are not to be considered unprotected by law.

Print ISBN: 978-3-433-03422-4
ePDF ISBN: 978-3-433-61183-8
oBook ISBN: 978-3-433-61182-1
ePub ISBN: 978-3-433-61184-5

Coverdesign: Petra Franke/Ernst & Sohn GmbH using a design by Sophie Bleifuß, Berlin, Germany
Typesetting: Straive, Chennai, India

Printing:
Binding:

Printed on acid-free paper.

To my wife, children and grandchildren

In remembrance of Professor A. de Grave, a civil engineer who introduced building physics as a new discipline at the University of Leuven, Belgium, in 1952.

Hugo Hens

Applied Building Physics

Ambient Conditions, Functional Demands, and Building Part Requirements

- content well structured combining theory with typical building engineering practice
- equally suitable as a textbook and for practitioners
- applicable independent of national or other standard requirement

As with all engineering sciences, Building Physics is oriented towards application, hence, after a first book on fundamentals this volume on Applied Building Physics discusses the heat, air, moisture performance metrics that affect building design, construction and performance.

Available as a package with Building Physics

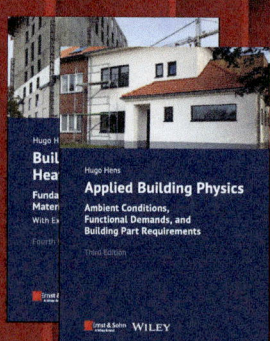

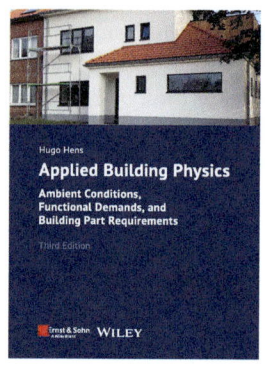

3. revised edition · 9 / 2023 ·
approx. 352 pages · approx. 187 figures ·
approx. 95 tables

Softcover
ISBN 978-3-433-03423-1
approx. **€ 69***

eBundle (Softcover + ePDF)
ISBN 978-3-433-03432-3
approx. **€ 99***

PACKAGE
Building Physics
+ Applied Building Physics
ISBN 978-3-433-03433-0
approx. **€ 119***

eBundle (Print + ePDF) PACKAGE
ISBN 978-3-433-03434-7
approx. **€ 179***

Available for pre-order.

ORDER
+49 (0)30 470 31-236
marketing@ernst-und-sohn.de
www.ernst-und-sohn.de/en/3423

* All book prices inclusive VAT.

Contents

Preface *xv*
About the Author *xix*
List of Units and Symbols *xxi*

0	**Introduction** *1*	
0.1	Subject of the Book *1*	
0.2	Building Physics? *1*	
0.2.1	Definition *1*	
0.2.2	Constraints *2*	
0.2.2.1	Comfort *2*	
0.2.2.2	Health and Well-being *3*	
0.2.2.3	Architecture and Materials *3*	
0.2.2.4	Economy *3*	
0.2.2.5	Sustainability *3*	
0.3	Importance? *4*	
0.4	History *5*	
0.4.1	In General *5*	
0.4.2	Applied Physics *5*	
0.4.2.1	Heat, Air, Moisture *5*	
0.4.2.2	Acoustics *8*	
0.4.2.3	Lighting *9*	
0.4.3	Indoor Air Quality and Thermal Comfort *9*	
0.4.4	Building Services *11*	
0.4.5	Building Design and Construction *11*	
0.4.6	Hall of Fame *12*	
0.4.7	Building Physics at the KULeuven and Other Universities in the Low Countries *13*	
	Further Reading *15*	
1	**Heat Transfer** *17*	
1.1	In General *17*	
1.1.1	Heat *17*	
1.1.1.1	What? *17*	
1.1.1.2	Sensible Heat *17*	

1.1.1.3	Latent Heat	18
1.1.2	Temperature	18
1.1.3	Why are Heat and Temperature so Compelling?	18
1.1.4	Some Definitions	19
1.2	Conduction	19
1.2.1	Conservation of Energy	19
1.2.2	Conduction Laws	20
1.2.2.1	First Law	20
1.2.2.2	Second Law	22
1.2.3	Thermal Conductivity	22
1.2.3.1	In General	22
1.2.3.2	Heat Transfer Modes Fixing the Property	22
1.2.4	Steady-State	26
1.2.4.1	What?	26
1.2.4.2	One Dimension, Flat Assemblies	27
1.2.4.3	Two Dimensions, Cylinder Symmetric	33
1.2.4.4	Two and Three Dimensions: Thermal Bridges	35
1.2.5	Non-steady-state	38
1.2.5.1	In General	38
1.2.5.2	Periodic Boundary Conditions, Flat Assemblies	39
1.2.5.3	Any Boundary Conditions, Flat Assemblies	48
1.2.5.4	Two and Three Dimensions: Thermal Bridges	52
1.3	Heat Exchange at Surfaces by Convection and Radiation	52
1.3.1	What?	52
1.3.2	Convection	53
1.3.2.1	In General	53
1.3.2.2	Typology	55
1.3.2.3	Quantifying the Convective Surface Film Coefficient	55
1.3.2.4	Values for the Convective Surface Film Coefficient	58
1.3.3	Radiation	63
1.3.3.1	In General	63
1.3.3.2	Definitions	63
1.3.3.3	Reflection, Absorption and Transmission	64
1.3.3.4	Radiant Surfaces	66
1.3.3.5	Simple Formulae	74
1.4	Building-related Applications	76
1.4.1	Surface Film Coefficients and Reference Temperatures	76
1.4.1.1	Methodology	76
1.4.1.2	Indoors	76
1.4.1.3	Outdoors	78
1.4.2	Steady-state, Flat Assemblies	80
1.4.2.1	Thermal Transmittance of Envelope Assemblies, Partitions and Party Walls	80
1.4.2.2	Average Thermal Transmittance of Envelope Parts in Parallel	83
1.4.2.3	Electrical Analogy	84

1.4.2.4	Thermal Resistance of Non-ventilated Cavities *84*
1.4.2.5	Interface Temperatures *86*
1.4.2.6	Effect of Ever Thicker Insulation Layers on the Thermal Transmittance *87*
1.4.2.7	Solar Transmittance *88*
1.4.3	Local Inside Surface Film Coefficients *91*
1.4.4	Steady-state: Two and Three Dimensions *92*
1.4.4.1	Pipes *92*
1.4.4.2	Floors on Grade *93*
1.4.4.3	Thermal Bridges *94*
1.4.4.4	Windows *98*
1.4.4.5	Building Envelopes *99*
1.4.5	Heat Balances *100*
1.4.6	Non-steady-state *101*
1.4.6.1	Periodic Boundary Conditions: Flat Assemblies *101*
1.4.6.2	Periodic Boundary Conditions: Spaces *101*
1.4.6.3	Any Boundary Conditions: Thermal Bridges *105*
	Problems and Solutions *106*
	Further Reading *118*

2	**Mass Transfer** *121*
2.1	In General *121*
2.1.1	Facts *121*
2.1.2	Definitions *122*
2.1.3	Saturation Degree Scale *123*
2.1.4	Air and Moisture Transfer *123*
2.1.5	Moisture Sources *125*
2.1.6	Air and Moisture in Relation to Durability *127*
2.1.7	Links with Energy Transfer *127*
2.1.8	Conservation of Mass *128*
2.2	Air *129*
2.2.1	In General *129*
2.2.2	Air Pressure Differentials *130*
2.2.2.1	Wind *130*
2.2.2.2	Stack Effect *130*
2.2.2.3	Fans *133*
2.2.3	Air Permeability and Air Permeances *133*
2.2.4	Airflow in Open-porous Materials *135*
2.2.4.1	The Conservation Law Adapted *135*
2.2.4.2	One Dimension: Flat Assemblies *138*
2.2.4.3	Two and Three Dimensions *139*
2.2.5	Airflow Through Assemblies with Air-open Layers, Leaky Joints, Leaks, Cavities, etc. *140*
2.2.6	Airflow at the Building Level *141*
2.2.6.1	Definitions *141*

2.2.6.2   Thermal Stack   *142*
2.2.6.3   Large Openings   *142*
2.2.6.4   The Conservation Law Applied   *143*
2.2.6.5   Applications   *145*
2.2.7   Combined Heat and Airflow Through Assemblies Composed of Open-porous Layers   *148*
2.2.7.1   Heat Balance   *148*
2.2.7.2   Steady-state: Flat Assemblies   *148*
2.2.7.3   Steady-state, Two and Three Dimensions   *152*
2.2.7.4   Non-steady-state, Flat Assemblies   *152*
2.2.7.5   Non-steady-state, Two and Three Dimensions   *153*
2.2.7.6   Air-permeable Layers, Joints and Leaks   *153*
2.2.7.7   Vented Cavities   *153*
2.3   Water Vapour   *156*
2.3.1   Water Vapour in the Air   *156*
2.3.1.1   In General   *156*
2.3.1.2   Quantities   *156*
2.3.1.3   Vapour Saturation Pressure   *157*
2.3.1.4   Relative Humidity   *157*
2.3.1.5   Changes of State in Humid Air   *161*
2.3.1.6   Enthalpy of Humid Air   *162*
2.3.1.7   Measuring Air Humidity   *162*
2.3.2   Vapour Balance in Spaces   *163*
2.3.3   Relative Humidity On Inside Surfaces   *165*
2.3.4   Vapour in Open-porous Materials   *168*
2.3.4.1   Different from Air?   *168*
2.3.4.2   Sorption/Desorption Isotherm   *168*
2.3.5   Vapour Transfer in the Air   *172*
2.3.6   Vapour Flow by Diffusion in Open-porous Materials and Building Assemblies   *174*
2.3.6.1   Flow Equation   *174*
2.3.6.2   Vapour Resistance Factor µ   *175*
2.3.6.3   Mass Conservation   *176*
2.3.6.4   Applicability of the <Equivalent> Diffusion Concept   *177*
2.3.6.5   Steady State: Flat Assemblies   *177*
2.3.6.6   Steady State: Two and Three Dimensions   *186*
2.3.6.7   Non-steady State   *187*
2.3.7   Vapour Flow by Diffusion and Moist Air Moving Through Open-porous Assemblies   *189*
2.3.7.1   In General   *189*
2.3.7.2   Isothermal, Single- and Multi-layered Assemblies   *190*
2.3.7.3   Non-isothermal, Single- and Multi-layered Assemblies   *191*
2.3.8   Surface Film Coefficients for Diffusion   *195*
2.3.8.1   Derivation   *195*
2.3.8.2   Applications   *198*

2.3.9	Evaluating Interstitial Condensation in Practice	201
2.3.9.1	Boundary Conditions Used	201
2.3.9.2	Calculation Sequence	203
2.3.9.3	Example	204
2.4	Moisture	209
2.4.1	In General	209
2.4.2	Water Flow in a Pore	209
2.4.2.1	Capillarity	209
2.4.2.2	Poiseuille's Law	211
2.4.2.3	Isothermal Water Flow in a Pore Contacting Water	212
2.4.2.4	Isothermal Water Flow in a Pore After Water Contact	218
2.4.2.5	Non-isothermal Water Transfer in a Pore After Water Contact	218
2.4.2.6	Remark	219
2.4.3	Vapour Flow in a Pore Containing Water Isles with Air Inclusions in Between	219
2.4.3.1	A Short Description	219
2.4.3.2	Isothermal	219
2.4.3.3	Non-isothermal	220
2.4.4	Moisture Flow in and Through Materials and Assemblies	221
2.4.4.1	Transport Equations	221
2.4.4.2	Moisture Permeability	223
2.4.4.3	Mass Conservation	223
2.4.4.4	Starting, Boundary and Contact Conditions	224
2.4.4.5	Remarks	224
2.4.5	Simple Moisture Flow Model	225
2.4.5.1	How to Do?	225
2.4.5.2	Applying the Simple Model	227
	Problems and Solutions	240
	Further Reading	261
3	**Heat, Air and Moisture Combined**	**265**
3.1	Why?	265
3.2	Material and Assembly Level	265
3.2.1	Assumptions	265
3.2.2	Solution	266
3.2.3	Conservation of Mass	266
3.2.4	Conservation of Energy	267
3.2.5	Flux Equations	270
3.2.5.1	Heat	270
3.2.5.2	Mass, Air	270
3.2.5.3	Mass, Moisture	270
3.2.5.4	Remark	271
3.2.6	Equations of State	271
3.2.6.1	Enthalpy and Vapour Saturation Pressure in Relation to Temperature	271

3.2.6.2	Relative Humidity in Relation to Moisture Content	*271*
3.2.6.3	Suction in Relation to Moisture Content	*271*
3.2.7	Start, Boundary and Contact Conditions	*271*
3.2.8	Two Examples of Simplified Models	*272*
3.2.8.1	Assemblies Composed of Non-Hygroscopic, Non-Capillary Materials	*272*
3.2.8.2	Assemblies Composed of Fine Porous, Hygroscopic Materials	*274*
3.3	Whole Building Level	*274*
3.3.1	In General	*274*
3.3.2	Balance Equations	*275*
3.3.2.1	Vapour	*275*
3.3.2.2	Air	*276*
3.3.2.3	Heat	*276*
3.3.2.4	Closing the Loop	*278*
3.3.3	Sorption Active Surfaces and Hygric Inertia	*279*
3.3.3.1	In General	*279*
3.3.3.2	Sorption-Active Thickness	*280*
3.3.3.3	Zone with One Sorption-Active Surface	*282*
3.3.3.4	Zone with Several Sorption-Active Surfaces	*283*
3.3.3.5	Harmonic Analysis	*284*
3.3.4	Consequences	*284*
	Problems and Solutions	*287*
	Further Reading	*300*
4	**Heat, Air and Moisture Material Property Values**	*303*
4.1	In General	*303*
4.2	Dry Air and Water	*304*
4.3	Thermal Properties	*305*
4.3.1	Definitions	*305*
4.3.2	Standard Values	*305*
4.3.2.1	Regardless of Being on the In- or on the Outside of the Thermal Insulation	*305*
4.3.2.2	Depending on Being on the In- or on the Outside of the Thermal Insulation	*309*
4.3.3	Surfaces, Radiant Properties	*316*
4.3.4	Measured Values	*317*
4.3.4.1	Thermal Conductivity, Test Methods	*317*
4.3.4.2	Test Results	*318*
4.4	Air Properties	*325*
4.4.1	Standard Values	*325*
4.4.2	Measured Values	*325*
4.4.2.1	Air Permeance, Test Method	*325*
4.4.2.2	Test Results	*327*
4.5	Moisture Properties	*336*
4.5.1	Standard Values	*336*

4.5.1.1	Building and Finishing Materials	*336*
4.5.1.2	Insulation Materials	*340*
4.5.2	Measured Values	*340*
4.5.2.1	Diffusion Resistance Factor (μ), Test Method	*340*
4.5.2.2	Test Results	*342*
	Further Reading	*357*

Postscript *359*

Index *361*

Bill Addis (Ed.)

Physical Models

Their historical and current use in civil and building engineering design

- the book summarizes the history of model testing by design and construction engineers in a single volume for the first time
- model testing is alongside knowledge of materials and structural behaviour a major driver in progress in civil and building engineering

The book traces the use of physical models by engineering designers from the eighteenth century, through their heyday in the 1950s-70s, to their current use alongside computer models. It argues that their use has been at least as important in the development of engineering as scientific theory has.

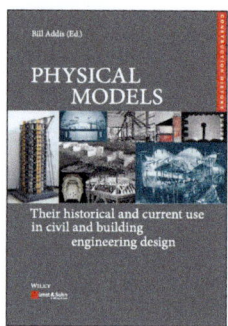

2020 · 1114 pages · 896 figures · 14 tables

Hardcover
ISBN 978-3-433-03257-2 € 149*

eBundle (Print + PDF)
ISBN 978-3-433-03305-0 € 249*

ORDER
+49 (0)30 470 31-236
marketing@ernst-und-sohn.de
www.ernst-und-sohn.de/en/3257

* All book prices inclusive VAT.

Preface

Until the energy crisis of 1973, building physics was a dormant beauty within building engineering, with seemingly limited applicability in practice. While soil mechanics, structural mechanics, construction materials, building itself and heating, ventilation, air conditioning (HVAC) were perceived as essential, designers only demanded advice on room acoustics, moisture tolerance, summer overheating or lighting when really needed or when in newly occupied buildings problems arose. Energy was no concern, while thermal comfort and indoor environmental quality were presumably guaranteed thanks to air infiltration, window opening and the HVAC system. 1973 and the energy crisis of 1979, persisting moisture problems, complaints about sick buildings, thermal, visual and olfactory discomfort, the move to more sustainability and, since the 1980s, global warming with today the quest for carbon neutrality changed this all. Besides, the pressure to diminish energy use and carbon emitted without degrading building usability more than activated the importance of a performance-based building and building part design and construction. As a result, building physics and related potentiality to quantify performances moved to the frontline of building innovation.

Like all engineering sciences, building physics is oriented towards application. This demands a sound knowledge of the basics in each of its branches: heat and mass transfer, acoustics, lighting, energy and indoor environmental quality. Advancing the basics on heat and mass transfer is the main objective of this volume, be it for mass flow limited to air, water vapour and moisture. In the introduction, building physics as a discipline is sketched and its history is given. The first chapter then concentrates on heat transport, with conduction, convection and radiation as main topics, followed by common concepts linked to and applications in the field of building and building part or assembly design and construction. The second chapter treats mass transport, with air, water vapour and moisture as the main topics. Also here, attention goes to the concepts and applications related to whole buildings and building parts. The third chapter discusses combined heat, air and moisture transport. All chapters end with exercises. In the fourth chapter, standard lists with heat, air and moisture material properties and measured data are given.

This content is the result of 38 years of teaching building physics to architectural, building and civil engineering students, that, coupled to more than 36 years of experience in building and building part performance research and more than 50 years of activity in consultancy and in curing hundreds of heat, air and moisture-related damage cases. When and where needed, information from international sources and literature has been consulted, which is why all chapters end with an extended list of references and further reading. The book uses SI units. It could be of help for undergraduate and graduate students in architectural and building engineering, although also students in mechanical engineering studying HVAC and practising building engineers, who want to refresh their knowledge, may benefit. Presumed is the reader has a sound knowledge of calculus and differential equations along with a background in physics, thermodynamics, hydraulics, construction materials and building design and construction.

Compared to the third edition published in 2017, the book has been reorganised, corrected, revised and expanded where appropriate for this fourth edition.

Acknowledgements

The book reflects the work of many, not only of the author. Therefore, we thank the thousands of students we had during the 38 years of teaching. They gave us the opportunity to test the content. Also, the book should not been written the way it is if not standing on the shoulders of those, who preceded it. Although we started our carrier as a structural engineer, our predecessor Professor Antoine de Grave planted the seeds that fed the interest in building physics. Bob Vos of TNO, the Netherlands, and Helmut Künzel of the Fraunhofer Institüt für Bauphysik, Germany, showed the importance of experimental work and field testing to understand whole building and building part or assembly performance, while Lars Erik Nevander of Lund University, Sweden, taught that solving problems in building physics does not always demands complex modelling, mainly because reality in building construction is much more complex than any model can simulate.

During the four decades at the Unit of Building Physics and Sustainable Construction within the Department of Civil Engineering of the KULeuven, several researchers, then PhD students, got involved. They all contributed by the topics chosen to the advancement of the research done at the unit. Most grateful I am to Gerrit Vermeir, my colleague from the start in 1975, professor emeritus now, to Staf Roels, Dirk Saelens, Hans Janssen and Bert Blocken, who succeeded me as professors at the unit.

The experience gained the first 4 years of my career as a structural engineer and building site supervisor for a medium-sized architectural office, as building assessor during some 50 years, and as operating agent of four IEA, EXCO on Energy in Buildings and Communities Annexes forced me to rethink the engineering-based performance approach each time again. The many ideas exchanged in Canada and the United States with Kumar Kumaran of NRC, Paul Fazio of Concordia University in Montreal, Bill Brown, William B. Rose of the University of Illinois in Urbana-Champaign, Joe Lstiburek of the Building Science Corporation, Anton Ten Wolde and those participating in ASHRAE TC 1.12 'Moisture management in buildings' and TC 4.4 'Building materials and building envelope performance' were also of great value.

Finally, I thank my family, my wife Lieve, who managed living together with a busy engineering professor, our three children, our children in law and our grand children.

March 2023
KU Leuven, Leuven, Belgium

Hugo S.L.C. Hens

About the Author

Dr. Ir. Hugo S.L.C. Hens is an emeritus professor of the University of Leuven (KU Leuven), Belgium. Until 1972, he worked as a structural engineer and site supervisor at a mid-sized architectural office. After the sudden death of his predecessor and promotor Professor A. de Grave in 1975 and after defending his PhD thesis, he stepwise built up the Unit of Building Physics at the Department of Civil Engineering.

He taught Building Physics from 1975 to 2003, performance-based building design from 1975 to 2005 and building services from 1975 to 1977 and 1990 to 2008. He authored and co-authored 68 peer-reviewed journal papers and 174 conference papers about the research done, has helped to manage hundreds of building damage cases and acted as coordinator of the CIB W40 working group on Heat and Mass Transfer in Buildings from 1983 to 1993. Between 1986 and 2008, he was operating agent of the Annexes 14, 24, 32 and 41 of the IEA EXCO on Energy in Buildings and Communities. He is a fellow of the American Society of Heating, Refrigeration and Air Conditioning Engineers (ASHRAE).

ECCS – European Convention for Constructional Steelwork (Ed.)

Design of Steel Plated Structures with Finite Elements

- design of steel bridges and other plated structures is increasingly FEM-based
- leading European steel design experts explain background and procedure
- examples, benchmarks and verifications support designers

The book deals with the practical design of welded plated steel structures with the finite element method and especially the proof of plate buckling resistance.

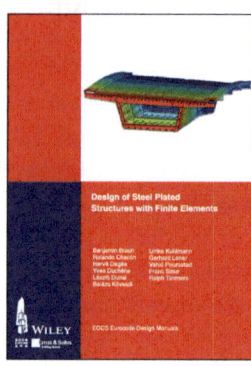

2023 · 162 pages · 137 figures · 17 tables

Softcover
ISBN 978-3-433-03416-3   € 55*

ORDER
+49 (0)30 470 31-236
marketing@ernst-und-sohn.de
www.ernst-und-sohn.de/en/3416

All book prices inclusive VAT.

List of Units and Symbols

Units

The book uses the SI system, internationally mandatory since 1977, with as base units the metre (m), the kilogram (kg), the second (s), the Kelvin (K), the ampere (A) and the candela. Derived units of importance when studying building physics are:

Unit of force	Newton (N)	$1\,N = 1\,kg.m.s^{-2}$
Unit of pressure	Pascal (Pa)	$1\,Pa = 1\,N/m^2 = 1\,kg.m^{-1}.s^{-2}$
Unit of energy	Joule (J)	$1\,J = 1\,N.m = 1\,kg.m^2.s^{-2}$
Unit of power	Watt (W)	$1\,W = 1\,J.s^{-1} = 1\,kg.m^2.s^{-3}$

Symbols

For the symbols, the ISO standards (International Standardization Organization) are followed. For quantities not included, the CIB W40 recommendations (International Council for Building Research, Studies and Documentation, Working Group 'Heat and Moisture Transfer in Buildings') and the list edited by Annex 24 of the IEA, EBC (International Energy Agency, Executive Committee on Energy in Buildings and Communities) apply.

Table 1 List with symbols and quantities.

Symbol	Meaning	Units
a	Acceleration	m/s^2
a	Thermal diffusivity	m^2/s
b	Thermal effusivity	$W/(m^2.K.s^{0.5})$
c	Specific heat capacity	$J/(kg.K)$
c	Concentration	kg/m^3, g/m^3
e	Emissivity	–
f	Specific free energy	J/kg
f	Temperature ratio	–
g	Specific free enthalpy	J/kg
g	Acceleration by gravity	m/s^2
g	Mass flux	$kg/(m^2.s)$
h	Height	m
h	Specific enthalpy	J/kg
h	Surface film coefficient for heat transfer	$W/(m^2.K)$
k	Mass-related permeability (mass may be moisture, air, salt..)	s
l	Length	m
l	Specific enthalpy of evaporation or melting	J/kg
m	Mass	kg
n	Ventilation rate	s^{-1}, h^{-1}
p	Partial pressure	Pa
q	Heat flux	W/m^2
r	Radius	m
s	Specific entropy	$J/(kg.K)$
t	Time	s
u	Specific latent energy	J/kg
v	Velocity	m/s
w	Moisture content	kg/m^3
x, y, z	Cartesian co-ordinates	m
A	Water sorption coefficient	$kg/(m^2.s^{0.5})$
A	Area	m^2
B	Water penetration coefficient	$m/s^{0.5}$
D	Diffusion coefficient	m^2/s
D	Moisture diffusivity	m^2/s
E	Irradiation	W/m^2
F	Free energy	J

(Continued)

Table 1 List with symbols and quantities. (Continued)

Symbol	Meaning	Units
G	Free enthalpy	J
G	Mass flow (mass = vapour, water, air, salt)	kg/s
H	Enthalpy	J
I	Radiation intensity	J/rad
K	Thermal moisture diffusion coefficient	kg/(m.s.K)
K	Mass permeance	s/m
K	Force	N
L	Luminosity	W/m^2
M	Emittance	W/m^2
P	Power	W
P	Thermal permeance	W/(m^2.K)
P	Total pressure	Pa
Q	Heat	J
R	Thermal resistance	m^2.K/W
R	Gas constant	J/(kg.K)
S	Entropy, saturation degree	J/K, -
T	Absolute temperature	K
T	Period (of a vibration or a wave)	s, days, etc.
U	Latent energy	J
U	Thermal transmittance	W/(m^2.K)
V	Volume	m^3
W	Air resistance	m/s
W	Work	J
X	Moisture ratio	kg/kg
Z	Diffusion resistance	m/s
α	Thermal expansion coefficient	K^{-1}
α	Absorptivity	–
β	Surface film coefficient for diffusion	s/m
β	Volumetric thermal expansion coefficient	K^{-1}
η	Dynamic viscosity	N.s/m^2
θ	Temperature	°C
λ	Thermal conductivity	W/(m.K)
λ	Wavelength	m
μ	Vapour resistance factor	–
ν	Kinematic viscosity	m^2/s
ρ	Density	kg/m^3

(Continued)

Table 1 List with symbols and quantities. (Continued)

Symbol	Meaning	Units
ρ	Reflectivity	–
σ	Surface tension	N/m
ω	Thermal pulsation	J/(m².K)
τ	Transmissivity	–
ϕ	Relative humidity	–
α, ϕ, Θ	Angle	rad
ξ	Specific moisture capacity	kg/kg per unit of moisture potential
Ψ	Porosity	–
Ψ	Volumetric moisture ratio	m³/m³
Φ	Heat flow	W

Table 2 List with currently used suffixes.

Symbol	Meaning	Symbol	Meaning
Indices			
A	Air	m	Moisture, maximal
c	Capillary, convection	r	Radiant, radiation
e	Outside, outdoors	sat	Saturation
h	Hygroscopic	s	Surface, area, suction
i	Inside, indoors	rs	Resulting
cr	Critical	v	Water vapour
CO_2, SO_2	Chemical symbol for gasses	w	Water
		ϕ	Relative humidity

Notation	Meaning
[], bold	Matrix, array, value of a complex number
Dash (ex...:$\bar{a}$)	Vector

0
Introduction

0.1 Subject of the Book

This is the first volume in a series of three:

- Building Physics: Heat, Air and Moisture, Fundamentals, Engineering Methods, Material Properties and Exercises
- Applied Building Physics: Ambient Conditions, Whole Building and Building Assembly Performance
- Performance-Based Building Design: from Below Grade over Floors, Walls, Roofs, and Windows to Finishes

Discussed are the physics governing the heat, air, moisture, also called the hygrothermal response of materials, building assemblies and whole buildings with added a chapter with tables and measured values concerning the heat, air, moisture properties of building, insulating and finishing materials. The second volume on Applied Building Physics in turn deals with the ambient conditions to be considered, the performance requirements at the whole building and the heat, air, moisture requirements and metrics at the building assembly's level. The third volume on 'Performance-Based Building Design: from Below Grade over Floors, Walls, Roofs, and Windows to Finishes' finally document the overall structural, building physics, fire safety, economics and sustainability-related performance metrics, which help realising high-quality buildings.

By the way, the notion 'Building Physics' is hardly used in the Anglo-Saxon world. 'Building science', the field is called there. A difference with Building Physics is that Building Science does not include acoustics and lighting and focuses more directly on practice-related issues.

0.2 Building Physics?

0.2.1 Definition

As an applied science, 'Building Physics' studies the hygrothermal, acoustical and visual performances at the material, building assembly, space, whole building and built environment level, the last then under the name 'Urban Physics'. The constraints faced are user demands related to overall comfort, healthiness and safety, several architectural restrictions, durability issues, economical demands and sustainability-related challenges with, given the reality of a global warming, energy use and decarbonisation as key concerns.

The term 'applied' indicates that Building Physics is oriented to the application with the theory as tool, not as purpose. Topics tackled in the heat, air, moisture subfield are airtightness, thermal insulation, transient thermal response, moisture tolerance, thermal bridging, salt transport, temperature and humidity-induced stresses and strains, net energy demand, gross energy demand, end energy use, ventilation, thermal comfort and indoor air quality.

In the building acoustics subfield, the topics discussed include acoustical comfort, the air- and structure-borne noise transmission through outer walls, floors, partitions, party walls, glazing and roofs, room acoustics and the abatement of installation and ambient noise. In the lighting subfield, the topics handled concern, daylighting, artificial lighting and the impact both have on human wellbeing and end energy use.

Urban physics finally looks among others to the thermal, acoustical, visual and wind-induced comfort issues outdoors, the wind and rain patterns on buildings in cities, the spread of air pollution, the heat island effect and the energy management at the city level.

0.2.2 Constraints

0.2.2.1 Comfort

Comfort can be defined as the state of mind that expresses satisfaction with the conditions in the direct environment. Attaining a comfortable situation there depends on the human need to feel thermally, acoustically and visually at ease: neither too cold nor too warm, not too noisy, no unacceptable contrasts in luminance, etc.

Thermal comfort engages the human physiology and psychology. As exothermal creatures with a core temperature of $\approx 37\,°C$ (310 K), humans have to lose heat to the environment under all circumstances, be it by conduction, convection, radiation, perspiration, transpiration or breathing. The air temperature, its gradient, the radiant temperature, the radiant asymmetry, the contact temperatures, the air velocity relative to the body, the air turbulence and the relative humidity (RH) in the direct environment are the parameters that determine how much heat will be exchanged. For a given activity and clothing, certain combinations will be quoted as comfortable, others as uncomfortable, although adaptation plays.

Acoustical comfort is strongly connected to mental awareness. Physically, young adults hear frequencies between 20 and 16 000 Hz. But, as humans scale sound intensity logarithmically with a better hearing for higher frequencies, a logarithmic quantity has been introduced to judge sound and noise: the decibel (dB), with 0 dB as audibility limit and 140 dB as pain threshold. Undesired noises produced by neighbours, traffic, industry and aircraft may give complaints and are often the cause of protracted disputes.

Visual comfort finally combines mental and physical facts. Physically, the eyes see electromagnetic waves having wavelengths between 0.38 and 0.78 μm with a maximum sensitivity around 0.58 μm, the yellow-green light, while the overall sensitivity adapts to the mean luminance, up to 10 000 times higher when dark than during daytime, although the eyes perceive this logarithmically. Besides, too large differences in brightness disturb and a well-adapted lighting creates cosiness.

0.2.2.2 Health and Well-being

Not only the absence of illness, but also no neuro-vegetative complaints, no psychological stress and no physical unease determines what's perceived as healthiness and well-being. Dust, fibres, (S)VOCs, radon, CO, viruses and bacteria, moulds and mites, too much noise, thermal discomfort, to large luminance contrasts, all are menacing factors.

0.2.2.3 Architecture and Materials

Applying the tools and knowledges building physics offers always faces architectural and material-linked constraints. While façade and roof form, aesthetics and the materials chosen shape buildings, at the same time their design must satisfy a huge set of performance metrics. Conflicting structural and physical issues often complicate solutions. Necessary thermal cuts may interfere with the strength and stiffness demands connections have to ensure. Being waterproof and vapour permeable at the same time is not always compatible. Necessary acoustic absorption could oppose vapour tightness. Some materials may not turn and stay wet, etc.

0.2.2.4 Economy

Not only the construction costs must respect the budget available but also the total present value, an economic parameter that adds the initial investment to the costs of end energy consumed, maintenance, upgrades and replacements over the usage period transposed to today, should preferentially be the lowest achievable. A building designed and constructed according to the performance metrics advanced by building physics and other engineering fields will generally generate a lower total present value than if done without such fitness for purpose approach.

0.2.2.5 Sustainability

From a human welfare point of view, it should be a blessing if the new and renovated buildings, that a worldwide growing population needs, could offer a better comfort and good indoor environmental quality (IEQ). However, related ever-growing energy need, if fossil based, will have worrying consequences in terms of global warming. At the same time, building use gives other solid, liquid and gaseous waste, whereby NO_2, if heating with fossil fuels, is of growing concern.

The pursuit for more sustainability is reflected in a growing use of life cycle inventory and analysis tools (LCIA) plus certification instruments. In LCIA, buildings are evaluated on their environmental impact 'cradle to cradle', i.e., from material and component production over construction and occupancy to demolition with related reuse of materials and components. Per step, all material, energy and water inflows and all polluting outflows are quantified and the impact on human well-being and environment assessed. Certification instruments instead focus on the fitness for purpose results that new and renovated buildings or urban environments guarantee.

0.3 Importance?

The necessity to create a comfortable indoors that protects people from the fickle of the weather gave birth to the knowledge field called Building Physics. Basic thereby is that an appropriate design should annihilate when needed the impact of the various loads on the building enclosure, take sun, rain, wind, noise, temperature, vapour and air pressure differentials, but use these when aiding comfort and well-being, while demanding as little energy as doable.

In earlier days, experience was guiding. Former generations only had a limited range of materials available – wood, straw, loam, brick, natural stone, lead, copper, cast iron, and blown glass – for which how to use them evolved over centuries. Standard solutions for roofs, roof edges and outer walls took shape. From the size and orientation of the windows to the overall layout, everything was conceived to limit the heating needs in winter and overheating in summer. Because outside the urban centres, deafening noisy sources were scarce, at the countryside sound was hardly a problem, while a lifestyle adapted to the seasons saved energy. Came the industrial revolution. New materials such as steel, reinforced concrete, prestressed concrete, nonferrous metals, synthetics, bitumen and insulation inundated the market. Advanced technologies turned existing materials into innovative products, take cast and floated glass, rolled metals, pressed bricks, etc. Structural mechanics allowed any span. Energy, first coal, later petroleum, natural gas and electricity, was so cheap that efficient use was no issue. Building exploded and turned into a demand/supply market. The result was mass construction, often of dubious quality.

The early twentieth century saw a 'modern school' of architects emerging, who experimented with alternative structural solutions, simple details and new materials. The buildings they designed were neither energy efficient nor of good building physical quality. Typical was the profuse use of steel, concrete and glass, all difficult materials from a heat, air, moisture point of view. Overhangs and façade reliefs were banned. As a consequence, their buildings suffered from obvious failures requiring premature restoration, what a sound knowledge of building physics could have prevented. Figure 0.1 shows the house Guiette, designed by Le Corbusier, built in 1926, before and after renovation in 1987. Just built, the outer walls and low-sloped roof lacked any thermal insulation, which in winter gave surface condensation against the ceiling in the sleeping rooms under the roof, with water dripping down in the beds. The house got damaged during World War 2 and was sloppy restored after, whereby the originally white stuccoed outer walls got covered with grey slates, see figure. The renovation saw the roofs insulated and against the outer walls the application of a white stuccoed outside insulation. Before, heating the whole to a temperature of 18 °C demanded ≈16 000 l of fuel a year. After, if fully heated some 6500 m^3

Figure 0.1 House Guiette designed by Le Corbusier, before and after restoration in 1987.

of natural gas a year should have been used. As the owners never heated the sleeping rooms and applied a day/night control, the real consumption touched ≈4000 m^3. In 2023, a correcting retrofit was finished.

Contrary to the trial and error of the past, for which building technology, the requirements mandated and architectural fashion is evolving too fast, today fit for purpose building requires a correct application of all building physics-related performance metrics.

0.4 History

0.4.1 In General

Building physics emerged at the crossroad between applied physics, comfort and health, building services, building design and building construction.

0.4.2 Applied Physics

0.4.2.1 Heat, Air, Moisture

Until the end of the nineteenth century, heat transmission was the most studied part. In 1822, Jean Fourier from France (Figure 0.2) introduced a property that

Figure 0.2 Left Jean Fourier (1768–1830), middle Jean-Claude Pèclet (1793–1857), right Hermann Rietschel (1847–1914).

Table 0.1 An early list with better insulating materials.

Material	λ value, W/(m.K)
Seagrass	0.14
Dry sand	0.58 + 0.68
Wood chips	0.07–0.093
Wood-wool/cement boards, thickness 2.5–3.5 cm	0.09
Straw boards	0.057
Straw, hay	0.047–0.06
Expanded cork boards	0.04–0.045

quantified the ability of a material to conduct heat, the 'thermal conductivity' having actually as symbol λ and as SI-units W/(m.K). Some Anglo-Saxon countries still use as symbol k and as units the PI-ones. Six years later, in 1828, Jean-Claude Pèclet from France (Figure 0.2) proposed the 'thermal transmittance' as a property quantifying the steady state heat flow by conduction through flat walls with then k as symbol, today U as symbol and as SI-units W/(m^2.K). However, the heat exchange between the surfaces on both sides and their environment was not included. For that, one had to wait till 1885, when Hermann Rietschel, professor at the TU-Berlin in Germany (Figure 0.2), introduced the concept of a surface film coefficient to quantify that transfer due to convection and radiation. This expanded the equation of the thermal transmittance from transmission face to face to transmission ambient to ambient through a flat wall, a step, which allowed estimating the heat loss of building enclosures.

In the early twentieth century, tables with λ-values of some materials appeared, see Table 0.1.

In 1921, an exhibition showing walls that were perceived as better insulating was organised in Germany, see Figure 0.3. All were brick-based.

In 1924, architect Bugge of Norway proved experimentally that a better insulation lowered the energy needed for heating, see Figure 0.4.

In 1952, a first edition of the German insulation standard DIN 4108 was published with $U \leq 1.6$ W/(m^2.K) as requirement for outer walls, a number equal to the thermal transmittance of a 36 cm thick brick wall, see the exhibition of 1921.

The quest for more energy efficiency, with a better thermal insulation of building envelopes as preferred measure, started end of 1973, when the Yom Kippur War and related stop in crude delivery by the Organization Petroleum Exporting Countries (OPEC) resulted in a sharp increase in oil prices (Figure 0.5).

The scientization of moisture transport started quite a time after A. Fick (Figure 0.6) advanced his diffusion law, stating that differences in gas concentration create gas fluxes from higher to lower gas concentration proportionally to the local gradient.

Figure 0.3 The walls exhibited.

Figure 0.4 The test huts used to prove the usability of thermal insulation for saving heating.

Figure 0.5 Price evolution of crude between 1950 and 2010.

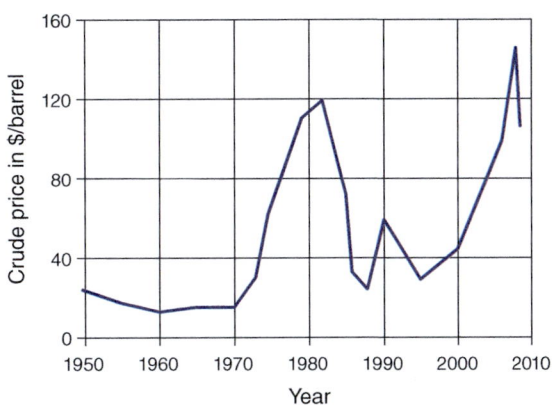

As diffusion could so be a cause of vapour movement in the air filling the pores in a material, the consequences, if happening when also heat moves through, became a subject of interest in the late 1930s, when Teesdale of the US Forest Products Laboratory published a study on 'Condensation in Walls and Attics'. In 1952, a paper written by J.S. Cammerer on 'Die Berechnung der Wasserdampf diffusion in der Wänden' (The calculation of Water Vapour Diffusion in Walls) appeared in 'Der Gesundheitsingenieur'. In 1958–1959, H. Glaser published four papers in 'Kältetechnik', advancing an upgraded calculation method to evaluate

Figure 0.6 Adolf Fick (1829–1901).

'interstitial' condensation by vapour diffusion in cold storage walls. In his book 'Wasserdampfdiffusion im Bauwesen' (Water vapour diffusions in buildings), in 1967, K. Seiffert applied Glaser's method on building assemblies. Due to the use of non-realistic winter-based constant boundary conditions, the vapour pressure curve touched saturation in nearly any assembly. The result was a vapour barrier mania. In reality, interstitial condensation by diffusion often combines a limited moisture deposit during the colder months with full drying when warmer.

Overlooked by these authors but a common cause of interstitial condensation in wall assemblies subjected to temperature differences is air exfiltration. Its impact was first noted in Canada, a country with a timber-frame tradition, when in 1961, A.G. Wilson of the NRC wrote 'One of the most important aspects of air leakage in relation to the performance of Canadian buildings is the extent to which it is responsible for serious condensation problems. Unfortunately, this is largely unrecognised in the design and construction of many buildings, and, even when failures develop, the source of moisture is often incorrectly identified. From the late 1960s on, researchers started studying combined heat, air, moisture transport and its consequences for moisture tolerance. Among them were O. Krischer, J.S. Cammerer and H. Künzel in Germany, A. De Vries, B.H. Vos and E. Tammes in the Netherlands, L.E. Nevander in Sweden, the author of this book in Belgium and A. Tveit in Norway.

0.4.2.2 Acoustics

In the early twentieth century, physicists got interested in noise control in buildings. In 1912, Berger submitted a Ph.D. thesis at the Technische Hochschule München with as title: 'Uber die Schalldurchlässigkeit' (About sound transmission). In 1920, Sabine published his reverberation time formula. Since, room acoustics became a favourite with studies on speech intelligibility, optimal reverberation time, etc. A decennium later, L. Cremer (Figure 0.7) forced a breakthrough in understanding airborne sound transmission through walls.

In his paper 'Theorie der Schalldämmung dünner Wände bei schrägem Einfall' (Theory of sound insulation of thin walls at oblique incidence), he recognised that coincidence between the sound waves in the air and the bending waves on a wall played a major role in degrading its sound insulation. Later, his studies about structure-borne noise transmission through floors forwarded floating screeds as a solution. K. Gösele and M. Heckl in turn linked building acoustics to building

Figure 0.7 Lothar Cremer (1905–1990).

construction by formulating rules on how to compose floors, walls, and roofs to get a correct air- and structure-borne sound attenuation. Meanwhile in the United States, in 1971, Beranek published his book 'Noise and Vibration Control', a reference for engineers who have to solve noise problems and a remake of his book 'Noise reduction', published in the 1960s.

0.4.2.3 Lighting

In 1931, a study was completed at the Universität Stuttgart, dealing with 'Der Einfluss der Besonnung auf Lage und Breite von Wohnstrassen'. (The influence of solar radiation on the location and width of residential streets.) Later, physicists used the radiation theory to calculate the illumination by artificial and natural light sources on surfaces and to study the surrounding luminance contrasts. End of the 1960s, the daylight factor was introduced as a quantity to evaluate the illumination indoors by natural light from outdoors. More recently, after the energy crises of the 1970s, the relation between artificial lighting and energy use surged as a topic. In the 1990s; the search for more energy-efficient lighting caused a move from light bulbs to LEDs.

0.4.3 Indoor Air Quality and Thermal Comfort

Already in the nineteenth century, engineers advanced healthy housing and urban hygiene as important issues. Max von Pettenkofer (Figure 0.8) was the first to evaluate the impact of ventilation on the CO_2 concentration indoors. He introduced the 1500 ppm acceptability limit. Also, the notion 'breathing material' is his, reflecting the link he found between more health complaints in stony than in brick dwellings and the measurements done proving that stony, not brick walls, blocked air passage. The true reason of course is the poorer thermal performance of stony walls, resulting in more complaints about mould, surface condensation and cold indoors.

Thermal comfort moved to the forefront of interest when in the 1930s research sponsored by the American Society of Heating and Ventilation Engineers (ASHVE), a predecessor of ASHRAE and conducted by C. Yaglou (Figure 0.9) lead to the notion 'operative temperature'. Originally, the definition did not include radiation, what changed after A. Missenard, a French engineer (Figure 0.9), by critically reviewing Yaglou's data detected the impact the radiant temperature had. In his book 'Thermal Comfort. Analysis and Applications in Environmental Engineering' published in 1970 by the Danish Technical Press in Copenhagen, the late P. O. Fanger (Figure 0.9) in turn firmly founded the relation between perceived thermal

Figure 0.8 Max von Pettenkofer (1818–1901).

Figure 0.9 left: Constantin Yaglou (1897–1960), middle: André Missenard (1901–1989), right: Per Ole Fanger (1934–2006).

comfort and all parameters having impact. Based on the human physiology, the heat exchanged between the clothed body and the environment and the differences in comfort perception between individuals, he developed a steady state thermal model for the active, clothed individual. Since, his 'Predicted Mean Vote (PMV) versus Predicted Percentage of Dissatisfied (PPD)' curve forms the kernel of all comfort standards worldwide. In 2004, P.O. Fanger received a doctorate honoris causa at the KULeuven for his lifetime achievements in the fields of thermal comfort and indoor air quality (IAQ).

After 1985, the adaptive model, being a refinement of Fanger's work, surged as a better approach to evaluate the thermal comfort perceived in naturally ventilated buildings.

Concerns about IAQ lead to cataloguing possible pollutants and related health risks. Over the years, with the move to fully air-conditioned buildings, the sick building syndrome (SBS) became more of a problem. It reinforced the need for an ever better understanding of the impact of IAQ on the well-being of people. Nonetheless, that 'better' was not always based on a sound interpretation of facts. Too often, discontentment with the job was overlooked. Also here, P.O. Fanger had an impact with his work on perceived indoor air quality linked to bad smell and the air enthalpy.

0.4.4 Building Services

In the nineteenth century, technicians were searching for methods to predict at the design stage the heating and cooling load by building use. The knowledge developed over the years, providing concepts such as the thermal transmittance of flat envelope assemblies, helped a lot. Quite early, organisations such as ASHVE and the Verein Deutsche Ingenieure (VDI) established technical committees dealing with the subject. An active member of ASHVE was W.H. Carrier (1876–1950) (Figure 0.10), in the United States, recognised as the father of chiller-based air conditioning. He was the first to publish a usable psychometric chart. For H. Rietschel (Figure 0.10), professor at the Technische Universität Berlin and author of a book on 'Heizung und Lüftungstechnik' (Heating and Ventilation Techniques), heat loss and gain through ventilation was one of his concerns. With others, he noted that well-designed ventilation systems were malfunctioning in cases where the envelope lacked air-tightness. This pushed the interest in unwanted air in- and exfiltration.

Vapour in the air became a worry once air-conditioning [heating, ventilating, air conditioning (HVAC)] became widespread. Before, humidity was seen as a troubling reality, mainly because of possible health effects. Due to the noisiness of many HVAC systems, sound attenuation surged as topic, while lighting did because HVAC engineers got contracts not only to design the heating, ventilating and air conditioning system in non-residential buildings but also the whole artificial lighting layout.

From 1973 till the early 1980s, the then-high energy prices pushed the development of energy-efficient building services to be used in energy-efficient premises. Since, a necessary moderation of the pending global warming issue has become the main motivator for that.

0.4.5 Building Design and Construction

The many complaints about bad acoustics and failing moisture tolerance, modernist architects were confronted with in the first half of the twentieth century, slowly gave building physics a boost to avoid these inconveniences, their 'state-of-the art' approach struggled with. The early 1930s saw an accrued concern about moisture issues in the United States, where peeling and blistering of paints on insulated timber-frame facades became a widespread problem. Insulation materials were new

Figure 0.10 left: Willis Carrier (1876–1950), right: Hermann Rietschel (1847–1914).

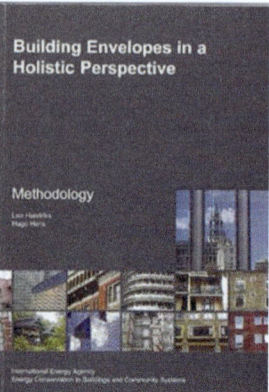

Figure 0.11 left: 'Freiland Versuchsstelle Holzkirchen', right: the Annex 32 report.

at that time. It motivated Teesdale to study interstitial condensation in timber-frame walls. Some years later, ventilated attics with insulation at ceiling level were tested by F. Rowley, professor in Mechanical Engineering at the University of Minnesota. This resulted in a series of instructions on the use of vapour retarders and the need for attic ventilation.

In Germany, from 1951 on, the 'Freiland Versuchsstelle Holzkirchen' (Figure 0.11, left) used building physics as a driver to upgrade the overall construction quality. When, after 1973, energy efficiency became a hot topic and insulation a necessity, the knowledge gathered on the test buildings there proved extremely useful for the construction of high-quality, well-insulated buildings and the development of better insulating glazing with lower solar and good visual transmittance.

In the 1990s, the need for better quality resulted in the performance rationale as proposed by IEA EBC Annex 32 'Integral Building Envelope Performance Assessment' in the report 'Building Envelopes in a Holistic Perspective' (Figure 0.11, right). The 'fitness for purpose' rationale advanced in this report inspired the high-performing building evaluation tools with classifications in terms of comfort, IAQ, safety, durability and sustainability, used today in several countries.

0.4.6 Hall of Fame

A hall of fame remembers the deceased, who helped advancing building physics and related sciences in a substantial way, be it as a teacher, researcher or practitioner. Proposing names of course is a challenge in a field so broad. Anyhow, following names deserve being part of it:

– Max von Pettenkofer, Germany	– Cornelis Kosten, the Netherlands
– Frank Rowley, USA	– Bob Vos, the Netherlands
– Josef Cammerer, Germany	– Lars Erik Nevander, Sweden
– Wallace Sabine, USA	– Per Ole Fanger, Denmark
– Lothar Cremer, Germany	––

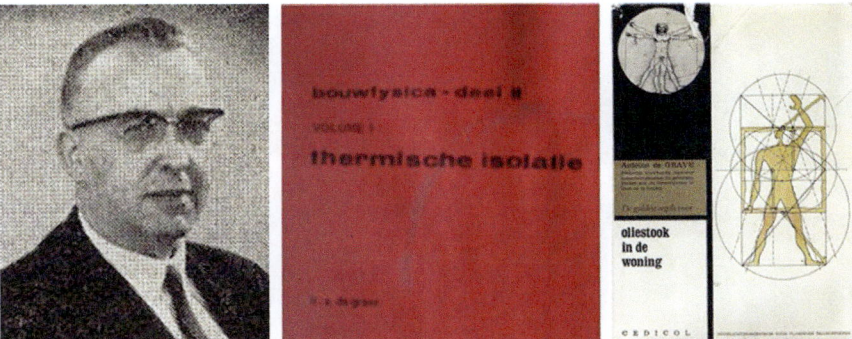

Figure 0.12 Antoine de Grave (1914–1975) and his two books.

0.4.7 Building Physics at the KULeuven and Other Universities in the Low Countries

At the KULeuven lecturing on building physics and building services started in 1952, when both topics became compulsory for students in architectural and optional for students in civil engineering. Nominated as professor was Antoine de Grave, a civil engineer, the head of the building department at the Ministry of Public Works (Figure 0.12). He taught both courses until 1975, when he passed away suddenly at the age of 61. The 2 books in Dutch he published, the first called 'Bouwfysica deel II, thermische isolatie' (building physics tome II, thermal insulation), and the second called 'Oliestook in de woning' (oil heating in dwellings) served as handbook for the lectures, he taught with burning enthusiasm.

Soon after 1975, the year that the Faculty of Engineering Sciences of the KULeuven asked the author of this book to immediately take over both courses, building physics became compulsory for civil engineering. End of 1975, together with Gerrit Vermeir, then researcher at the Acoustics Laboratory at the Department of Physics and from 1992 on professor in building acoustics, a working group on building physics was established in the then 'Department of Construction', with an objective of doing research and consultancy in that field.

In 1978, the working group became the 'Laboratory of Building Physics' to turn into the 'Unit of Building Physics' at the Department of Civil Engineering in 1990, since 2020 called the 'Unit of Building Physics and Sustainable Construction'. Over the years, the research topics tackled included the hygrothermal properties of building and insulating materials, the heat, air, moisture performance of well-insulated building assemblies, end energy use in buildings, indoor environmental quality in buildings, air- and structure-borne noise attenuation and room acoustics, while from 1990 on, teaching also included 'Performance-based building design'. In 1996, a test building was inaugurated allowing on-site research on highly insulated façade and roof assemblies (Figure 0.13).

In 2002, the unit moved to a completely new, well-equipped laboratory building, that houses all test setups needed to measure the hygrothermal properties of materials and the heat, air, moisture response of building assemblies under controlled

Figure 0.13 The test building.

environmental conditions. A main objective of the research and consultancy done between the 1970s and the 2000s was: upgrading the quality of the built environment in cooperation with the building sector and the building industry. In 2013, urban physics was added as a discipline.

In 2008, after the author of this book became emeritus, first Staf Roels and Dirk Saelens, in 2010 Hans Janssen and in 2013 Bert Blocken took over. Staf Roels is teaching 'Performance-based building design', Dirk Saelens 'Building services, energy use and indoor environmental quality', Hans Janssen the basics of 'Heat, air and moisture transport' and Bert Blocken, until 2023 full professor at the Technical University Eindhoven (TU/e) and part-time professor at the KULeuven, 'Urban Physics'. All are former PhD-students at the unit. In 2011, first Arne Dijckmans, later Edwin Reynders succeeded Gerrit Vermeir, who became emeritus then, for Building Acoustics and Lighting. All continue the research activities, started in 1975.

Ghent University (UGent) waited until 1999 to nominate Arnold Janssens, until then a post-doc researcher at the KULeuven Unit of Building Physics, as full-time professor in building physics. Since, he did a tremendous job in teaching, organising and heading research within the fields covered by building physics.

At the Vrije Universiteit Brussel (Free University Brussels) (VUB) a part-time professorship was offered to F. Descamps, again a former researcher at the KULeuven Unit of Building Physics and actually manager of an engineering office, specialised in building physics.

In the Netherlands, before World War 2, professor Zwikker gave lectures at the Technical University Delft (TU-Delft), in which he analysed buildings from a physical point of view. However, only from 1955 on, a course entitled 'Building Physics' was offered, having professor C. Kosten of the Applied Physics Faculty at the TU as chair holder. In 1963, professor Verhoeven took over. In 1969, the Technical University Eindhoven (TU/e) was founded. Professor P. De Lange became holder there of the chair of Building Physics at the Faculty of the Built Environment. Since, many professors got involved, among them J.A. Wisse, Nico Hendriks, Martin de Wit, Jan Hensen and Olaf Adan.

Further Reading

Beranek, L. (ed.) (1971). *Noise and Vibration Control*. New York: McGraw-Hill.

BRE (2010). BREEAM bespoke 2010.

CIB-W40 (1975). Quantities, Symbols and Units for the description of heat and moisture transfer in Buildings, conversion factors, IBBC-TNP, report n° BI-75-59/03.8.12.

De Freitas, V.P. and Barreira, E. (2012). *Heat, Air and Moisture Transfer Terminology, Parameters and Concepts*, vol. 369. CIB publication, 52 p.

Donaldson, B. and Nagengast, B. (1994). *Heat and Cold: Mastering the Great Indoors*, 339. Atlanta: ASHRAE Publication.

Hens H. (2008). Building Physics: from a dormant beauty to a key field in building engineering. *Proceedings of the Building Physics Symposium*, Leuven.

Hendriks, L. and Hens, H. (2000). *Building Envelopes in a Holistic Perspective, IEA-ECBCS Annex 32*. Leuven: ACCO.

ISO 1000 (1981). SI units and recommendations for the use of their multiples and of certain other units (ISO-International Standardisation Committee).

Kumaran, K. (1996). *Task 3: Material Properties, Final Report IEA EXCO ECBCS Annex 24*, 135. Leuven: ACCO.

Künzel, H. (2001). *Bauphysik Geschichte und Geschichten, Fraunhofer IRB-Verlag*. Stuttgart 143 p.

Northwood, T. (ed.) (1977). *Architectural Acoustics, Dowden*. Stroudsburg, PA: Hutchinson & Ross, Inc.

Rose, W. (2003). The rise of the diffusion paradigm in the US. In: *Research in Building Physics*, 327–334. A.A. Balkema Publishers.

USGBC (2008). LEED 2009 for New Construction and Major Renovations, Washington, DC, 200371.

Winkler Prins Technische Encyclopedie, deel 2 (1976). *Article on Building Physics*, 157–159. Amsterdam: Uitgeverij Elsevier.

Daniel Mugnier, Daniel Neyer, Stephen D. White (eds.)

The Solar Cooling Design Guide

Case Studies of Successful Solar Air Conditioning Design

- presents detailed guidance to prevent errors in the design process
- gives indicative economic analysis of the solar cooling plants

This book presents results of research initiated by the International Energy Agency's Solar Heating and Cooling Program and conducted by leading experts. It provides cutting-edge information on the design of solar air conditioning plants.

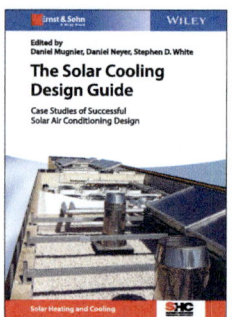

2017 · 158 pages · 100 figures
Hardcover
ISBN 978-3-433-03125-4   € 79*

ORDER
+49 (0)30 470 31-236
marketing@ernst-und-sohn.de
www.ernst-und-sohn.de/3125

* All book prices inclusive VAT.

1

Heat Transfer

1.1 In General

1.1.1 Heat

1.1.1.1 What?

A first description of what heat is comes from thermodynamics, a discipline that considers everything as a system surrounded by an environment, with which energy is exchanged. Can be the system: a material, a building assembly, a building, a heating, ventilating, air Conditioning (HVAC) installation, even a whole city. Energy transmitted as work between both looks purposeful and organized, energy transmitted as heat diffuse and chaotic. A second description of what heat is resides in particle physics, where heat refers to the statistically distributed kinetic energy of atoms and free electrons. Whatever, heat is the least noble, most diffuse form of energy, to which each nobler form degrades, see the second law of thermodynamics.

1.1.1.2 Sensible Heat

Sensible heat is unambiguously linked to what people feel, the temperature, see below. Transferring sensible heat, be it by conduction, convection and radiation, demands differences in temperature.

Conduction refers to the heat flow in a medium induced by its vibrating atoms, whose spheres of influence collide, and the movement of free electrons. Heat transmitted between solids at different temperature in ideal contact with each other and between points in these is straightforwardly conduction-based. This is also the case between gases and liquids and contacting surfaces. According to the second law of thermodynamics, conduction always moves direction lower temperatures. It needs a medium and does not induce macroscopic movement.

Convection in turn is the result of macroscopic movement in liquids and gases, in which temperature differences exist or that touch colder or warmer surfaces. As well external forces, differences in density or both together are inducing such movements and fix the type of convection generated: forced, natural or mixed. Convection needs a medium.

Radiation finally concerns heat transferred between surfaces due to them emitting and absorbing electromagnetic waves. Above 0 K, each surface radiates. When two or more surfaces at different temperature see each other, the result is heat exchanged between them. Radiation does not need a medium, while the laws governing it are very different from these shaping conduction and convection.

1.1.1.3 Latent Heat

Latent heat is linked to the changes of state between solid, liquid and gaseous. A release, for example liquid evaporating, or a deposit, for example liquid turning solid, require no differences in temperature, although both will impact the amounts of heat moving. Water evaporating absorbs the sensible heat of evaporation, so creates a heat sink. When then the water vapour so formed moves to a colder spot, where it condenses, the sensible heat of evaporation is emitted again, creating a heat source. Such sources and sinks not only impact the temperature profile in materials and assemblies, but they also have quite some impact on the sensible heat transferred.

1.1.2 Temperature

The temperature, described above as what people feel, mirrors the heat quality. Higher temperatures stay for more quality. Thanks to the related increase in kinetic energy of atoms and free electrons, the result is more exergy, which represents the potential to convert more heat into work via a cyclic process. Instead, lower temperatures and related decrease in the kinetic energy of atoms and free electrons stay for less exergy. Heightening the temperature of a system requires warming it, lowering the temperature of a system cooling it. Like any potential, temperature is a scalar. Sensing it is not a problem but measuring it is. Happily, many material properties depend on temperature, which allows its indirect quantification. A mercury thermometer allows scaling by using the volumetric expansion of mercury when heated and its volumetric contraction when cooled. In a Pt100 thermometer, the change of the electrical resistance of a platinum wire with temperature is used, while for thermocouples the varying contact potential between two metals is.

The SI system uses two temperature scales, one empiric, the degree Celsius (°C) with θ as symbol, and the other thermodynamic, the degree Kelvin (K) with T as symbol. 0 °C coincides with the triple point of water, and 100 °C with its boiling point at 1 Atmosphere. 0 K instead stands for the absolute zero, and 273.15 K for the triple point of water. Temperature differences are given in K, temperatures in °C or in K with as relation between both:

$$T = \theta + 273.15$$

Instead of degree Celsius (°C), the USA still uses degree Fahrenheit (°F). The link between the two is:

$$°F = 32 + 9/5 °C$$

1.1.3 Why are Heat and Temperature so Compelling?

That heat is an issue in buildings, follows from the human demand for thermal comfort. In cold and temperate climates, the comfort temperatures required demand heating during the colder months. In most cases, the heat sources still used are oil and natural gas. Their overall share in delivering the end energy needed causes such CO_2 release, that since the second half of the 1970s, energy efficiency became imperative. A prime opportunity to realize is by minimizing the heat loss through the building envelope. Knowing which envelope properties have a decisive influence on that, became a necessity for designers and builders.

Which temperatures are considered as being important depends on the situation and possible consequences. In winter, inside surface temperatures close to the air temperature indoors felt as comfortable upgrade thermal comfort. Instead, those much lower not only degrade the thermal comfort but also increase mould and surface condensation risk, both perceived as triggering healthiness. Also, too high summer temperatures indoors clash with thermal comfort and IAQ, while high temperature differences across outer assemblies increase the air and moisture movement in them, the thermal stresses experienced and crack risk. Large temperature gradients also favour the displacement of dissolved salts, while high temperatures accelerate the chemical breakdown of synthetics. Furthermore, too many temperature fluctuations from above to below freezing may damage wet, frost-sensitive porous materials. Whether all these effects will remain controllable, depends on how building assemblies are designed and built.

1.1.4 Some Definitions

Amount of heat, symbol Q, unit [J]

Quantifies the energy exchanged as heat. As heat is a scalar, the amount also is.

Heat flow, symbol Φ, unit [J/s] = [W]

Stands for the heat exchanged per unit of time. Heat flow is a measure for 'power', thus, a scalar.

Heat flux, symbol **q**, unit [W/m²]

Quantifies the heat exchanged per unit of time through a unit surface normal to the flow direction. The flux so is a vector with same direction as the surface vector. Its components in Cartesian coordinates are q_x, q_y, q_z, in polar coordinates q_R, q_ϕ, q_Θ.

Solving a heat transfer problem now means determining the scalar temperature field (T) and the vectorial heat fluxes field (**q**). Computing both so requires a scalar and a vector equation.

1.2 Conduction

1.2.1 Conservation of Energy

A first relation between the heat flux (**q**) and temperature (T) follows from the conservation of energy axiom. In case an infinitely small material volume is the system and what is around the environment, then, without mass displacement, the energy balance between both writes as:

$$d\Phi + d\Psi = dU + dW \tag{1.1}$$

with $d\Phi$ the resulting heat flow between system and environment, $d\Psi$ the heat dissipated uniformly in the system, dU the change in the system's internal energy and dW the labour exchanged with the environment, all per unit of time. Dissipation could include the heat produced by an exothermic reaction, the heat absorbed by an endothermic reaction, the Joule effect by an electric current passing through, latent heat released or absorbed, etc. The labour exchanged equals:

$$dW = P d (dV) = P d^2 V$$

with P the pressure exerted in Pa. The conservation balance so states that the heat exchanged ($=d\Phi$), released or absorbed, modifies the internal energy in this infinitely small material volume, while causing a labour exchange with the environment. If isobaric, the balance reshuffles to

$$d(U + PdV) = dQ + dE$$

with $U + PdV$ the enthalpy (H). The resulting heat flow, the change in enthalpy and the heat dissipated now write as:

$$d\Phi = -\text{div}(\mathbf{q})dV \quad dH = \left|\frac{\partial(\rho c_p T)}{\partial t}\right| dV \quad d\Psi = \Phi' dV$$

with c_p the specific heat capacity at constant pressure of the material (J/(kg·K)), ρ its density (kg/m³) and Φ' the heat dissipated per unit of time and volume, positive if a source, negative if a sink. The three turn the conservation equation into:

$$\left(\text{div}(\mathbf{q}) + \Phi' + \frac{\partial(\rho c_p T)}{\partial t}\right) dV = 0 \tag{1.2}$$

For solids and liquids, the specific heat capacity hardly depends on the change of state. So, one value, symbol c, can be used with the product ρc equal to the volumetric specific heat capacity. For gases, the value varies with the change of state, giving as relation between the specific heat capacity at constant pressure (c_p) and at constant volume (c_v):

$$c_p = c_v + R$$

with R the specific gas constant (in Pa·m³/(kg·K)). Because conservation of energy now holds for any infinitely small material volume, the relation between heat flux ($\mathbf{q}$) and temperature (T) so becomes:

$$\text{div } \mathbf{q} = -\frac{\partial(\rho c T)}{\partial t} - \Phi' \tag{1.3}$$

1.2.2 Conduction Laws

1.2.2.1 First Law

The name 'first law' is given to the empirical vector equation between heat flux and temperature, advanced by the French physicist Fourier:

$$\mathbf{q} = -\lambda \text{ grad } T = -\lambda \text{ grad } \theta \tag{1.4}$$

It states that the conductive heat flux in a point somewhere in a solid, liquid or gas varies proportionally to the temperature gradient there. The proportionality value λ figures as a material property called the 'thermal conductivity' with as units W/(m·K). It expresses the ability of a medium to conduct heat. The minus sign indicates that the flux and the temperature gradient, which as vector goes from colder to warmer, oppose each other. Thermodynamics in fact learns that, if not forced externally, heat always moves direction colder. Otherwise, the entropy would decrease without energy input, which is impossible. Following observation supports the first law. With the surfaces of equal temperature, called isotherms, drawn in a construction detail and the heat fluxes visualized by tracing the lines of equal flux, called isoflux lines, seen is that the last develop perpendicular to the isotherms, come closer and break up more where the isotherms do (Figure 1.1).

At the same time, in each material, the fluxes remain proportional to their thermal conductivity, which is often assumed to be a scalar and constant, even though for building and insulating materials its value depends on temperature, moisture content, sometimes the material's thickness and its age, while for anisotropic materials it becomes a tensor. Often, the term 'apparent' is therefore added to the supposedly constant value.

In right-angled Cartesian coordinates [x, y, z], the heat flux along the three axes writes as:

$$q_x = \mathbf{q}_x \mathbf{u}_x = -\lambda \frac{\partial T}{\partial x} \quad q_y = \mathbf{q}_y \mathbf{u}_y = -\lambda \frac{\partial T}{\partial y} \quad q_z = \mathbf{q}_z \mathbf{u}_z = -\lambda \frac{\partial T}{\partial z}$$

In case °C instead of K is used, the heat flow across a surface dA with direction n equals:

$$d\Phi_n = \mathbf{q} d\mathbf{A}_n = -\lambda \frac{\partial \theta}{\partial n} dA_n u_n^2 = -\lambda \frac{\partial \theta}{\partial n} dA_n$$

Along each of the three axes, this gives as heat flow:

$$d\Phi_x = -\lambda \frac{\partial \theta}{\partial X} dA_x \quad d\Phi_y = -\lambda \frac{\partial \theta}{\partial y} dA_y \quad d\Phi_z = -\lambda \frac{\partial \theta}{\partial Z} dA_z$$

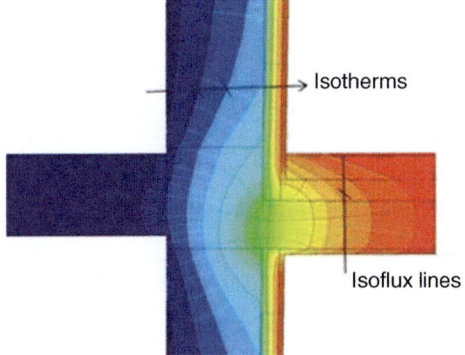

Figure 1.1 Floor with balcony traversing an outer wall insulated inside, lines of equal temperature and equal heat flux (the isotherms and isoflux lines).

1.2.2.2 Second Law

The second law is embedding the conduction equation in the conservation of energy one:

$$\operatorname{div}(\lambda \, \mathbf{grad} \, T) = \frac{\partial(\rho c T)}{\partial t} - \Phi' \tag{1.5}$$

This scalar relation allows calculating temperature fields. In case the thermal conductivity and the volumetric specific heat capacity are constant, the equation simplifies to what is called Fourier's second law:

$$\nabla^2 T = \left(\frac{\rho c}{\lambda}\right) \frac{\partial T}{\partial t} - \frac{\Phi'}{\lambda} \tag{1.6}$$

with ∇^2 the Laplace operator, in Cartesian coordinates equal to:

$$\nabla^2 = \frac{\partial^2}{\partial x^2} + \frac{\partial^2}{\partial y^2} + \frac{\partial^2}{\partial z^2} \tag{1.7}$$

After a deeper digging in what the thermal conductivity represents at the material level, further discussion will focus on applying both laws to building assemblies.

1.2.3 Thermal Conductivity

1.2.3.1 In General

As mentioned, the property λ is not a constant. Its value in fact depends on the way heat is transferred in a material, which causes the already mentioned dependence on temperature, moisture content, sometimes thickness and age. This demands a closer look to how λ gets a value.

1.2.3.2 Heat Transfer Modes Fixing the Property

In dry porous materials four modes coexist (Figure 1.2): conduction along the material matrix, conduction through the pore gas, convection in the pore gas and radiation in each pore between its walls. When humid, two modes add: conduction in the water film against the pore walls and a latent heat exchange by evaporation at the warmer part of these walls, diffusion of the vapour so formed through the pore

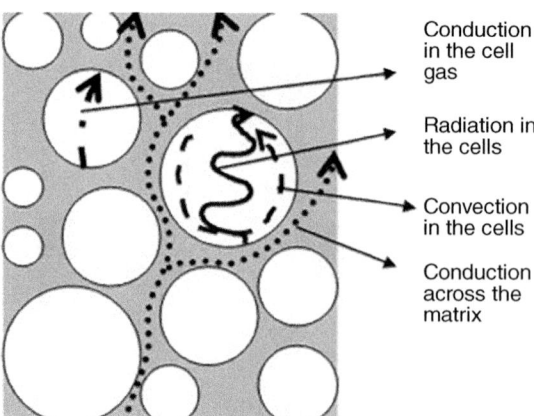

Figure 1.2 Heat flow modes in a porous material.

Conduction in the cell gas

Radiation in the cells

Convection in the cells

Conduction across the matrix

and condensation against the colder part of the pore walls. The value of a material's thermal conductivity depends on factors linked to these four or six modes.

If only conduction along the matrix and through the pore gas intervened, the thermal conductivity should equal:

$$\lambda_c = \lambda_M(1 - \Psi) + \lambda_G \frac{2\Psi}{1 + \Psi} \quad (1.8)$$

with λ_M the thermal conductivity of the matrix material, λ_G the thermal conductivity of the gas in the pores and Ψ the material's total porosity, equal to:

$$\Psi = (\rho_s - \rho)/\rho_s \quad (1.9)$$

with ρ the density and ρ_s the specific density of the matrix material, both in kg/m³. Accordingly, the thermal conductivity should decrease the higher the total porosity. Or, a more porous material with lower density (ρ), must see its thermal conductivity drop, a fact experiments prove, see Figure 1.3. Material matrixes with lower thermal conductivity or gases in the pores insulating better than stagnant air should show a same trend. If pores could be vacuumed, the gas-related thermal conductivity should even turn zero ($\lambda_G \approx 0$).

In case other gases than air are filling the pores of a highly porous material, diffusion of air in, diffusion of part of these gases out or of being adsorbed by the material matrix will slowly increase the λ-value. In the first weeks after manufacturing, this increase obeys:

$$\lambda_c = \lambda_c(0) + C_1 \sqrt{t}$$

where initially C_1 changes inversely proportional to the diffusion resistance factor μ of the matrix material and proportionally to the temperature with exponent $n < 1$ (T^n, T in K). Later the increase slows down to:

$$\lambda_c = \lambda_c(0) + (\lambda_c(\infty) - \lambda_c(0))[1 - \exp(C_2 t)]$$

with C_2 depending on the temperature and the diffusion resistance factor μ of the matrix material the same way as C_1 does, be it with values slackening the pace. Or, to

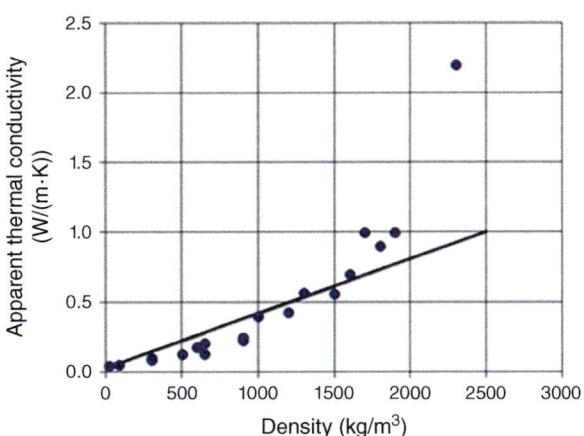

Figure 1.3 Thermal conductivity versus density. The full line stands for glass, from cellular to massive, $\lambda_M = 1\,W/(m \cdot K)$, the dots for measured values for other materials.

more or less permanently store another gas than air in the pores, the matrix should be as vapour retarding as possible. An alternative for thermally insulating materials consists of covering the fresh boards with a vapour-tight lining. Or, to act as thermal insulation, materials need a very low density with, if the matrix is enough vapour retarding, a gas filling the pores that retards heat better than air does.

However, in larger pores, convection may develop. Its impact is quantified by multiplying the thermal conductivity of the pore gas with its convection-related Nusselt number ($X \geq 1$):

$$\lambda_c = \lambda_M(1 - \Psi) + X\lambda_G \frac{2\Psi}{1 + \Psi} \tag{1.10}$$

As a result, more heat will flow through the material than without convection. Or, looking to insulation materials, their pores should be small enough to keep the Nusselt number 1. Thus, an insulation material should not only be very porous, but it should also have small enough pores.

But additionally, in a cell where the walls that are not perfectly reflective and at different temperature radiant heat is exchanged giving a radiant term (λ_R) completing the λ_c-value just given:

$$\lambda = \lambda_m(1 - \Psi) + X\lambda_G \frac{2\Psi}{1 + \Psi} + F_{RC} \underbrace{\frac{4C_b T_m^3 d}{100^4 \left(\frac{1}{e_1} + \frac{1}{e_2} + n\frac{1 + \rho - \tau}{1 - \rho + \tau} - 1\right)}}_{\lambda_R} \tag{1.11}$$

with:

$$F_{RC} = 1 + 100^4 \frac{\lambda_c(1/e_1 + 1/e_2 - 1)}{4C_b T_m^3 d} \left[\frac{\Delta\theta_1 + \Delta\theta_n}{\Delta\theta/n} - 1\right]$$

In case of insulation materials, in F_{RC}, the factor that corrects the thermal conductivity for the interaction between radiation, convection and conduction, is n standing for the number of pore walls, assumed parallel to the board's two faces, filling its thickness d. $\Delta\theta$ is the temperature difference between both faces, while $\Delta\theta_1$ and $\Delta\theta_n$ are the temperature differences between the linings that may cover both faces and the first pore wall encountered. In the λ_R formula, ρ and τ are the long wave reflectivity and transmissivity of these parallel pore walls, while for thermal insulation materials e_1 and e_2 are the long wave emissivity, side material of the linings covering both faces. As the radiant term λ_R is proportional to the third power of the temperature and the thermal conductivities of the matrix and the pore gas are temperature dependent, the overall relation with temperature may be rewritten as:

$$\lambda = \lambda_o + a_1 \theta^n + a_2 \theta^3 \tag{1.12}$$

with $0 < n < 1$. Between -20 and $50\,°C$, that equation becomes $\pm$ linear:

$$\lambda = \lambda_o + a_R \theta = \lambda_o \left(1 + a'_R \theta\right) \tag{1.13}$$

The lighter a highly porous material, meaning the thinner the pore walls or the larger the pores, the higher a_R and the more temperature dependant its thermal conductivity. Because few large than small pores fit into a given layer thickness and because thin pore walls show a higher long wave transmissivity than thicker ones, radiation gains importance in equally heavy materials having as well small or large pores. But the thermal conductivity must increase with a board's thickness. In fact, with the ratio between thickness (d) and mean pore width (d_P) replacing the number of pore walls, thickness appears in the numerator of λ_R but stays hidden in the denominator. This way, the radiation term (λ_R) shifts from 0 for a zero thickness to an asymptotic value ($\lambda_{R\infty}$) when infinitely thick:

$$\lambda_{R\infty} \approx \frac{4FC_b T_m^3 d_P}{100^4 \left[\frac{1+\rho-\tau}{1-\rho+\tau}\right]} \tag{1.14}$$

That asymptote increases for larger pores and pore walls transmitting more radiation, thus for materials with lower density. Or, radiant exchange prevents the thermal conductivity to continually drop with lower density. In fact, once below a certain value, further decrease demands larger pores or thinner pore walls. In both cases, radiation increases, turning the thermal conductivity into a sum of a monotonously decreasing conductive and an increasing radiant part. Or, surely for thermally insulating materials with given thickness, the λ-value must at some density pass a minimum (Figure 1.4):

$$\lambda = b_1 + b_2\rho + b_3/\rho \tag{1.15}$$

In a moist material, heat conduction in the adsorbed water layers and condensed water islands adds. For open porous materials, the result is a linear relation between thermal conductivity and moisture ratio, be it that for very porous insulating materials, the relation, now with the volumetric moisture ratio, is rather parabolic:

$$\lambda = \lambda_d(1 + a_X X) \quad \lambda = \lambda_d(1 + a_\Psi \Psi + b_\Psi \Psi^2) \tag{1.16}$$

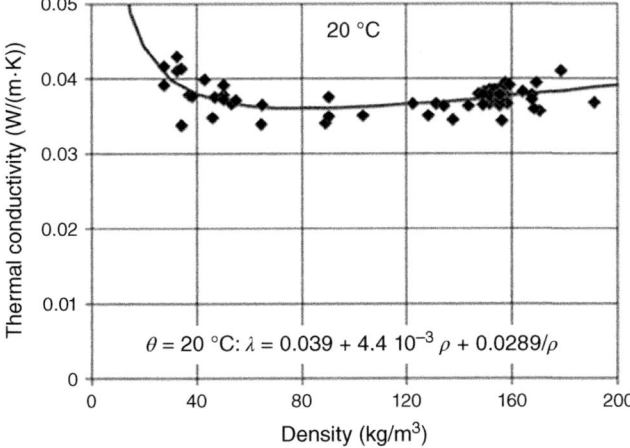

Figure 1.4 Mineral wool, thermal conductivity versus density.

In both, λ_d is the thermal conductivity when dry. Of course, porous materials will also see latent heat exchanged, adding as heat flux:

$$q_v = l_b g_v = -\frac{l_b}{\mu N} \frac{dp_{sat}}{d\theta} \text{grad}\theta \tag{1.17}$$

with l_b the heat of evaporation. As a result, the thermal conductivity of a moist material becomes:

$$\lambda = \lambda_d(1 + a_X X)\left(1 + \frac{l_b}{\mu N \lambda_d(1 + a_X X)} \frac{dp_{sat}}{d\theta}\right) \tag{1.18}$$

Due to the latent heat exchange, temperature affects the thermal conductivity of wet materials more than radiation does. The impact quickly increases with a lower vapour resistance factor of the matrix. Fibrous insulation materials such as mineral wool suffer the most. Evaporation at the warm side is giving a sudden boost to their thermal conductivity (Figure 1.5).

For simplicity reasons, in what follows, except if otherwise said, the 'apparent' thermal conductivity is considered constant.

1.2.4 Steady-State

1.2.4.1 What?

In steady state, the temperatures, the heat fluxes, the material properties and energy dissipation, if any, remain time independent. If so, the derivative of the temperature over time turns zero:

$$\partial T / \partial t = 0$$

With the temperature in °C, Fourier's second law then simplifies to:

$$\nabla^2 \theta = -\Phi'/\lambda \tag{1.19}$$

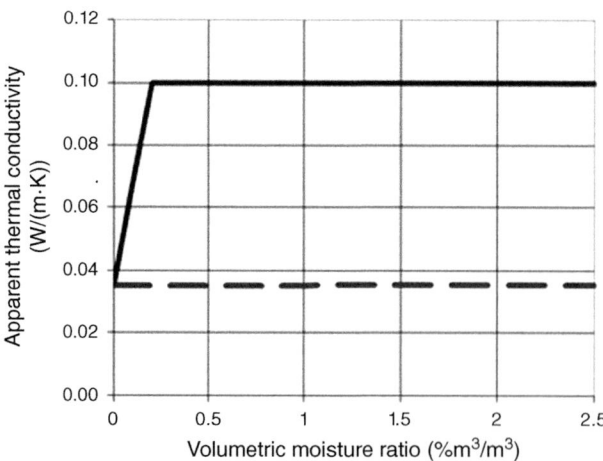

Figure 1.5 Mineral wool, thermal conductivity versus volumetric moisture ratio, a full line for the warm, a dotted for the cold side initially moist.

1.2.4.2 One Dimension, Flat Assemblies

1.2.4.2.1 *In General*

With the temperature change perpendicular to both end faces of a flat assembly, the second law further reduces to:

$$\frac{d^2\theta}{dx^2} = -\frac{\Phi'}{\lambda} \tag{1.20}$$

Without dissipation, this becomes:

$$d^2\theta/dx^2 = 0$$

with as solution: $\theta = C_1 x + C_2$ where C_1 and C_2 are the integration constants, fixed by the boundary conditions. This really simple equation governs heat conduction across flat assemblies with both sides at constant but different temperature. Most buildings are composed of flat assemblies, take low-slope roofs, sloped roof pitches, outer walls, floors, party walls, partitions, glass surfaces, etc. Their cross-section is either single- or multi-layered.

1.2.4.2.2 *Single-Layered Walls*

Supposedly known are the thermal conductivity and the thickness (d) of the material used. The boundary conditions are $x = 0: \theta = \theta_{s1}; x = d: \theta = \theta_{s2}$ with θ_{s1} and θ_{s2} the surface temperatures of both end faces. If end face s_1 is colder than end face s_2, the integration constants become:

$$C_2 = \theta_{s1} \quad C_1 = (\theta_{s2} - \theta_{s1})/d$$

This gives as temperature course through the assembly:

$$\theta = \frac{\theta_{s2} - \theta_{s1}}{d} x + \theta_{s1} \tag{1.21}$$

Or, in steady state and without heat source or sink, the temperatures in a single-layer assembly vary linearly from the warmer to the colder end face (Figure 1.6).

The heat flux becomes:

$$\mathbf{q} = -\lambda\, \mathbf{grad}\, \theta = -\lambda \frac{d\theta}{dx} = -\lambda \frac{\theta_{s1} - \theta_{s2}}{d} \tag{1.22}$$

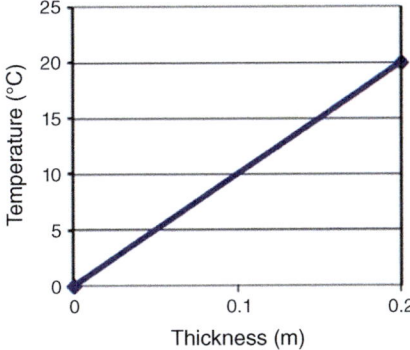

Figure 1.6 Single-layer assembly, temperatures.

Or, in absolute terms:

$$q = \lambda(\theta_{s2} - \theta_{s1})/d \tag{1.23}$$

The constant flux so develops proportionally to the thermal conductivity of the material and the difference in surface temperature of both end faces but inversely proportional to the assembly's thickness. With the difference and thickness given, a lower thermal conductivity reduces the heat flux, meaning less heat lost or gained. Materials with very low thermal conductivity are therefore named '(thermally) insulating'. Reshuffling the heat flux equation gives:

$$q = \frac{\Delta\theta}{d/\lambda} \tag{1.24}$$

with d/λ the thermal resistance of such flat, single-layer assembly, symbol R, units m²·K/W.

The higher R, the lower the heat flux for a given difference in surface temperatures or, the better insulating the assembly. That higher requires either a thicker assembly or the use of a better insulating material with same thickness. The inverse of the thermal resistance is called the thermal conductance, symbol P, units W/(m²·K), a quantity telling how much heat per unit of time and surface passes through an assembly at 1 °C difference between both end faces.

1.2.4.2.3 Multi-Layered Assemblies

Multi-layer assemblies, as most floors, walls and roofs in buildings are, consist of two or more plan-parallel material layers. An example is the cavity wall of Figure 1.7 with concrete block inside leaf, cavity fill and concrete block veneer.

In steady state and without local heat sources and sinks, the heat flux through must be the same in all layers. With the temperature θ_{s1} on end face 1 higher than θ_{s2} on end face 2, the thermal conductivities and thicknesses of all layers known and the contact resistances in between negligible, the constant heat flux **q** must equal:

Layer 1 $\quad q = \lambda_1 \dfrac{\theta_1 - \theta_{s1}}{d_1}$

Figure 1.7 Filled cavity wall.

Layer 2 $q = \lambda_2 \dfrac{\theta_2 - \theta_1}{d_2}$

Layer $n-1$ $q = \lambda_{n-1} \dfrac{\theta_{n-1} - \theta_{n-2}}{d_{n-1}}$

Layer n $q = \lambda_n \dfrac{\theta_{s2} - \theta_{n-1}}{d_n}$

with $\theta_1, \theta_2, \ldots, \theta_{n-1}$ the unknown interface temperatures. Rearrangement and summing give:

$$q \dfrac{d_1}{\lambda_1} = \theta_1 - \theta_{s1}$$

$$+ q \dfrac{d_2}{\lambda_2} = \theta_2 - \theta_1$$

$$+ \cdots$$

$$\dfrac{+ q \dfrac{d_n}{\lambda_n} = \theta_{s2} - \theta_{n-1}}{q \sum_{i=1}^{n} \left(\dfrac{d_i}{\lambda_i} \right) = \theta_{s2} - \theta_{s1}} \qquad (1.25)$$

or:

$$q = \dfrac{\theta_{s2} - \theta_{s1}}{\sum_{i=1}^{n}(d_i/\lambda_i)}$$

$\sum(d_i/\lambda_i)$, symbol R_T, units m²K/W, is called the total thermal resistance of the multi-layer assembly, while the ratios d_i/λ_i, symbol R_i, same units, represent the thermal resistances of the separate layers. The higher the total thermal resistance, the lower the steady state heat flux and the better insulating the assembly. A high total thermal resistance R_T requires the inclusion of a sufficiently thick insulation layer. After the energy crises of 1973 and 1979, lowering the net heating demand by insulating the multi-layer assemblies, building envelopes are composed of, finally became a legal requirement. Since the 2000s, limiting the CO_2 release when heating with fossil fuels, more, moving to carbon neutrality added extra pressure to excellently insulate the envelope assemblies in new construction and, if doable, in renovation.

The layers that make up an assembly fix its total thermal resistance. Being the sum of the thermal resistances of all, the commutation rule applies, meaning layer sequence has no impact on that value. So, inside insulation should not differ from outside insulation. Of course, from an overall building physics point of view, this is untrue: a same total thermal resistance yes, but a diverging overall heat, air and moisture performance.

With the surface temperatures on both end faces known, the temperatures in all consecutive interfaces follow from reshuffling layer per layer the heat flux equation:

$$\theta_1 = \theta_{s1} + q \dfrac{d_1}{\lambda_1} = \theta_{s1} + (\theta_{s2} - \theta_{s1}) \dfrac{R_1}{R_T}$$

$$\theta_2 = \theta_1 + q\frac{d_2}{\lambda_2} = \theta_1 + qR_2 = \theta_{s1} + (\theta_{s2} - \theta_{s1})\frac{(R_1 + R_2)}{R_T}$$

$$\theta_{n-1} = \theta_{n-2} + q\frac{d_{n-1}}{\lambda_{n-1}} = \theta_{s1} + (\theta_{s2} - \theta_{s1})\frac{\sum_{i=1}^{n-1} R_i}{R_T}$$

Writing $\theta_i = \theta_{i-1} + qd_i/\lambda_i$ as $(\theta_i - \theta_{i-1})/d_i = q/\lambda_i$ underlines that the temperature gradient in a layer is inversely proportional to its thermal conductivity. Hence, gradients are large in insulating layers and small in conductive layers. Each layer equation can be rewritten now as:

$$\theta_x = \theta_{s1} + qR_{s1}^x \quad (1.26)$$

with R_{s1}^x the thermal resistance between end face s1 and the considered interface x. If calculating should start at end face s2, that equation becomes:

$$\theta_x = \theta_{s2} - qR_{s2}^x \quad (1.27)$$

In an axis system with the thermal resistance R as apsis and the temperature θ as ordinate, both equations give a straight line between the temperatures on both end faces $((0, \theta_{s1})$ and $(R, \theta_{s2}))$ having the heat flux as slope. Or, a multi-layer assembly looks in this axis system as if single-layered. To construct the temperature line, the assembly is therefore redrawn with each layer as thick as its thermal resistance. Then, the surface temperature θ_{s1} on end face s1 and the surface temperature θ_{s2} on end face s2 are marked and the line linking both traced. The temperature versus thickness curve then follows from transposing the intersections between that temperature line and the successive interfaces to the same interface on the thickness graph and linking the consecutive temperatures with line segments, see Figure 1.8.

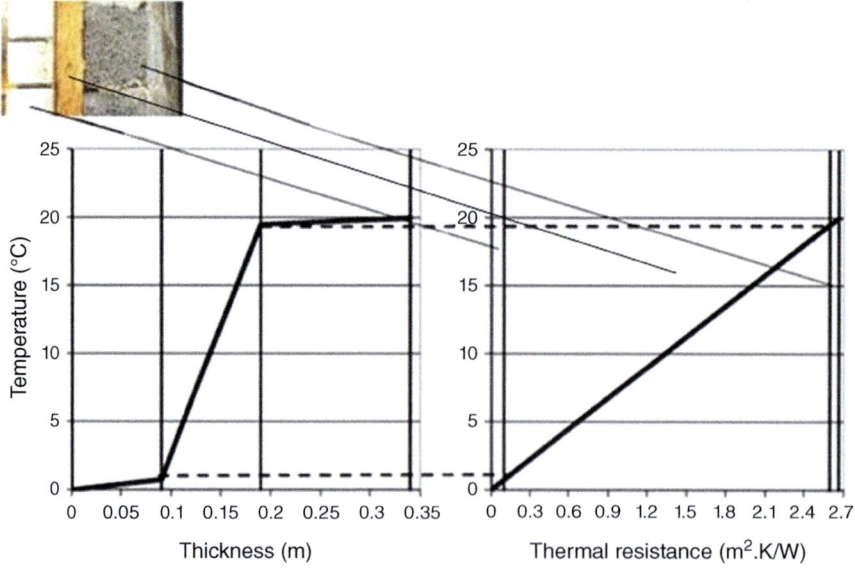

Figure 1.8 Temperatures in a composite assembly, graphical construction.

For single- and multi-layer assemblies, the heat flow through follows from the product of the heat flux with their surface (A) in m^2:

$$\Phi = qA \tag{1.28}$$

1.2.4.2.4 Electrical Analogy

The current passing through a series connection of electrical resistors (R_{ei}) subjected to a voltage differential ΔV is

$$i = \Delta V / \sum R_{ei} \tag{1.29}$$

Or; voltage so to say replaces the temperature, the current the heat flux and the resistors the thermal resistances. This allows converting heat conduction problems into an electrical analogy.

1.2.4.2.5 Special Cases

The first concerns a single-layer assembly, of which the thermal conductivity along its thickness (x) changes with the local temperature or with moisture content. Should its value depend linearly on temperature ($\lambda = \lambda_o + a\theta$), then the heat flux through becomes ($x = 0, \theta = \theta_{s1}; x = d, \theta = \theta_{s2}$):

$$\int_{\theta_{s1}}^{\theta_{s2}} (\lambda_o + a\theta) \, d\theta = \int_0^d q \, dx$$

This gives as solution:

$$\lambda_o(\theta_{s2} - \theta_{s1}) + \frac{a\left(\theta_{s2}^2 - \theta_{s1}^2\right)}{2} = qd \quad \text{or} \quad q = \lambda(\theta_m)\frac{\theta_{s2} - \theta_{s1}}{d} \tag{1.30}$$

with $\lambda(\theta_m)$ the thermal conductivity at the single-layer's average temperature. The thermal resistance so is $d/\lambda(\theta_m)$, while the temperature course turns into a parabolic relation:

$$a\theta^2/2 + \lambda_o\theta = qx + C$$

For an assembly with locally varying moisture content $w(x)$ (Figure 1.9) making the thermal conductivity a function of $w(x)$, thus of x, the heat flux through is ($x = 0: \theta = \theta_{s1}; x = d: \theta = \theta_{s2}$):

$$q \int_0^d \frac{dx}{\lambda(x)} = \int_{\theta_{s1}}^{\theta_{s2}} d\theta$$

Is the moisture content distributed in a way the thermal conductivity increases linearly with x ($\lambda = \lambda_o + ax$), solving the integrals then gives:

$$q = \frac{\theta_{s2} - \theta_{s1}}{\frac{1}{a} \ln\left(\frac{\lambda_o + a\,d}{\lambda_o}\right)} \tag{1.31}$$

Again, the denominator stands for the thermal resistance. The temperature curve becomes:

$$\theta_x = \theta_{s1} + q\left[\frac{1}{a} \ln\left(\frac{\lambda_o + ax}{\lambda_o}\right)\right] \tag{1.32}$$

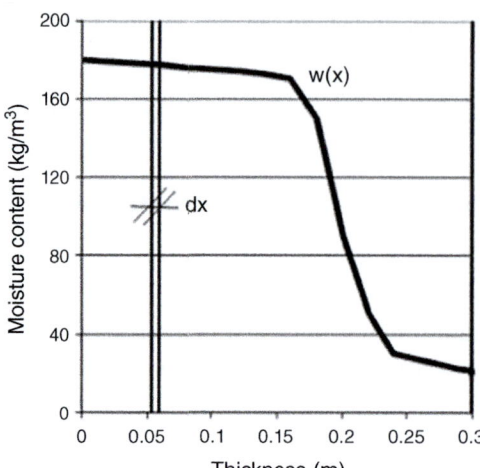

Figure 1.9 Moisture profile turning thermal conductivity into a function of the ordinate x, representing the thickness.

The second applies to a single-layer assembly dissipating or absorbing Φ' joule of heat per unit of time and volume. If spread uniformly over its thickness, the steady state balance becomes:

$$\frac{d^2\theta}{dx^2} = -\frac{\Phi'}{\lambda}$$

With as boundary conditions $x = 0$: $\theta = \theta_{s1}$; $x = d$: $\theta = \theta_{s2}$ ($\theta_{s1} < \theta_{s2}$), the solution so is:

$$\theta = -\frac{\Phi'}{2\lambda}x^2 + C_1 x + C_2 \tag{1.33}$$

This parabolic temperature curve turns convex when heat is dissipated and concave when heat is absorbed by the assembly (Figure 1.10).

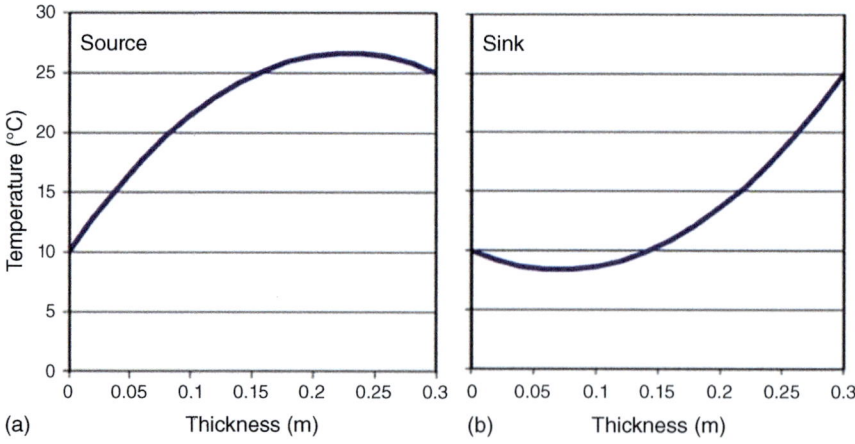

Figure 1.10 A single-layered assembly acting as uniformly distributed heat source (a) or sink (b), temperature curves.

Related heat flux becomes:

$$q = -\lambda \frac{d\theta}{dx} = \Phi' x - C_1 \lambda \qquad (1.34)$$

Opposite to no heat dissipated, the flux yet changes along the assembly's thickness. The boundary conditions give as integration constant:

$$C_1 = \frac{\theta_{s2} - \theta_{s1}}{d} + \frac{\Phi d'}{2\lambda} \quad C_2 = \theta_{s1}$$

The third concerns a multi-layer assembly with locally a heat source or sink, for example due to condensation in or drying from an interface x. The heat dissipated or absorbed per s is q', while the temperature θ_{s1} on one end face is higher than the temperature θ_{s2} on the other. A steady state heat balance for interface x then allows calculating the overall temperature curve. Assuming that heat flows from both end faces to interface x, related fluxes become:

From end face s1 to x: $q_{s1}^x = \dfrac{\theta_{s1} - \theta_x}{R_{s1}^x}$

From end face s2 to x: $q_{s2}^x = \dfrac{\theta_{s2} - \theta_x}{R_{s2}^x}$

In both equations, θ_x is the unknown temperature in x. Setting the difference between the two equals to the heat released (+) or absorbed (−) per unit of time gives for θ_x:

$$\theta_x = \frac{R_{s2}^x \theta_{s1} + R_{s1}^x \theta_{s2} + q' R_{s1}^x R_{s2}^x}{R_{s1}^x + R_{s2}^x} \qquad (1.35)$$

Including this result in both flux equations gives:

$$q_{s1}^x = \left(\frac{\theta_{s1} - \theta_{s2}}{R_T}\right) - \frac{q' R_{s2}^x}{R_T} \quad q_{s2}^x = -\left[\left(\frac{\theta_{s1} - \theta_{s2}}{R_T}\right) + \frac{q' R_{s1}^x}{R_T}\right] \qquad (1.36)$$

with R_T the total thermal resistance of the assembly: $R_T = R_{s1}^x + R_{s2}^x$. A heat source so gives a lower in- and higher outgoing flux, a sink the inverse. With the thermal resistance as apsis, the temperature curve consists of two straight lines with different slope, linking the temperatures on both end faces to the one in interface x (Figure 1.11).

1.2.4.3 Two Dimensions, Cylinder Symmetric

In cylindric coordinates, pipes behave as if one-dimensional. Considered is a hanged heating pipe with inside radius r_1 and outside radius r_2. Of interest is the heat loss per running metre the pipe's temperature and the insulation efficiency. The inner face has as temperature θ_{s1}, and the outer as temperature θ_{s2} (Figure 1.12).

In steady state and without dissipation other than from the liquid or gas transported, the heat flow passing each concentric cylinder in the pipe must be identical. With the centre as origin, the flow per running metre so must equal:

$$\Phi = -\lambda (2\pi r) d\theta / dr = C^t$$

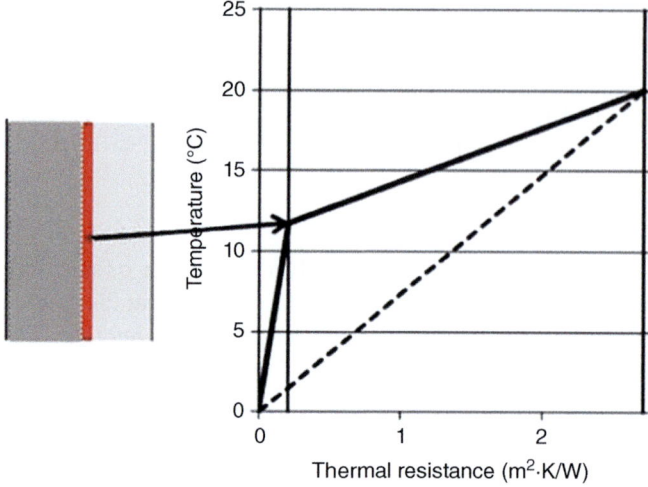

Figure 1.11 Two-layer assembly, heat source in the interface between the two.

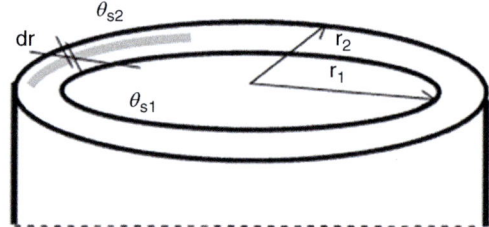

Figure 1.12 The pipe problem.

Integration gives:

$$\Phi \int_{r_1}^{r_2} \frac{dr}{r} = -2\pi\lambda \int_{\theta_{s1}}^{\theta_{s2}} d\theta \quad \text{or} \quad \Phi = \frac{\theta_{s1} - \theta_{s2}}{\left(\dfrac{\ln(r_2/r_1)}{2\pi\lambda}\right)} \tag{1.37}$$

The denominator, called the thermal resistance of the pipe per running metre units m·K/W, can be seen as the equivalent of the thermal resistance of a flat assembly. For a composite pipe, a same reasoning as for a multi-layer flat assembly gives as heat flow per running metre:

$$\Phi = \frac{\theta_{s1} - \theta_{s2}}{\sum_{i=1}^{n}\left[\dfrac{\ln(r_{i+1}/r_i)}{2\pi\lambda_i}\right]} \tag{1.38}$$

The temperatures follow from:

$$\theta_{i+1} = \theta_{s1} + \Phi \sum_{i=1}^{i}\left[\dfrac{\ln(r_{i+1}/r_i)}{2\pi\lambda_i}\right] \tag{1.39}$$

1.2.4.4 Two and Three Dimensions: Thermal Bridges

When looking in detail to outer walls, roofs, floors and partitions, the assumption 'flat' does not apply everywhere. What about lintels above windows? What about window reveals? What about junctions between two outer walls? What about corners between two outer walls and a low-slope roof (Figure 1.13)? All act as thermal bridges. Even both end faces of a flat assembly could be non-isothermal, turning its thermal picture from one to two or three dimensionally.

Studying the steady state heat transfer through, now requires a return to the second law:

$$\frac{\partial^2 \theta}{\partial x^2} + \frac{\partial^2 \theta}{\partial y^2} + \frac{\partial^2 \theta}{\partial z^2} = \pm \frac{\Phi'}{\lambda}$$

or, without heat dissipation:

$$\frac{\partial^2 \theta}{\partial x^2} + \frac{\partial^2 \theta}{\partial y^2} + \frac{\partial^2 \theta}{\partial z^2} = 0$$

For some very simple combinations - one material, simple geometry, simple boundary conditions - this partial differential equation can be solved analytically. However, in the majority of building-related cases, the use of control volumes (CVM) is preferred. Many building details consist of rectangular material volumes, which are easily meshed in cube or beam-like control volumes. In steady state and if three-dimensional, the heat flow from the six adjacent volumes to the central must be 0 then (Figure 1.14). If meshing gives control volumes whose sides coincide with the interfaces between layers, all are material-homogenous, but then the calculations will not give the interface temperatures. Preference therefore goes to control volumes with the centre on interfaces between materials. Related lack of homogeneity is compensated by the fact the calculations give the interface temperatures then.

In case all control volumes are cubic, then the distances to consider along the x-, y- and z-axis around each equal the one between its centre and that of the six adjacent control volumes. If additionally, all lay in the same material, then, with the named distance called the mesh width a, the heat flow from the adjacent control volume $(l-1, m, n)$ to the central (l, m, n) equals:

$$\Phi_{l-1,m,n}^{l,m,n} = \lambda \frac{(\theta_{l-1,m,n} - \theta_{l,m,n}) a^2}{a} = a\lambda(\theta_{l-1,m,n} - \theta_{l,m,n})$$

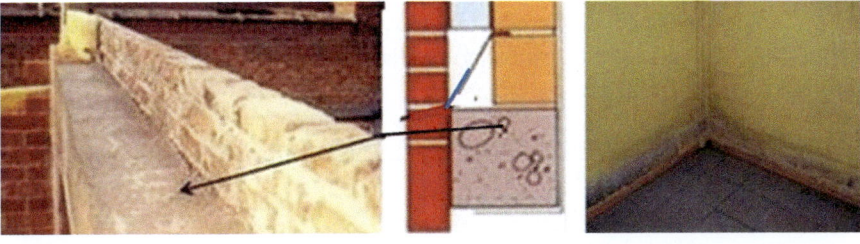

Figure 1.13 Lintel above a window and corner between two outside walls and the floor.

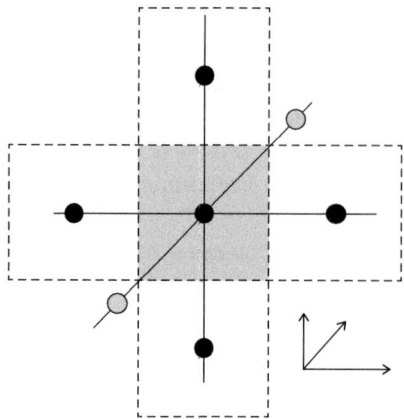

Figure 1.14 Central and adjacent control volumes.

For the other five, the flows become:

$$\Phi^{l,m,n}_{l+1,m,n} = \lambda \frac{(\theta_{l+1,m,n} - \theta_{l,m,n})\, a^2}{a} = a\lambda(\theta_{l+1,m,n} - \theta_{l,m,n})$$

$$\Phi^{l,m,n}_{l,m-1,n} = \lambda \frac{(\theta_{l,m-1,n} - \theta_{l,m,n})\, a^2}{a} = a\lambda(\theta_{l,m-1,n} - \theta_{l,m,n})$$

$$\Phi^{l,m,n}_{l,m+1,n} = \lambda \frac{(\theta_{l,m+1,n} - \theta_{l,m,n})\, a^2}{a} = a\lambda(\theta_{l,m+1,n} - \theta_{l,m,n})$$

$$\Phi^{l,m,n}_{l,m,n-1} = \lambda \frac{(\theta_{l,m,n-1} - \theta_{l,m,n})\, a^2}{a} = a\lambda(\theta_{l,m,n-1} - \theta_{l,m,n})$$

$$\Phi^{l,m,n}_{l,m,n+1} = \lambda \frac{(\theta_{l,m,n+1} - \theta_{l,m,n})\, a^2}{a} = a\lambda(\theta_{l,m,n+1} - \theta_{l,m,n})$$

Adding the six and sum zero gives:

$$\theta_{l-1,m,n} + \theta_{l+1,m,n} + \theta_{l,m-1,n} + \theta_{l,m+1,n} + \theta_{l,m,n-1} + \theta_{l,m,n+1} - 6\,\theta_{l,m,n} = 0$$

Unknown in this linear equation are the temperatures in the central and six adjacent control volumes. In two dimensions, each control volume has four adjacent, giving as sum:

$$\theta_{l-1,m} + \theta_{l+1,m} + \theta_{l,m-1} + \theta_{l,m+1} - 4\,\theta_{l,m} = 0$$

When the central control volume bridges an interface between two materials with thermal conductivity λ_1 and λ_2 parallel to the [x, y] plane, then, for a mesh width 'a' along the three axes, the heat flow from an adjacent control volume $(l-1, m, n)$ to the central (l, m, n) changes to:

$$\Phi^{l,m,n}_{l-1,m,n} = \lambda_1 \frac{(\theta_{l-1,m,n} - \theta_{l,m,n})\, a^2}{2a} + \lambda_2 \frac{(\theta_{l-1,m,n} - \theta_{l,m,n})\, a^2}{2a}$$

or:

$$\Phi^{l,m,n}_{l-1,m,n} = \frac{a(\lambda_1 + \lambda_2)(\theta_{l-1,m,n} - \theta_{l,m,n})}{2}$$

For the other five, that flow becomes:

$$\Phi_{l+1,m,n}^{l,m,n} = \frac{a(\lambda_1 + \lambda_2)(\theta_{l+1,m,n} - \theta_{l,m,n})}{2}$$

$$\Phi_{l,m-1,n}^{l,m,n} = \frac{a(\lambda_1 + \lambda_2)(\theta_{l,m-1,n} - \theta_{l,m,n})}{2}$$

$$\Phi_{l,m+1,n}^{l,m,n} = \frac{a(\lambda_1 + \lambda_2)(\theta_{l,m+1,n} - \theta_{l,m,n})}{2}$$

$$\Phi_{l,m,n-1}^{l,m,n} = a\lambda_1(\theta_{l,m,n-1} - \theta_{l,m,n})$$

$$\Phi_{l,m,n+1}^{l,m,n} = a\lambda_2(\theta_{l,m,n+1} - \theta_{l,m,n})$$

Adding and sum zero gives:

$$\frac{(\lambda_1 + \lambda_2)(\theta_{l-1,m,n} + \theta_{l+1,m,n} + \theta_{l,m-1,n} + \theta_{l,m+1,n})}{2}$$
$$+ \lambda_2 \theta_{l,m,n-1} + \lambda_1 \theta_{l,m,n+1} - 3(\lambda_1 + \lambda_2)\theta_{l,m,n} = 0$$

Again, the result is a linear equation with: the temperatures in the central and the six adjacent control volumes unknown. In two dimensions, the result is:

$$\frac{(\lambda_1 + \lambda_2)(\theta_{l-1,m} - \theta_{l+1,m})}{2} + \lambda_2 \theta_{l,m-1} + \lambda_1 \theta_{l,m+1} - 2(\lambda_1 + \lambda_2)\theta_{l,m,n} = 0$$

In case the central control volume lays on the intersection between three materials with thermal conductivity λ_1, λ_2 and λ_3 and interfaces parallel to the [x, y] and [y, z]-plane, the sum changes to:

$$(\lambda_2 + \lambda_3)\frac{\theta_{l-1,m,n}}{2} + (\lambda_2 + \lambda_3)\frac{\theta_{l+1,m,n}}{2} + \lambda_3 \theta_{l,m-1,n} + (\lambda_1 + \lambda_2)\frac{\theta_{l,m+1,n}}{2}$$
$$+ (\lambda_1 + \lambda_2 + \lambda_3)\frac{\theta_{l,m,n-1} - \theta_{l,m,n+1}}{4} - (3\lambda_1 + 3\lambda_2 + 6\lambda_3)\frac{\theta_{l,m,n}}{2} = 0$$

Again, a linear equation with: seven unknowns. Two dimensions give:

$$\lambda_3 \theta_{l-1,m} + (\lambda_1 + \lambda_2)\frac{\theta_{l+1,m}}{2} + (\lambda_2 + \lambda_3)\frac{\theta_{l,m-1}}{2}$$
$$+ (\lambda_1 + \lambda_3)\frac{\theta_{l,m+1}}{2} - (\lambda_1 + \lambda_2 + \lambda_3)\theta_{l,m} = 0$$

All other cases are solved the same way. In three dimensions, a control volume could contain up to eight materials, in two up to four. All generate a linear equation with seven or five unknowns. For p control volumes, the result is a system of p equations with p unknown temperatures, in which the known temperatures figure as boundary condition. Solving gives the temperature distribution in the building detail considered. Once known, the above equations allow calculating the heat fluxes along the three axes between the control volume centres.

Generalizing the algorithm begins with calling the surface-linked thermal conductance between adjacent control volumes P_s, units W/K. If both consist of the same material, then P_s becomes:

$$P_s = (\lambda/d)\,A$$

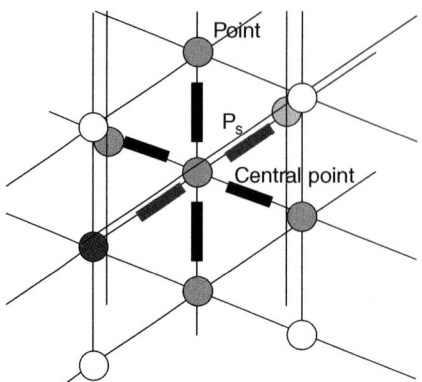

Figure 1.15 Summing up the six thermal conductance's between the central point considered and the six neighbour ones.

If the two belong to different materials, the resulting surface-linked conductance P_s turns into a serial, a parallel or a serial/parallel circuit of conductance, to be calculated in advance.

$$\Phi_{1-l,m,n}^{l,m,n} = P_{s1-l,m,n}^{l,m,n}(\theta_{1-l,m,n} - \theta_{l,m,n}) \tag{1.40}$$

Summing up the related heat flows gives per central point (Figure 1.15):

$$\sum_{\substack{i=l,m,n \\ j=\pm 1}} [P_{s,i+j}\theta_{i+j}] - \theta_{l,m,n} \sum_{\substack{i=l,m,n \\ j=\pm 1}} P_{s,i+j} = 0 \tag{1.41}$$

After transferring the known temperatures to the right, the system representing the two or three dimensional detail transposed into p control volumes converts to:

$$[P_s]_{p,p}[\theta]_p = [P_{s,i,j,k}\theta_{i,j,k}]_p$$

with $[P_s]_{p,p}$ the p rows, p columns matrix of the conductance's, $[\theta]_p$ the column matrix of the p unknown temperatures and $[P_{s,i,j,k}\theta_{i,j,k}]_p$ the column matrix of the known temperatures. Solving gives the temperature distribution in the detail, which then allows to calculate the heat flow passing through.

The accuracy of a CVM calculation depends on how fine is meshed. The finer, the more the solution will approximate reality. An infinitely fine mesh should produce the exact solution, but then, a system with an infinite number of equations has to be solved, which would last infinitely. Therefore, a compromise between accuracy and processing time includes making the mesh less fine where small and finer where large temperature gradients are expected. Today, powerful software packages to calculate the 2D or 3D heat transfer through building details are available. In former times, before computer software took over, electrical analogies were used.

1.2.5 Non-steady-state

1.2.5.1 In General

Opposite to what was assumed till now, due to varying boundary conditions, varying material properties and/or the presence of varying internal heat sources or sinks the temperatures in and heat fluxes through assemblies are mostly time dependent.

If nonetheless the material properties can be considered constant and no heat sources or sinks intervene, the second law becomes:

$$\nabla^2 \theta = \frac{\rho c}{\lambda} \frac{\partial \theta}{\partial t}$$

The temperatures and heat fluxes on both end faces of an assembly now may vary periodically or suddenly, periodically when linked to the air temperature outdoors, which fluctuates on a yearly, n days, daily, hourly or even a shorter time basis, suddenly when warming up a space (Figure 1.16).

1.2.5.2 Periodic Boundary Conditions, Flat Assemblies
1.2.5.2.1 Methodology

Setting $\lambda/(\rho c) = a$ reduces the second law to:

$$a \frac{\partial^2 \theta}{\partial x^2} = \frac{\partial \theta}{\partial t}$$

The ratio a, units m²/s, is called the thermal diffusivity of a material, a characteristic indicating how easily a local temperature change spreads in it. The higher a, the faster this happens.

A high thermal diffusivity requires a light material with high thermal conductivity or a heavy material with low volumetric heat capacity. None of the two exists. Light materials have a low thermal conductivity, and heavy materials have a high volumetric heat capacity. Very different materials can so have quite similar thermal diffusivities. An exception is the metals.

Substituting the thickness x in the equation by the layer's thermal resistance $R (= x/\lambda)$, which means multiplying the left and right with λ^2, gives as temperature and heat flux:

$$\frac{\partial^2 \theta}{\partial R^2} = \rho c \lambda \frac{\partial \theta}{\partial t} \qquad q = -\partial \theta / \partial R \tag{1.42}$$

Assumed now is that the temperatures on both end faces of a flat layer fluctuate with period T. Transposing these fluctuations into a Fourier series gives:

$$\theta_s = \frac{B_{s0}}{2} + \sum_{n=1}^{\infty} \left[A_{sn} \sin\left(\frac{2n\pi t}{T}\right) + B_{sn} \cos\left(\frac{2n\pi t}{T}\right) \right]$$

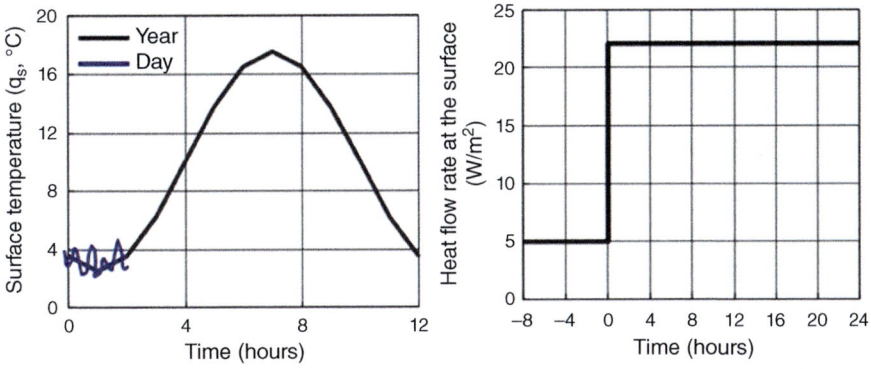

Figure 1.16 A periodic change and a sudden jump called a step change.

with:

$$A_{sn} = \frac{2}{T}\int_0^T \theta_s(t) \sin\left(\frac{2n\pi t}{T}\right) dt \qquad B_{sn} = \frac{2}{T}\int_0^T \theta_s(t) \cos\left(\frac{2n\pi t}{T}\right) dt$$

The value $B_{s0}/2$ represents the average temperature on the end face considered over the period T, while the values $A_{s1}, A_{s2}, \ldots, A_{sn}, B_{s1}, B_{s2}, \ldots, B_{sn}$ stay for the harmonics of the 1st, 2nd, ..., nth order. Rewriting the series in complex form using Euler's formulas for $\sin(x)$ and $\cos(x)$ gives:

$$\text{Euler}: \sin(x) = \frac{\exp(ix) - \exp(-ix)}{2i}$$

$$\cos(x) = \frac{\exp(ix) + \exp(-ix)}{2} \qquad x = 2n\pi t/T$$

$$\text{Result}: \theta_s(t) = \frac{1}{2}\sum_{n=-\infty}^{\infty}\left[\alpha_{sn} \exp\left(\frac{2in\pi t}{T}\right)\right]$$

$$= \frac{1}{2}\sum_{n=-\infty}^{\infty}\left[(A_{sn} + iB_{sn}) \exp\left(\frac{2in\pi t}{T}\right)\right] \tag{1.43}$$

α_{sn} being the nth complex temperature having as amplitude and phase shift:

Amplitude	Phase shift
$\sqrt{B_{sn}^2 + A_{sn}^2}$	$\operatorname{a\,tan}(-A_{sn}/B_{sn})$

The amplitude tells how large half the temperature variation is, and the phase shift how long the time delay is compared to a cosine with period $T/(2n\pi)$ (in radians). For the monthly mean air temperature outdoors at Uccle of Figure 1.17, the thick proxy temperature line has as equation:

$$\theta_e = 9.45 + 7.18 \cos\left(\frac{2\pi t}{365.25} - (-2.828)\right)$$

with t the time in days.

Included the leap years, a year counts 365.25 days. The annual average temperature (θ_{em}) so is 9.45 °C the annual amplitude (θ_{e1}) 7.18 °C, and the phase shift in radians −2.828. The time-dependent term resembles a rotating vector with value 7.18 starting −2.828 radians away from the real axis. Conversion into a Fourier series gives:

$$\theta_e = \frac{B_0}{2} + \theta_{e1}\left[A_1 \sin\left(\frac{2\pi t}{365.25}\right) + B_1 \cos\left(\frac{2\pi t}{365.25}\right)\right]$$

or, with $B_0 = 18.9$, $A_1 = \theta_{e1}\sin(\varphi_{e1}) = -2.21$ and $B_1 = \theta_{e1}\cos(\varphi_{e1}) = -6.83$:

$$\theta_e = 9.45 + -2.21 \sin\left(\frac{2\pi t}{365.25}\right) - 6.83 \cos\left(\frac{2\pi t}{365.25}\right)$$

Related complex value is $\alpha_{e1} = -6.83 - i2.21$, a number with amplitude 7.18 °C and phase shift −2.828 radians. As shown in Figure 1.17, the averages over the period 1991–2020 give different values, proving the impact of global warming. The annual average moved from 9.45 to 10.93 °C.

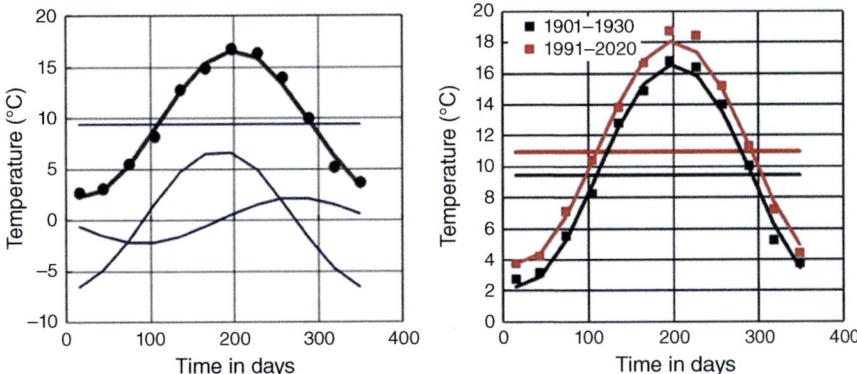

Figure 1.17 Uccle, monthly mean temperature, left: for the period 1901–1930. The black circles are the averages, the thick line gives the result of a Fourier analysis with one harmonic. The thin lines show the annual mean, the sine and the cosine term; right: result for the period 1991–2020. Global warming plays.

Taken is a material layer now, of which one end face endures a periodic temperature and heat flux change from time zero on. The response will be twofold, first a transient that slowly dies and secondly a lasting periodic change. As the layer can neither compress nor extend thermal signals, the periodic passing across must contain the same harmonics as the signal on the end face but the temperature and heat flux amplitudes will dampen and gradually run behind with as result an increasing phase shift during the traverse. Or, the temperature and heat flux in the layer will behave as being complex. The solution so must be a Fourier series with the thermal resistance as independent variable:

$$\theta(R,t) = \frac{1}{2}\sum_{n=-\infty}^{\infty}\left[\alpha_n(R)\exp\left(\frac{2in\pi t}{T}\right)\right]$$

$$q(t) = -\frac{d\theta(R,t)}{dR} = \frac{1}{2}\sum_{n=-\infty}^{\infty}\left[\alpha'_{sn}(R)\exp\left(\frac{2in\pi t}{T}\right)\right]$$

The accent on α'_{sn} reminds that, mathematically, the complex heat flux is the first derivative of the complex temperature to the thermal resistance, showing as amplitude and phase shift:

Amplitude	Phase shift
$\sqrt{B'^2_{sn} + A'^2_{sn}}$	$\mathrm{bgtg}\left(-A'_{sn}/B'_{sn}\right)$

1.2.5.2.2 Single-layered

Inserting the equation for the complex temperature into Fourier's second law gives:

$$\sum_{n=-\infty}^{\infty}\left\{\left[\frac{d^2\alpha_n(R)}{dR^2} - \frac{2\rho c\lambda in\pi}{T}\alpha_n(R)\right]\exp\left(\frac{2in\pi t}{T}\right)\right\} = 0 \qquad (1.44)$$

A sum zero requires all coefficients of the time exponentials being zero, or:

$$\frac{d^2 \alpha_n(R)}{dR^2} - \frac{2\rho c \lambda in\pi}{T} \alpha_n(R) = 0 \quad (1.45)$$

The outcome so is $2\infty + 1$ second-order differential equations with: the complex temperature as dependent variable and the integer n moving from $-\infty$ over 0 to $+\infty$. As the solutions for n positive and n negative mirror each other, only $\infty + 1$ equations must be solved. All have as solution:

$$\alpha_n(R) = C_1 \exp(\omega_n R) + C_2 \exp(-\omega_n R), \ 0 \leq n \leq \infty \text{ and } \omega_n^2 = \frac{2\rho c \lambda in\pi}{T} \quad (1.46)$$

The term ω is called the thermal pulsation. Or, Euler's formulas give as complex temperatures and complex heat fluxes in a flat single-layer assembly:

$$\alpha_n(R) = (C_1 - C_2)\sinh(\omega_n R) + (C_1 + C_2)\cosh(\omega_n R)$$

$$\alpha'_n(R) = \frac{d\alpha}{dR} = \omega_n[(C_1 - C_2)\cosh(\omega_n R) + (C_1 + C_2)\sinh(\omega_n R)] \quad (1.47)$$

The integration constants $(C_1 - C_2)$ and $(C_1 + C_2)$ follow from the boundary conditions. These state that the complex temperature $\alpha_{sn}(0)$ of and complex heat flux $\alpha'_{sn}(0)$ on end face $R = 0$ must equal:

$$\alpha_{sn}(0) = (C_1 - C_2) 0 + (C_1 + C_2) = C_1 + C_2$$

$$\alpha'_{sn}(0) = \omega_n[(C_1 - C_2) + (C_1 + C_2) 0] = \omega_n(C_1 - C_2) \quad (1.48)$$

This converts the temperature and heat flux equations into:

$$\alpha_n(R) = \alpha_{sn}(0)\cosh(\omega_n R) + \alpha'_{sn}(0)\frac{\sinh(\omega_n R)}{\omega_n}$$

$$\alpha'_n(R) = \alpha_{sn}(0) \omega_n \sinh(\omega_n R) + \alpha'_{sn}(0)\cosh(\omega_n R)$$

Two equations mean two unknowns, whose inclusion in their Fourier series gives the time dependency. Of interest is the link between the complex values on both end faces. Is for envelopes $R = 0$ the inner and $R = R_T$ the outer end face, for partitions R_T is the inner end face at the other side. Anyhow, for $R = R_T$ the system becomes:

$$[A_{sn}(R_T)] = [W_n][A_{sn}(0)] \quad (1.49)$$

with $[A_{sn}(0)]$ a column matrix of the unknown complex temperature and heat flux on the inner end face, $[A_{sn}(R_T)]$ a column matrix of the complex temperature and heat flux on the other end face and $[W_n]$ the system matrix of the nth harmonic, depending on the assembly's thickness, its material properties, the base period and the harmonic considered. $[W_n]$ so replaces the thermal resistance, although containing additional information.

$n = 0$ Then, $\alpha_{so}/2$ and $\alpha'_{so}/2$ represent the average temperature and heat flux on the inner end face, while the thermal pulsation then becomes:

$$\omega_0^2(n = 0) = 2i\rho c \lambda 0\pi/T = 0$$

This makes the complex surface temperature and heat flux on the other end face equal to:

$$\alpha_{so}(R_T) = \alpha_{so}(0) + \alpha'_{so}(0)\frac{0}{0} \quad \alpha'_{so}(R_T) = \alpha_{so}(0) 0 + \alpha'_{so}(0) = \alpha'_{so}(0)$$

Or, the in- and outgoing average heat fluxes then are the same. As this holds on each plan-parallel plane between 0 and R_T in the single-layer assembly, $n = 0$ means steady state. The ratio 0/0 in the temperature equation can be clarified using l' Hospital's rule:

$$\lim_{\omega_0 \to 0}[\sin h(\omega_o R_T)/\omega_o] = \lim_{\omega_0 \to 0}[R_T \cos h(\omega_o R_T)] = R_T$$

which gives:

$$\alpha_{so}(R_T) = \alpha_{so}(0) + \alpha'_{so}(0)\,R_T$$

a result, holding all over the assembly. The average temperature curve in the $[R,\theta]$-axis system so is a straight line with the heat flux as slope. This extends the concept steady state to the average over a long enough period of time.

$n > 0$ A first case is the temperature on the inner end face remaining constant. The complex temperature on the other end face then becomes:

$$\alpha_{sn}(R_T) = \alpha'_{sn}(0)\frac{\sin h(\omega_n R_T)}{\omega_n} \quad \text{or} \quad \frac{\alpha_{sn}(R_T)}{\alpha'_{sn}(0)} = \frac{\sin h(\omega_n R_T)}{\omega_n} \qquad (1.50)$$

The function on the right now stands for the ratio between the complex surface temperature on the other and the complex heat flux on the inner face. In steady state, this ratio is the thermal resistance. Therefore, that function is called the dynamic thermal resistance for the nth harmonic, symbol D_q^n, units m²·K/W, with as amplitude and time shift (φ_q^n):

$$[D_q^n] = [\sin h(\omega_n R_T)/\omega_n] \quad \phi_\theta^n = \arg[\sin h(\omega_n R_T)/\omega_n]$$

The assumption made looks theory. However, the case refers to spaces at constant temperature. For the period infinitely long, the amplitude and phase shift of the dynamic thermal resistance become:

$$[D_q^n] = \lim_{n \to 0}\left[\frac{\sin h(0)}{0}\right] = \lim_{n \to 0}[R_T \cos h(0)] = R_T \quad \phi_q^n = \lim_{n \to 0}\left[\arg\frac{\sin h(0)}{0}\right] = \infty$$

Or, its value equals the steady state one then. Logic as infinitely long means steady state. For a really fast oscillation with period ≈ 0 s, the amplitude and phase shift of the dynamic thermal resistance change to:

$$[D_q^n] = \lim_{n \to \infty}\left[\frac{\sin h(\infty)}{\infty}\right] = \lim_{n \to \infty}[R_T \cos h(\infty)] = \infty \quad \phi_q^n = \lim_{n \to \infty}\left[\arg\frac{\sin h(\infty)}{\infty}\right] = 0$$

Or, ever faster oscillations push its value to infinite, meaning the assembly dampens the signal completely. The dynamic thermal resistance so is always larger than the steady state one, meaning that imposing high thermal resistances suffices to get a high dynamic value.

A second case is the heat flux on the inner end face constant. As no complex heat fluxes exist there then $(a'_{sn}(0) = 0)$, the complex temperature on the other end face becomes:

$$\alpha_{sn}(R_T) = \alpha_{sn}(0)\cos h(\omega_n R_T) \quad \text{or}: \quad \frac{\alpha_{sn} R_T}{\alpha_{sn}(0)} = \cos h(\omega_n R_T) \qquad (1.51)$$

The function on the right so stands for the ratio between the complex temperatures on the other and the inner end face. It is called the temperature damping for the nth harmonic, symbol D_θ^n with as amplitude and time shift (φ_θ^n):

$$[D_\theta^n] = [\cos h(\omega_n R_T)] \quad \varphi_\theta^n = \arg[\cos h(\omega_n R_T)]$$

At first glance, that quantity looks fictional, the more because there's no steady state equivalent. Anyhow, it shows the ability of an assembly to moderate the impact at the inner end face of a temperature change at the other. In cases with few heat gains indoors and a restricted ventilation, an enclosure with high temperature damping might so keep the indoor temperature quite stable. This gives it a performance status. For the period infinitely long, the amplitude and phase of the temperature damping become:

$$[D_\theta^n] = \lim_{n \to 0}[\cos h(0)] = 1 \quad \varphi_\theta^n = \lim_{n \to 0}[\arg[\cos h(0)]] = \infty$$

Or, it then nears 1. In fact, in steady state, without heat gains and losses, the temperatures at both end faces must remain the same. For a period nearing 0 s, the amplitude and phase of the temperature damping become:

$$[D_\theta^n] = \lim_{n \to \infty}[\cos h(\infty)] = \infty \quad \varphi_\theta^n = \lim_{n \to \infty}[\arg[\cos h(\infty)]] = 0$$

Or, ever faster fluctuations push its value to infinite.

A third case is the temperature on the outer or the end face at the other side constant. The complex temperature of the inner end face then converts to:

$$0 = D_\theta^n \, \alpha_{sn}(0) + D_q^n \, \alpha'_{sn}(0) \quad \text{or}: \quad \frac{\alpha'_{sn}(0)}{\alpha_{sn}(0)} = -\frac{D_\theta^n}{D_q^n} = -\omega_n \cot gh(\omega_n R) \quad (1.52)$$

The function on the right, equal to the ratio between the complex heat flux and temperature on the inner end face, thus to the ratio between the temperature damping (D_θ^n) and the dynamic thermal resistance (D_q^n), is called the admittance for the nth harmonic, symbol Ad^n, units W/(m²·K), with as amplitude and time shift (φ_{Ad}^n):

$$[Ad^n] = [-\omega_n \cot gh(\omega_n R_T)] \quad \varphi_{Ad}^n = \varphi_\theta^n - \varphi_q^n$$

Again, the quantity looks fictional. However, it indicates how easily single-layer assemblies pick up heat when the temperature on the inner end face fluctuates. The higher its amplitude, the more effective heat is stored. Or, the assessment 'capacitive' points to envelope assemblies or partitions having a high admittance. The thermal pulsation ω_n now also writes as:

$$\omega_n = \sqrt{i} \, \sqrt{\rho c \lambda} \, \sqrt{2 n \pi / T}$$

This shows that a high admittance requires a high value of the square root of the product of the volumetric heat capacity with the thermal conductivity, a quantity called the contact coefficient or effusivity of the material, symbol b, units J/(m²·s$^{-1/2}$·K). The higher its effusivity, take stony materials compared to timber, the more active the material as heat storage medium. For a period infinitely long, the amplitude and phase of the admittance become:

$$[Ad^n] = \lim_{n \to 0}\left(\frac{D_\theta^n}{D_q^n}\right) = \frac{1}{R_T} \quad \varphi_{As}^n = \lim_{n \to 0}\left[\arg\left(-\frac{D_\theta^n}{D_q^n}\right)\right] = \infty$$

Or, ever slower fluctuations push the admittance to the thermal conductance (P). Logic, as in steady state, the heat flux entering or leaving an assembly equals $P\Delta\theta_s$. For a period nearing $0\,s$, the amplitude and phase of the admittance touch:

$$[Ad^n] = \lim_{n\to\infty}\left(\frac{D^n_\theta}{D^n_q}\right) = \infty \qquad \phi^n_{As} = \lim_{n\to\infty}\left[\arg\left(\frac{D^n_\theta}{D^n_q}\right)\right] = 0$$

Or, ever faster fluctuations push the admittance to infinite.

With the dynamic thermal resistance and the temperature damping known, the system matrix $[W_n]$, see above, can be rewritten as:

$$[W_n] = \begin{bmatrix} D^n_\theta & D^n_q \\ \omega^2_n D^n_q & D^n_\theta \end{bmatrix}$$

A transposition to real numbers first requires reformulating the thermal pulsation:

$$\omega_n = \sqrt{i}\,b\sqrt{\frac{2n\pi}{T}} = \frac{(1+i)\,b\sqrt{\frac{2n\pi}{T}}}{\sqrt{2}} = (1+i)\,b\sqrt{\frac{n\pi}{T}}$$

The conversion of $\sqrt{i}$ is based on $(1+i)^2 = 2i$ or $\sqrt{i} = (1+i)/\sqrt{2}$. The product $\omega_n R$ so becomes:

$$\omega_n R = \frac{(1+i)\,b\cdot\sqrt{\frac{n\pi}{T}}}{\lambda} = (1+i)\cdot\sqrt{\frac{n\pi}{aT}}$$

with x stretching along the single-layer assembly's thickness and $a = \lambda^2/b^2$ its thermal diffusivity. Setting $X_n = x\sqrt{n\pi/aT}$ allows rewriting the thermal pulsation as $(1+i)X_n/R$, giving for the temperature damping D^n_θ:

$$D^n_\theta = \cosh(\omega_n R) = \cosh[(1+i)X_n] = \cosh(X_n)\cosh(iX_n) + \sinh(X_n)\sinh(iX_n)$$

or, with $\cosh(iX_n) = \cos(X_n)$ and $\sinh(iX_n) = i\sin(X_n)$:

$$\cosh(\omega_n R) = \cosh(X_n)\cos(X_n) + i\,\sinh(X_n)\sin(X_n)$$

Analogously, the dynamic thermal resistance D^n_q and admittance Ad^n of the single-layer assembly turn into:

$$D^n_q = \frac{\sinh(\omega_n R)}{\omega_n} = \frac{R}{2X_n}\left\{\begin{array}{l}[\sinh(X_n)\cos(X_n) + \cosh(X_n)\sin(X_n)] \\ +i\,[\cosh(X_n)\sin(X_n) - \sinh(X_n)\cos(X_n)]\end{array}\right\}$$

$$Ad^n = \omega_n \sinh(\omega_n R) = \frac{X_n}{R}\left\{\begin{array}{l}[\sinh(X_n)\cos(X_n) - \cosh(X_n)\sin(X_n)] \\ +i\,[\cosh(X_n)\sin(X_n) + \sinh(X_n)\cos(X_n)]\end{array}\right\}$$

The temperature damping, the dynamic thermal resistance and the term $\omega^2_n D^n_q$ in the system matrix can be rewritten as:

$$D^n_\theta = \cosh(\omega_n R) = G_{n1} + iG_{n2}$$

$$D^n_q = \frac{\sinh(\omega_n R)}{\omega_n} = R\,(G_{n3} + iG_{n4})$$

$$\omega^2_n D^n_q = \omega_n \sinh(\omega_n R) = \frac{G_{n5} + iG_{n6}}{R}$$

The G-functions in these three formulas equal:

$$G_{n1} = \cosh(X_n)\cos(X_n) \quad G_{n2} = \sinh(X_n)\sin(X_n)$$

$$G_{n3} = [\sinh(X_n)\cos(X_n) + \cosh(X_n)\sin(X_n)]/(2X_n)$$

$$G_{n4} = [\cosh(X_n)\sin(X_n) - \sinh(X_n)\cos(X_n)]/(2X_n)$$

$$G_{n5} = X_n[\sinh(X_n)\cos(X_n) - \cosh(X_n)\sin(X_n)]$$

$$G_{n6} = X_n[\cosh(X_n)\sin(X_n) + \sinh(X_n)\cos(X_n)]$$

The amplitude and phase shift of the temperature damping, the dynamic thermal resistance and the admittance so become:

Amplitude	Phase shift
$[D_q^n] = R_T\sqrt{G_{n3}^2 + G_{n4}^2}$	$\varphi_q^n = \mathrm{bgtg}(G_{n4}/G_{n3})$
$[D_\theta^n] = \sqrt{G_{n1}^2 + G_{n2}^2}$	$\varphi_\theta^n = \mathrm{bgtg}(G_{n2}/G_{n1})$
$[Ad^n] = [D_\theta^n]/[D_q^n]$	$\varphi_{Ad}^n = \varphi_\theta^n - \varphi_q^n$

With all this, per harmonic n the complex 2×2 matrix converts into following real 4×4 matrix:

$$W_n = \begin{bmatrix} G_{n1} & G_{n2} & R_T G_{n3} & R_T G_{n4} \\ -G_{n2} & G_{n1} & -R_T G_{n4} & R_T G_{n3} \\ G_{n5}/R_T & G_{n6}/R_T & G_{n1} & G_{n2} \\ -G_{n6}/R_T & G_{n5}/R_T & -G_{n2} & G_{n1} \end{bmatrix}$$

What concerns the phase shifts, Figure 1.18 shows which quarter convenes with which sign of the G-functions.

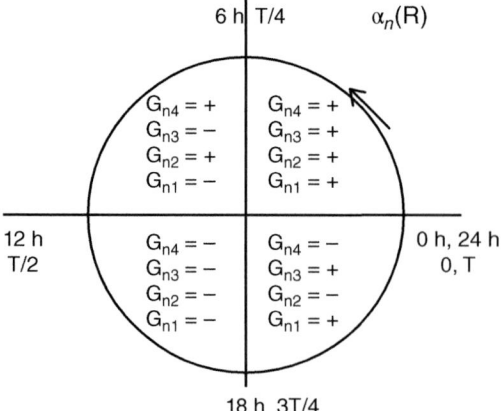

Figure 1.18 Phase shift: sign of the G-functions.

1.2.5.2.3 Multi-layered

Multi-layer assemblies behave as a serial composition of flat layers, of which each has its own layer matrix $[W_{n,i}]$. The system matrix $[W_{nT}]$ of the whole then becomes:

$$[A_{sn}(R_T)] = W_{nT}[A_{sn}(0)] \tag{1.53}$$

with $[A_{sn}(R_T)]$ and $[A_{sn}(0)]$ the column matrixes of the nth complex temperature and heat flux on both end faces, see Figure 1.19.

The relation between the complex temperatures and heat fluxes at the interface j between layer j and the one adjacent layer and $j+1$ between layer j and the other adjacent layer is:

$$[A_{n,j+1}] = W_{n,j}[A_{n,j}]$$

For the inner end face coinciding with R = 0, the inner end face, the series of layer matrices for an assembly counting m layers becomes:

$$[A_{n,1}] = W_{n,1}[A_{s,n}(0)]$$
$$[A_{n,2}] = W_{n,2}[A_{n,1}]$$
$$\ldots$$
$$[A_{n,m-1}] = W_{n,m-1}[A_{n,m-2}]$$
$$[A_{s,n}(R_T)] = W_{n,m}[A_{n,m-1}]$$

Transposing now each preceding equality in the next gives:

$$[A_{s,n}(R_T)] = W_{n,m}W_{n,m-1}\ldots W_{n,2}W_{n,1}[A_{s,n}(0)]$$

Or, the system matrix of a multi-layer assembly equals:

$$[W_{nT}] = \prod_{j=1}^{m}[W_{n,j}] \tag{1.54}$$

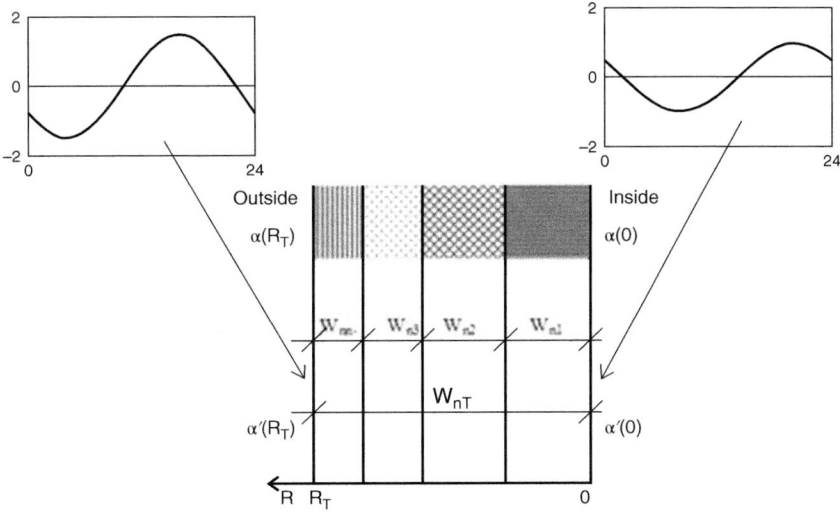

Figure 1.19 Composite assembly, system matrix.

Multiplying each next layer matrix with the product of all preceding layer matrices starts with the layer at the inner end face to end with the layer at the other end face. As the commutation property does not apply for a product of matrices, the layer sequence matters. Or, contrary to the thermal resistance, the transient response of a multi-layer assembly changes with how that sequence looks! Of course, each layer matrix keeps the same formulation as for a single-layer assembly. If two of the complex temperatures and heat fluxes at the end faces are known, then the other two follow from:

$$[A_{sn}(R_T)] = [W_{nT}][A_{sn}(0)]$$

To quantify the complex temperatures of and heat fluxes at all interfaces, the layer equations must be ascended or descended in the correct order.

Besides, to get the complex temperature and heat flux course in a single-layer assembly, it should be divided in m layers of thickness Δx and layer matrix $[W_{\Delta x}]$, giving as system matrix $[W_n] = (W_{\Delta x})^m$. Further calculations then are as for multi-layer.

1.2.5.3 Any Boundary Conditions, Flat Assemblies
1.2.5.3.1 Convolution

Convoluting starts from Dirac impulses (Figure 1.20), whereby the temperature or heat flux in a single- or multi-layer assembly remains 0, unless for an infinitely short time step dt, when at one of the end faces it jumps to 1.

If this is the temperature θ_{s1} on end face s1, then after a time also the heat flux $q_{s1}(t)$ there and the temperature $\theta_{s2}(t)$ and heat flux $q_{s2}(t)$ on the other end face s2 will vary a little. In case the materials used have constant properties, these changes $q_{s1}(t)$, $\theta_{s2}(t)$ and $q_{s2}(t)$, called 'response factors', are written as:

$$I_{\theta_{s1} q_{s1}} \quad I_{\theta_{s1} \theta_{s2}} \quad I_{\theta_{s1} q_{s2}}$$

Analogously, for the heat flux on end face s1 jumping to 1 or the temperature or heat flux on end face s2 jumping to 1, related response factors look:

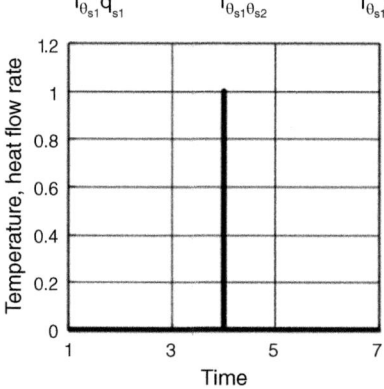

Figure 1.20 Dirac impulse.

Heat flux, end face 1	$I_{q_{s1}\theta_{s1}}$	$I_{q_{s1}\theta_{s2}}$	$I_{q_{s1}q_{s2}}$
Temp., end face 2	$I_{\theta_{s2}q_{s1}}$	$I_{\theta_{s2}\theta_{s1}}$	$I_{\theta_{s1}q_{s2}}$
Heat flux, end face 2	$I_{q_{s2}q_{s1}}$	$I_{q_{s2}q_{s1}}$	$I_{q_{s2}q_{s2}}$

For multi-layer assemblies, the response factors depend on the layer sequence. As a consequence, their values change with on which end face both jump: s1 or s2.

If the temperature or heat flux on end face s1 shows a jump with value θ_o or q_o, then, with the response factors known, the temperature and heat flux change on end face s2 and the heat flux or temperature on end face s1 become:

Pulse θ_o on end face s1	Pulse q_o on end face s1
$q_{s1} = \theta_o I_{\theta_{s1}q_{s1}}$	$\theta_{s1} = q_o I_{q_{s1}\theta_{s1}}$
$\theta_{s2} = \theta_o I_{\theta_{s1}\theta_{s2}}$	$\theta_{s2} = q_o I_{q_{s1}\theta_{s2}}$
$q_{s2} = \theta_o I_{\theta_{s1}q_{s2}}$	$q_{s2} = q_o I_{q_{s1}q_{s2}}$

A jump on end face s2 gives analogous relations. Any signal $\theta_{s1}(t)$ on end face s1 can be seen now as an addition of jumps $\theta_{s1}(t)\Delta t$, giving as temperature (θ_{s2}) on end face s2 (Figure 1.21):

$t = 0 \quad \theta_{s2}(0) = 0$

$t = \Delta t \quad \theta_{s2}(\Delta t) = \theta_{s1}(t = 0) I_{\theta_{s1}\theta_{s2}}(t = \Delta t)$

$t = 2\Delta t \quad \theta_{s2}(2\Delta t) = \theta_{s1}(t = 0) I_{\theta_{s1}\theta_{s2}}(t = 2\Delta t) + \theta_{s1}(t = \Delta t) I_{\theta_{s1}\theta_{s2}}(t = \Delta t)$

$t = 3\Delta t \quad \theta_{s2}(3\Delta t) = \theta_{s1}(t = 0) I_{\theta_{s1}\theta_{s2}}(t = 3\Delta t) + \theta_{s1}(t = \Delta t) I_{\theta_{s1}\theta_{s2}}(t = 2\Delta t)$
$+ \theta_{s1}(t = 2\Delta t) I_{\theta_{s1}\theta_{s2}}(t = \Delta t)$

...

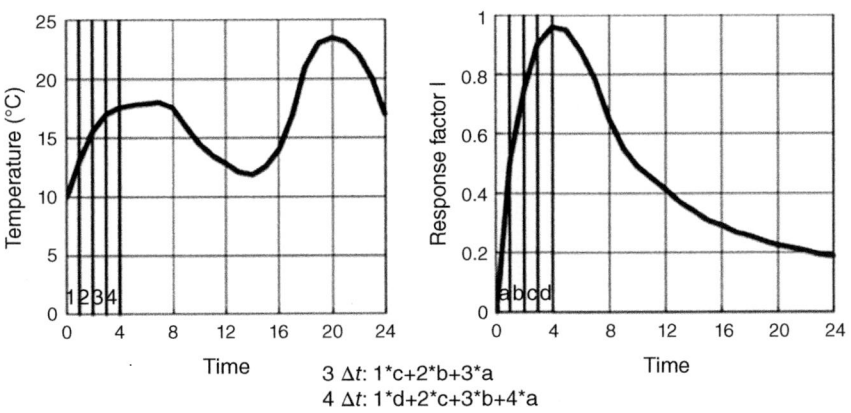

3 Δt: 1*c+2*b+3*a
4 Δt: 1*d+2*c+3*b+4*a

Figure 1.21 Convolution, principle.

$$t = n\Delta t \quad \theta_{s2}(n\Delta t) = \theta_{s1}(t=0)\, I_{\theta_{s1}\theta_{s2}}(t=n\Delta t)$$
$$+ \theta_{s1}(t=\Delta t) I_{\theta_{s1}\theta_{s2}}(t=(n-1)\Delta t)$$
$$+ \theta_{s1}(t=2\Delta t) I_{\theta_{s1}\theta_{s2}}(t=(n-2)\Delta t) + \cdots$$

$$\text{or}: \theta_{s2}(n\Delta t) = \sum_{j=0}^{n-1} \theta_{s1}(j\overrightarrow{\Delta t})\, I_{\theta_{s1}\theta_{s2}}((n-j)\Delta t)$$

Written as an integral with the time steps infinitely small:

$$\theta_{s2}(t) = \int_0^t \theta_{s1}(\tau)\, I_{\theta_{s1}\theta_{s2}}(t-\tau)\, d\tau$$

Defined so is the convolution integral of the temperature θ_{s2} on end face s2 for a signal $\theta_{s1}(t)$ on end face s1. The signal $\theta_{si}(\tau)$ has to be scanned clockwise, and the response factor $I_{\theta_{s1}\theta_{s2}}(t-\tau)$ counterclockwise. A same approach, be it with the correct response factors, holds for the heat fluxes on both end faces, more, for all situations with randomly changing end face conditions. Response factors and convolution integrals can only be calculated numerically.

1.2.5.3.2 Temperature Change in a Semi-infinitely Thick Layer

At time 0, the surface temperature of a uniformly warm, semi-infinitely thick layer suddenly jumps from θ_{so} to $\theta_{so} + \Delta\theta_{so}$ (Figure 1.22). Solving Fourier's second law, in this case, requires a separation of variables with as initial condition $t=0$, $\theta = \theta_{s,o}$ for $0 \leq x \leq \infty$ and as boundary condition on the surface: $t \geq 0$, $\theta_s = \theta_{so} + \Delta\theta_{so}$.

The solution is:

$$\theta(x,t) = \theta_{so} + \Delta\theta_{so}\left(\frac{2}{\sqrt{\pi}}\int_{q=\frac{x}{2\sqrt{at}}}^{\infty} \exp(-q^2)\, dq\right) \quad (1.55)$$

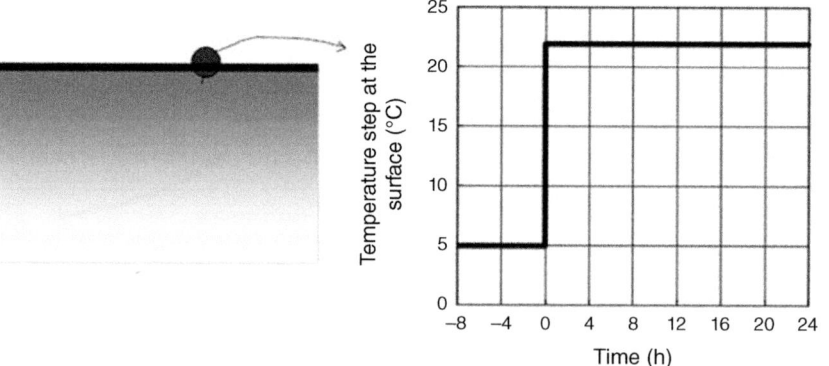

Figure 1.22 Temperature jump at the surface of a uniformly warm semi-infinitely thick layer.

The term between brackets stands for the inverse error function with 'a' the thermal diffusivity of the medium. The heat flux ($x = 0$) on the surface in turn equals:

$$q = -\lambda \frac{d\theta}{dx} = -\lambda \left[\frac{2\Delta\theta_{s0}}{2\sqrt{\pi at}} \exp\left(-\frac{x^2}{4at}\right) \right]_{x=0} = -\frac{\Delta\theta_{s0}\, b}{\sqrt{\pi t}} \tag{1.56}$$

with b the contact coefficient of the medium. Applying the definition of response factor to this equation gives:

$$I_{\theta_{s1} q_{s1}} = -b/\sqrt{\pi t}$$

Semi-infinite mediums do not exist, but the soil and really thick material layers are close. In any case, the heat flux equation illustrates what the contact coefficient is doing. A high value means fast heating with increasing and fast cooling with decreasing surface temperature, and a low value both going slowly. Materials with high contact coefficient so act as storage volumes. If in passive solar buildings walls and floors cannot temporarily store a large part of the solar gains, the indoors could become far too warm. Or, partitions and floors in such buildings should consist of heavy materials and be at least as thick as structurally needed. The heat in- or outflow per m² from time 0 on in such semi-infinite medium equals:

$$Q = \int_0^t q\, dt = \frac{2b\, \Delta\theta_{s0}}{\sqrt{\pi}} \sqrt{t} = A_q \sqrt{t} \tag{1.57}$$

with A_q the heat absorption coefficient, units J/(m²·K·s$^{1/2}$).

In case two materials, one at temperature θ_1, the other at temperature θ_2, contact each other, the heat flux there will equal ($\theta_1 > \theta_2$):

In material 1 : $q_{s1} = (\theta_1 - \theta_c) b_1 / \sqrt{\pi t}$

In material 2 : $q_{s2} = (\theta_c - \theta_2) b_2 / \sqrt{\pi t}$

As both fluxes must have the same absolute value, the contact temperature θ_c will be:

$$\theta_c = \frac{b_1 \theta_1 + b_2 \theta_2}{b_1 + b_2} \tag{1.58}$$

Its value so depends on the temperature and contact coefficient of the materials contacting each other, a reality showing what people experience when touching materials. Those with high contact coefficient feel cold, and those with low contact coefficient comfortably warm. Indeed, in the first case, the skin temperature, 32–33°C, will move direction surface temperature of the material, while in the second case, the surface temperature of the material will near the skin temperature. Concrete and aluminium so feel unpleasant, wood pleasant. The term cold and warm materials is often used. Chair finishes should so best have a low contact coefficient.

1.2.5.4 Two and Three Dimensions: Thermal Bridges

To quantify heat transmitted two or three dimensionally, Fourier's second law for transient conduction in 2 or 3 dimensions, applies. Since analytical solutions fail, CVM is used. The principles were explained. Now, additionally, the heat capacity $\rho c \Delta V$ of each control volume plays. Without dissipation, the resulting heat flow per time step from any adjacent control volume to the central equals the change in heat during that step stored in or released by the central one:

$$\left(\sum \Phi_m\right) \Delta t = \rho c \, \Delta V \, \Delta \theta$$

In it, Φ_m is the weighted average heat flow during time Δt between adjacent and central control volume:

$$\Phi_m = p\Phi_{t+\Delta t} + (1-p)\Phi_t \quad (1 > p > 0)$$

The balance equation so rewrites as ($i = l, m, n$ and $j = \pm 1$):

$$p\left[\sum (P_{s,i+j}\theta_{i+j}) - \theta_{l,m,n} \sum P_{s,i+j}\right]^{(2)} + (1-p)$$

$$\cdot \left[\sum (P_{s,i+j}\theta_{i+j}) - \theta_{l,m,n} \sum P_{s,i+j}\right]^{(1)} = \rho c \, \Delta V \frac{\theta_{l,m,n}^{(2)} - \theta_{l,m,n}^{(1)}}{\Delta t}$$

In it, $p = 0$ means forward, $p = 1$ backward and $p = 0.5$ average differencing, called the Cranck–Nicholson scheme that after rearrangement looks:

$$\left[\sum (P_{s,i+j}\theta_{i+j}) - \theta_{l,m,n}\left(\sum P_{s,i+j} + \frac{2\rho c \, \Delta V}{\Delta t}\right)\right]^{(2)}$$

$$= \left[-\sum (P_{s,i+j}\theta_{i+j}) + \theta_{l,m,n}\left(\sum P_{s,i+j} - \frac{2\rho c \, \Delta V}{\Delta t}\right)\right]^{(1)}$$

In it, the actual temperatures (2) are the unknowns, the preceding (1) the known ones. Each time step so gives a system with as many equations as control volumes with unknown temperature. With the initial and boundary conditions known, the system so found is solved time step wise. A question of course is: which meshes and time steps to use? Best choice is to refine the meshes in materials with high thermal diffusivity and to shorten the time step where mesh density is higher, the faster the boundary conditions change and/or the higher the desired accuracy.

1.3 Heat Exchange at Surfaces by Convection and Radiation

1.3.1 What?

Till now, the temperatures on the end faces of an assembly were assumed to be known. However, in most cases, it's the air temperature at both sides, sometimes the heat flux at a surface, that is known. In fact, weather stations measure the air temperature outdoors, while logging the temperature in the air indoors is simpler than logging surface temperatures. Therefore, the focus now moves to the heat transfer ambient to ambient through assemblies. Of course, the key is to know how heat

1.3 Heat Exchange at Surfaces by Convection and Radiation

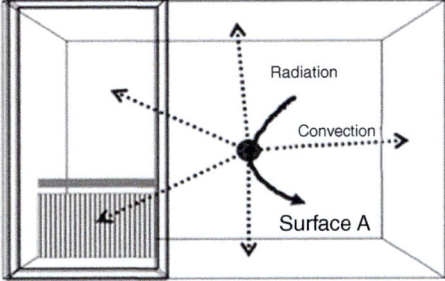

Figure 1.23 Heat exchange at surface A by convective exchange with the room air and radiation to and from all other surfaces.

from the ambient reaches a surface. Two modes intervene: convection between the air and the surface called A in Figure 1.23 and radiant exchanges with all faces that surface sees.

1.3.2 Convection

1.3.2.1 In General

Here, the term 'convection' refers to the heat exchanged between a liquid or a gas and a surface they are contacting. Related heat flux and flow are commonly written as:

$$q_c = h_c(\theta_{fl} - \theta_s) \quad \Phi_c = h_c(\theta_{fl} - \theta_s)A \tag{1.59}$$

with θ_{fl}, the temperature in the said liquid or gas close to the surface, θ_s the surface temperature and h_c a property called the convective surface film coefficient, units W/(m²·K). Both equations are called 'Newton's law'. It links convection linearly to the driving temperature difference $(\theta_{fl} - \theta_s)$. The law anyhow rather serves as definition of the surface film coefficient, which reflects the complexity of the heat and mass flow in a liquid or gas touching a surface. Are fixing the mass flow: the scalar conservation of mass and vectorial conservation of momentum equations:

$$\text{Mass (scalar)} \quad \text{div}(\rho \mathbf{v}) = 0$$

$$\text{Momentum (Navier–Stokes, vector)} \quad \frac{d(\rho \mathbf{v})}{dt} = \rho \mathbf{g} - \mathbf{grad}\, P + \mu \nabla^2 \mathbf{v}$$

In both, ρ is density of the fluid, μ its dynamic viscosity, P the total pressure and $\rho \mathbf{g}$ the gravity gradient. Unknowns are the three velocity components v_x, v_y, v_z and the total pressure. When the flow turns turbulent, how to shape turbulency adds. Often used for that is the (k,ε)-approach, with k the turbulent kinetic energy and ε its dissipation. Two additional scalar equations so add, one for k and one for ε. In addition, estimating the three velocity components in the x, y, and z direction turns the vectorial momentum equation in three scalar equations, so that quantifying turbulent flow requires six scalar equations with six unknowns. Constant properties, negligible kinetic energy and hardly any friction in turn allow to write the conservation of energy in a moving fluid with thermal diffusivity a as:

$$a\nabla^2 \theta = \frac{d\theta}{dt} \tag{1.60}$$

In the Navier–Stokes and in this equation, d/dt is the total derivative:

$$\frac{d}{dt} = \frac{\partial}{\partial x}\frac{\partial x}{\partial t} + \frac{\partial}{\partial y}\frac{\partial y}{\partial t} + \frac{\partial}{\partial z}\frac{\partial z}{\partial t} + \frac{\partial}{\partial t} = \underbrace{\frac{\partial}{\partial x}v_x}_{(1)} + \underbrace{\frac{\partial}{\partial y}v_y}_{(2)} + \underbrace{\frac{\partial}{\partial z}v_z}_{(3)} + \frac{\partial}{\partial t}$$

In the conservation of energy equation, the terms (1), (2) and (3) multiplied with $\rho c\theta$ give the enthalpy, the θ °C warm liquid or gas displace at a velocity with components v_x, v_y, v_z. Temperature θ figures as seventh unknown. All information about convection follows from solving that system of seven scalar partial differential equations equations. With the temperature field known, then, due to convection becoming conduction in the laminar boundary layer against the surface, the heat flux to it follows from Fourier's first law, giving as surface film coefficient:

$$h_c = -\lambda_{fl}\frac{(\text{grad }\theta)_s}{\theta_{fl} - \theta_s} \tag{1.61}$$

In it, $(\text{grad }\theta)_s$ is the temperature gradient in the laminar boundary layer and θ_{fl} the temperature in the undisturbed liquid or gas. Although Computerized Fluid Dynamics (CFD) is mostly used, solving laminar natural convection along a semi-infinite vertical surface and laminar forced convection along a semi-infinite horizontal surface is analytically doable. In both cases, the velocity and temperature equations look similar, which with air as gas gives:

$$\frac{\partial v_x}{\partial x} + \frac{\partial v_y}{\partial y} = 0 \quad v_x\frac{\partial v_x}{\partial x} + v_y\frac{\partial v_y}{\partial y} = v_{fl}\frac{\partial^2 v_x}{\partial y^2} \quad v_x\frac{\partial \theta}{\partial x} + v_y\frac{\partial \theta}{\partial y} = a_{fl}\frac{\partial^2 \theta}{\partial y^2}$$

For the thermal diffusivity equal to the kinematic viscosity, the three equations even become identical. The so coinciding heat and mass flow patterns give as surface film coefficient:

$$h_{c,x} = \frac{0.664\,\lambda_{fl}}{x}\sqrt{\frac{v_\infty x}{v_{fl}}}$$

with x the distance to the surface's free edge and v_∞ the velocity of the undisturbed air. The result is a convective surface film coefficient that changes along the surface. However, any trial to calculate analytically the convective heat transfer along walls in a room, along the envelope outdoors or along any other surface is doomed to fail. Happily, even in complex situations, convection depends on all parameters and properties fixing the heat and mass flow:

Fluid properties	Flow parameters
Thermal conductivity (λ_{fl})	Geometry
Density (ρ_{fl})	Surface roughness
Specific heat capacity (c_{fl})	Temperature difference ($\theta_{fl} - \theta_s$)
Kinematic viscosity (μ_{fl}/ρ_{fl})	Nature and direction of the flow
Volumetric expansion coefficient	Velocity components v_x, v_y, v_z

1.3.2.2 Typology
1.3.2.2.1 Driving Forces
In case differences in air density due to temperature and vapour concentration gradients act as driving force, natural convection with a flow pattern mainly reflecting the temperature field in the air is the result, typically the case indoors. If instead an imposed pressure difference acts as driving force, then forced convection is a fact with a flow pattern independent of temperature, usually the case outdoors. Both convection types can coincide, giving mixed convection. Outdoors, a natural component adds when close to windless. Indoors, opening and closing doors may add forced convection peaks to natural convection.

1.3.2.2.2 Flow Types
Mass flows can be laminar, turbulent or transient. If laminar, streamlines never cross and particle velocities and the overall flow velocity are the same. In turbulent flow, a chaotic momentum field creates whirling eddies with particle velocities different from the overall flow velocity. Turbulent kinetic energy (k) builds up in the eddies while turbulent dissipation (ε) continuously dampens their motion. Describing turbulent flow is hard, even with CFD and a fractional eddy approach. Transient flow in turn fills the gap between both. Small disturbances along a flow path may induce switches from laminar to turbulent, after which the eddies formed fade away and the flow turns laminar again. In between, it is transient. Turbulency favours heat transfer the most, because in laminar flow, heat conduction can only develop perpendicular to the flow direction. So:

flow → Driving force ↓	Laminar	Transient	Turbulent
Density differences	X	X	X
Density and pressures	X	X	X
Pressures	X	X	X

1.3.2.3 Quantifying the Convective Surface Film Coefficient
1.3.2.3.1 Analytically
The few solvable cases learn that the convective surface film coefficient changes from spot to spot along a surface. For heat exchange, normal is to use a surface averaged value:

$$h_c = \frac{1}{A} \int h_{cA} dA$$

If surface temperatures on specific spots are what matters, then the value there of the convective surface film coefficient should be used.

1.3.2.3.2 Numerically
Although CFD helps to calculate h_c as the results in Table 1.1 and Figure 1.24 show, it requires assuming a velocity distribution in the boundary layer.

Table 1.1 Convective surface film coefficient (h_e) against the envelope of a detached cube-shaped building: CFD-based correlations for a wind speed <15 m/s and 10 °C temperature difference with the air all around.

Emmel et al. (2007)	Surface to wind angle (°)	h_e, W/(m²·K) (v_w = wind speed)
Outer walls	0	$5.15 v_w^{0.31}$
	45	$3.34 v_w^{0.34}$
	90	$4.78 v_w^{0.71}$
	135	$4.05 v_w^{0.77}$
	180	$3.54 v_w^{0.76}$
Roofs	0	$5.11 v_w^{0.78}$
	45	$4.6 v_w^{0.79}$
	90	$3.76 v_w^{0.85}$

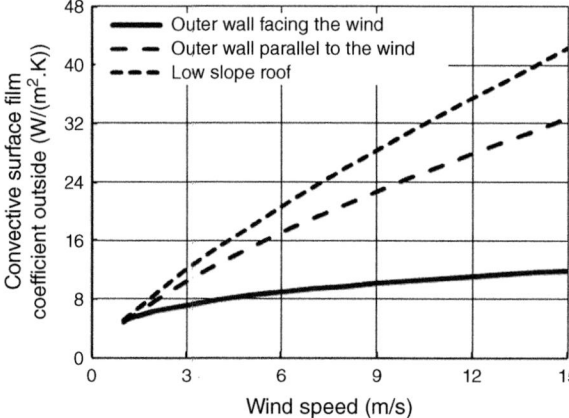

Figure 1.24 Cube-shaped building, CFD-based outside convective surface film coefficient.

1.3.2.3.3 Dimensionally

Many information on convection comes from experiments that help determining which dimensionless ratios between fluid properties, geometric data and kinematic characteristics require a same value in the experiments as in reality to allow extrapolation of the test results to reality. These ratios either follow from the differential equations or from Buckingham's π-theorem, stating that if a problem depends on n single-valued physical properties with p basic dimensions, then n–p dimensionless ratios will fix the solution.

Describing forced convection so involves seven physical properties (L, λ_{fl}, v_{fl}, ρ_{fl}, μ_{fl}, c_{fl}, h_c (see above)) having four basic dimensions (representative length L, time t, mass M, temperature θ). So, three dimensionless ratios (π_1, π_2, π_3) are needed for a solution. They follow from turning $\pi = L^a \lambda_{fl}^b v_{fl}^c \rho_{fl}^d \mu_{fl}^e c_{fl}^f h_c^g$ into an equation containing the three ratios, written as $\pi_1 = f(\pi_2, \pi_3)$:

1.3 Heat Exchange at Surfaces by Convection and Radiation

$$\pi = [L]^a \left[\frac{ML}{t^3\theta}\right]^b \left[\frac{L}{t}\right]^c \left[\frac{M}{L^3}\right]^d \left[\frac{M}{Lt}\right]^e \left[\frac{L^2}{t^2\theta}\right]^f \left[\frac{M}{t^3\theta}\right]^g$$

As the three must be dimensionless, the sum of the exponents per dimension should be 0, or:

for M		b			+ D		+ e			+ g	= 0
for L	a	+ b	+ c		− 3d		− e		+ 2f		= 0
for t		− 3b	− c				− e		− 2f	− 3g	= 0
for θ		− b							− f	− g	= 0

The three so follow from recalculating the π-function three times, first for $g=1$, $c=0$, $d=0$, then for $g=0$, $a=1$, $f=0$ and finally for $g=0$, $e=1$, $c=0$:

Solution 1 $a = 1, b = -1, e = 0, f = 0$ or $\pi_1 = \dfrac{h_c L}{\lambda_{fl}}$

Solution 2 $b = 0, c = 1, d = 1, e = -1$ or $\pi_2 = \dfrac{\rho_{fl} v_{fl} L}{\mu_{fl}}$

Solution 3 $a = 0, b = -1, d = 0, f = 1$ or $\pi_3 = \dfrac{c_{fl} \mu_{fl}}{\lambda_{fl}}$

Natural convection in turn demands four dimensionless ratios, of which two combine to one. The three for forced convection are called the Reynolds, Nusselt and Prandl number, while the three for natural convection are called the Grasshof, Nusselt and Prandl number:

Reynolds : $\text{Re} = v_{fl} L / \nu_{fl}$ $(= \pi_2)$ (1.62)

with v_{fl} the fluid's velocity, ν_{fl} the fluid's kinematic viscosity and L the characteristic length fixing the geometry (the hydraulic diameter for pipes, the dimension in the flow direction for walls and a calculated length for more complex cases). Reynolds so stands for the ratio between the inertia force and viscous friction. If low, friction gains and the result is laminar flow. If high, inertia wins and the result is turbulent flow. Or, the number fixes the flow type: laminar for $\text{Re} \leq 2000$, turbulent for $\text{Re} \geq 20\,000$, transient for $2000 < \text{Re} < 20\,000$.

Nusselt : $\text{Nu} = h_c L / \lambda_{fl}$ $(= \pi_1)$ (1.63)

with λ_{fl} the thermal conductivity of the fluid. The right-hand side can be rewritten as:

$$h_c L / \lambda_{fl} = (\text{grad } \theta)_s / [(\theta_{fl} - \theta_s)/L]$$

Or, Nusselt represents the ratio between the temperature gradient against a surface in the fluid and the mean temperature gradient along the characteristic length. A high value stands for the first large and the second small, the case with high fluid velocities. Physically, the number confirms that conduction governs the heat transfer against a surface. In fact, even for turbulent flow, a laminar boundary layer remains, however with a thickness that shrinks the higher the fluid velocity. The number underlines the importance of convection compared to conduction.

Prandl : $\text{Pr} = \nu_{fl} / a_{fl}$ $(= \pi_3)$ (1.64)

This number combines the heat and mass transfer by rating two analogous quantities: the thermal diffusivity, which determines how easy a local temperature change is spreading in the fluid, and the kinematic viscosity v, which indicates how easy a local velocity change does it.

$$\text{Grasshof:} \quad \text{Gr} = \beta_{fl} g L^3 \Delta\theta / v_{fl}^2 \tag{1.65}$$

with β_{fl} the volumetric expansion coefficient, g the acceleration by gravity and $\Delta\theta$ the representative temperature difference. The number replaces Reynolds in natural convection as the velocity then is the result of mainly temperature-induced density differences (βg), which simultaneously change the temperature field. This is why the terms L^3 and v_{fl}^2 replace L and v_{fl}, in the Reynolds number.

$$\text{Rayleigh:} \quad \text{Ra} = \text{GrPr} \tag{1.66}$$

This number replaces the product GrPr in the many natural convection formulae.

All experimental, numerical and analytical expressions for the convective surface film coefficient are now written as:

Natural convection	Mixed convection	Forced convection
$\text{Nu} = c(\text{Ra})_n$	$\text{Nu} = F(\text{Re}/\text{Gr}^{1/2}, \text{Pr})$	$\text{Nu} = F(\text{Re}, \text{Pr})$

The coefficient c, the exponent n and $F(\ )$ differ among geometries, flow type and direction.

1.3.2.4 Values for the Convective Surface Film Coefficient
1.3.2.4.1 Flat Surfaces

For natural convection, L is the height for vertical surfaces, the side for square horizontal surfaces and the average of length and width for rectangular horizontal surfaces. The mean temperature between surface and undisturbed liquid or gas in turn fixes the value of the properties intervening. The relations advanced are:

	Conditions	Functions
Vertical surfaces	$\text{Ra}_L \leq 10^9$	$\text{Nu}_L = 0.56\, \text{Ra}_L^{1/4}$
	$\text{Ra}_L > 10^9$	$\text{Nu}_L = 0.025\, \text{Ra}_L^{2/5}$
Horizontal surfaces		
Heat flow upwards	$10^5 < \text{Ra}_L \leq 2 \cdot 10^7$	$\text{Nu}_L = 0.56\, \text{Ra}_L^{1/4}$
	$2 \cdot 10^7 < \text{Ra}_L < 3 \cdot 10^{10}$	$\text{Nu}_L = 0.138\, \text{Ra}_L^{1/3}$
Heat flow downwards	$3 \cdot 10^5 < \text{Ra}_L < 10^{10}$	$\text{Nu}_L = 0.27\, \text{Ra}^{1/4}$

For forced convection, they are:

	Conditions	Functions
Laminar flow	$Pr > 0.1$, $Re_L < 5 \cdot 10^5$	$Nu_L = 0.644\, Re_L^{1/2} Pr^{1/3}$
Turbulent flow	$Pr > 0.5$, $Re_L > 5 \cdot 10^5$	$Nu_L = 0.036\, Pr^{1/3} \left(Re_L^{4/5} - 23\,200\right)$

Turning to buildings, air at atmospheric pressure and the ambient temperature are what counts. This simplifies the equations. For natural convection, the temperature difference $\Delta\theta$ between surface and undisturbed air is driving, as reflected by the relation heading the table:

$h_c = a(\Delta\theta/L)^b$	Conditions	a	b	L
Vertical surfaces	$10^{-4} < L^3 \Delta T \leq 7$	1.4	1/4	Height
	$7 < L^3 \Delta T \leq 10^3$	1.3	1/3	1
Horizontal surfaces				
Heat flow upwards	$10^{-4} < L^3 \Delta T \leq 0.14$	1.3	1/4	Eq. side
	$0.14 < L^3 \Delta T \leq 200$	1.5	1/3	1
Heat flow downwards	$2 \cdot 10^{-4} < L^3 \Delta T \leq 200$	0.6	1/4	1

In buildings with HVAC, the convective surface film coefficients for mixed convection are often related to the air change rate n, a number telling how many times an hour the air in a space is exchanged (ach). For rectangular spaces:

	Configuration	h_c (W/(m²·K))
Walls	Forced convection, air diffusers	$-0.199 + 0.18 n^{0.8}$
Floor	at the ceiling, room isothermal	$0.159 + 0.116 n^{0.8}$
Ceiling		$-0.166 + 0.484 n^{0.8}$
Walls	Forced convection, air diffusers	$-110 + 0.132 n^{0.8}$
Floor	in the walls, room isothermal	$0.704 + 0.168 n^{0.8}$
Ceiling		$0.064 + 0.00444 n^{0.8}$

For forced convection outdoors, wind is the main actor, giving as relations:

Wind speed	Relation	Remarks
$v \leq 5\,\text{m/s}$	$h_c = 5.6 + 3.9\,v$	For $v \leq 5$ m/s natural convection still intervenes, therefore the constant 5.6
$v > 5\,\text{m/s}$	$h_c = 7.2\,v^{0.78}$	

That the value increases with wind speed, follows from a related drop in boundary layer thickness.

All relations given apply for air flowing along freestanding flat surfaces. Angles between two and corners between three surfaces disturb the flow. Moreover, if surfaces form a room, the overall flow pattern must satisfy the continuity equation. All this makes convection so complex that standards give constant average values:

EN-standard	Heat loss	Surface temperatures
Natural convection (=indoors)		
Vertical surfaces	3.5	2.5
Horizontal surfaces:		
Heat upwards (↑)	5.5	2.5
Heat downwards (↓)	1.2	1.2
Forced convection (=outdoors)	19.0	19.0

The reference temperature in indoor spaces is the air temperature, 1.7 m above the floor's centre. Outdoors, it's the air temperature measured by the nearest weather station. When calculating surface temperatures using local convective surface film coefficients, the reference should be the air temperature at the same spot just outside the boundary layer. Large temperature differences, complex geometries or inner surfaces hided by furniture require a more correct calculation. In such cases, the tabled formulae or data from literature should be used.

1.3.2.4.2 Cavities

The name cavity refers to an air layer between material layers with a width, small compared to its length or height, see Figure 1.25. The convective heat fluxes against the warm (1) and cold cavity face (2) equal:

$$q_{c1} = h_{c1}(\theta_{s1} - \theta_c) \qquad q_{c2} = h_{c2}(\theta_c - \theta_{s2})$$

In both, θ_c is the gas or air temperature in the centre of the cavity. If unvented, both fluxes must on average be equal, giving as mean value:

$$q_c = \frac{h_{c1} h_{c2}}{h_{c1} + h_{c2}} (\theta_{s1} - \theta_{s2}) \qquad (1.67)$$

Figure 1.25 Wall with cavity between veneer and inside leaf.

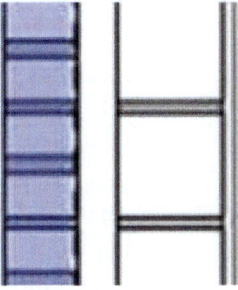

Replacing both surface film coefficients by one value h_c, simplifies the formula to:

$$q_c = (h_c/2)(\theta_{s1} - \theta_{s2})$$

In reality, convection in a cavity includes conduction, reason why this relation is often rewritten as if the last dominates, be it with the λ_{fl}-value multiplied with the Nusselt number Nu:

$$q + q_c = (\lambda_{fl} \text{Nu}) \Delta \theta_s / d = h'_c \, \Delta \theta_s \qquad (1.68)$$

Here, d is the cavity width (m) and $\Delta \theta_s$ the temperature difference between both cavity faces (°C). In infinitely long horizontal cavities, circular eddies, called Bénard cells, develop, while in infinitely high vertical cavities, air rotation does. In both cases, Nu to the width d writes as:

$$\text{Nu}_d = \max \left[1, 1 + \frac{m \, \text{Ra}_d^r}{\text{Ra}_d + n} \right] \quad (10^2 \leq \text{Ra}_d \leq 10^8) \qquad (1.69)$$

In it, temperature difference between both cavity faces acts as reference for the Rayleigh number Ra_d, while cavity width (d) act as characteristic length and m, r and n have as values:

	m	n	r
Horizontal cavity			
Heat flow downwards	0		
Heat flow upwards	0.07	3200	1.33
Vertical and tilted cavity with slope above 45°			
	0.024	10 100	1.39
Tilted cavity with slope below 45°			
Heat flow upwards	0.043	4100	1.36
Heat flow downwards	0.025	13 000	1.36

Ra_d lower than 100 means pure conduction, $\text{Nu}_d = 1$. Convection in finite cavities diverges strongly from convection in infinite ones. With d cavity width, H cavity height and L cavity length, all in m, with ^{lam} the superscript for laminar, ^{turb} the superscript for turbulent and ^{transient} the superscript for transient, Nu_d becomes:

	Nu_d
	Vertical cavity
Rayleigh number upper limit value (Ra_{max}) for the applicability of Nu_d depending on the ratio H/d: $H/d =$ 5 20 40 80 100 $Ra_{max} = 10^8$ 2.10^6 2.10^5 3.10^4 $1.2\,10^4$	$\max\left(Nu_d^{lam}, Nu_d^{turb}, Nu_d^{transient}\right)$, with $Nu_d^{lam} = 0.242\left(\dfrac{Ra_d d}{H}\right)^{0.273}$ $Nu_d^{turb} = 0.0605\,Ra_d^{0.33}$ $Nu_d^{transient} = \left[1 + \left(\dfrac{0.104\,Ra_d^{0.293}}{1 + (6310/Ra_d)^{1.36}}\right)^3\right]^{0.33}$
	Horizontal cavity
Heat flow upwards $Ra_d \leq 1708$	1
$Ra_d > 1708$	$\max\left[1.1537\,d^2\left(\dfrac{\Delta\theta}{L}\right)^{1/4}\right]$
Heat flow downwards	1
	Tilted cavity: see literature

1.3.2.4.3 Pipes

Experimental and semi-experimental work has been done on convection between a pipe's outer surface and the fluid or gas in the ambient around (Figure 1.26).

The result is a series of formulae. For natural convection:

Vertical pipe	$Ra_d \leq 10^9$	$Nu_L = 0.555\,Ra_d^{1/4}$
	$Ra_d > 10^9$	$Nu_L = 0.021\,Ra_d^{2/5}$
Horizontal pipe	$Ra_d \leq 10^9$	$Nu_L = 0.530\,Ra_d^{1/4}$

Figure 1.26 Convection around an insulated pipe.

For forced convection:

$Re_d < 500$	$Nu_d = 0.43 + 0.48\, Re_d^{1/2}$
$Re_d > 500$	$Nu_L = 0.46 + 0.00128\, Re_d$

In these equations, the characteristic length (d) is the pipe's outer diameter, while all properties are linked to the average temperature between the undisturbed fluid or gas around (θ_{fl}) and the pipe's outer surface (θ_s).

1.3.3 Radiation

1.3.3.1 In General

Radiation differs fundamentally from conduction and convection. In fact, what displaces heat is neither the kinetic energy of atoms and free electrons in a material nor the macroscopic movements in a gas or a liquid but the electromagnetic waves surfaces exchange. Each surface warmer than 0 K emits and absorbs electromagnetic waves, with those absorbed agitating the atoms and electrons in the material closest to that surface, that so is warming up.

Electromagnetic waves span a wide range of wavelengths (λ), with ultraviolet (UV), visible light (L) and infrared (IR) ($\lambda = 10^{-7}$ to 10^{-3} m) as the thermally active ones. Aside, the wavelength stands for the ratio between the speed of the photons propagated in m/s and the frequency in Hz, 1 Hz being 1 wavelength passing per second:

$\lambda^{(1)} \leq 10^{-6}\, \mu m$	Cosmic radiation
$10^{-6} < \lambda \leq 10^{-4}\, \mu m$	Gamma rays
$10^{-4} < \lambda \leq 10^{-2}\, \mu m$	X-rays
$10^{-2} < \lambda \leq 0.38\, \mu m$	UV
$0.38 < \lambda \leq 0.76\, \mu m$	Visible light
$0.76 < \lambda \leq 10^3\, \mu m$	IR
$\lambda > 10^3\, \mu m$	Radio waves

Thermal radiation does not demand a medium. In fact, the transfer is most unhindered in vacuum, where it moves with a speed equal to 299 792.5 km/s. How much radiation material surfaces emit, depends on their nature and their temperature. A net heat exchange between surfaces anyhow requires them being at a different temperature.

1.3.3.2 Definitions

Table 1.2 outlines the way thermal radiation can be quantified. The related, so-called spectral values are given by the quantity considered per wavelength. A single wavelength emitted is giving monochromatic radiation, different wavelengths emitted together are giving coloured radiation.

Table 1.2 Radiant heat transfer, variables.

Variable	Definition (units)	Equations
Radiant heat Q_R	The heat emitted or received under the form of electromagnetic waves. Q_R is a scalar, units J	
Radiant heat flow Φ_R	The radiant heat per unit of time. Φ_R is a scalar, units W	$\dfrac{dQ_R}{dt}$
Radiant heat flux q_R	The radiant heat flow per unit of surface. As a surface emits radiation to and receives radiation from all directions, q_R, is a scalar, units W/m². The term irradiation, symbol E, is used for the incoming radiant heat flux, the term emittance, symbol M, for the one emitted	$\dfrac{d^2 Q_R}{dA\,dt}$
Radiation intensity I	The radiant energy emitted in a specific direction. The intensity is a vector, units W/(m²·rad), with $d\omega$ the elementary angle in the direction considered	$\dfrac{dq_R}{d\omega}$ or $\dfrac{d^2\Phi_R}{dA\,d\omega}$
Luminosity L	The ratio between the radiant heat flow rate in a direction ϕ and the apparent surface, seen from that direction. The luminosity is a vector, units W/(m²·rad). It describes how a receiving surface sees one emitting.	$\dfrac{d^2\Phi_R}{\cos(\phi)\,dA\,d\omega}$

1.3.3.3 Reflection, Absorption and Transmission

When a radiant flux (q_{Ri}) emitted by a surface at temperature T touches another, a part is absorbed (q_{Ra}), a part is reflected (q_{Rr}) and, if transparent, a part is transmitted (q_{Rt}):

$$\alpha = q_{Ra}/q_{Ri} \quad \rho = q_{Rr}/q_{Ri} \quad \tau = q_{Rt}/q_{Ri} \tag{1.70}$$

In there, α, ρ and τ are the average absorptivity, reflectivity and transmissivity at the temperature of the surface touched. Conservation of energy now states that the sum of these three must be 1, or:

$$\alpha + \rho + \tau = 1$$

Sum 1 however does not hold for radiation at different temperatures.

A distinction must also be made between diffuse and specular reflection. The last obeys the laws governing optics: incident and reflected beam in the same plane perpendicular to the surface and the angles of incidence ϕ_i and reflection ϕ_r equal (Figure 1.27). Most surfaces however reflect incident radiation diffuse, in all directions.

The reflectivity in direction α can be defined opposite the radiant intensity touching the surface under an angle ϕ ($I_{Ri\phi}$):

$$\rho_\varphi = I_{Rr\alpha}/I_{RiRi\varphi}$$

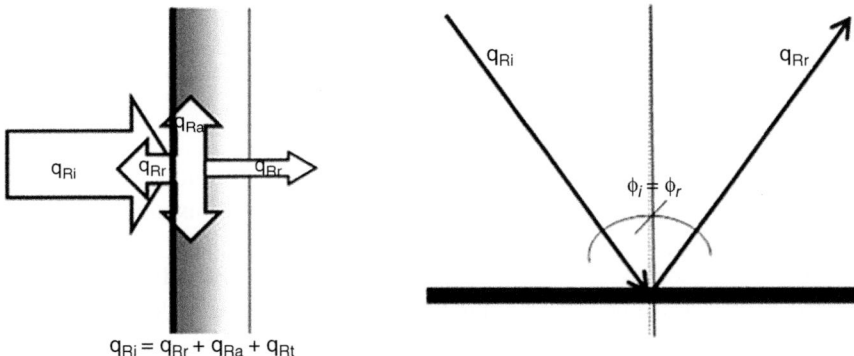

Figure 1.27 Reflection, absorption and transmission at a surface, specular reflection.

In it, $I_{Rr\alpha}$ is the reflected intensity in direction α. For specular reflecting surfaces, the reflectivity is:

$$\rho_\varphi = (I_{Rr}/I_{Ri})_\varphi$$

with ρ_ϕ function of the angle of incidence.

Most building and insulating materials are opaque for thermal radiation ($\tau = 0$). What comes in is absorbed in a thin surface layer, 10^{-6} m thick for metals and 10^{-4} m thick for other materials. Because of that, the term absorbing surface is commonly used. Instead, most gases, liquids and some solids such as glass and many synthetics are selectively transparent but also show selective absorption in their mass depending on their extinction coefficient (a), (Figure 1.28):

$$\frac{dq_R}{q_R} = -a\,dx \tag{1.71}$$

For such material layer with thickness d, the transmitted and absorbed radiant fluxes so become:

$$q_{Rt} = q_{Ri} \exp(-ad) \quad q_{Ra} = q_{Ri} - q_{Rt} = q_{Ri}[1 - \exp(-ad)]$$

with q_{Ri} the incoming radiant flux. Absorptivity and transmissivity consequently equal:

$$\alpha = 1 - \exp(-ad) \quad \tau = \exp(-ad)$$

Figure 1.28 Transparent materials, the way they absorb radiation passing through.

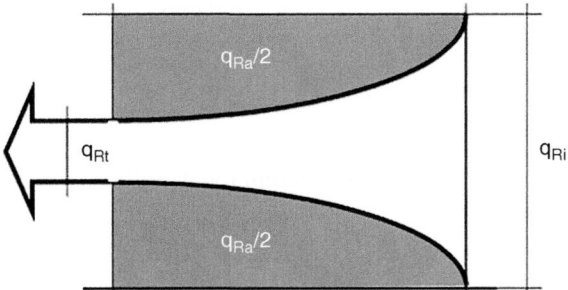

For an irradiated surface of a transparent material, the specular reflectivity so looks:

$$\rho = \frac{I_r}{I_i} = \left[\frac{n_1 \cos(\phi_i) - n_2 \cos(\phi_t)}{n_1 \cos(\phi_i) + n_2 \cos(\phi_t)}\right]^2$$

with n_1 the refractive index of the gas crossed, n_2 the refractive index of the material, ϕ_i the angle of incidence in the gas and ϕ_t the angle of transmittance in the transparent material.

As mentioned, absorptivity, reflectivity, and transmissivity vary with temperature, thus, with wavelength, albeit for the radiation intensity also the angle of incidence matters. This dependency anyhow can change quickly. Take glass. Its transmissivity for visible light is large whereas for low temperature IR it nears zero with the absorptivity then passing 0.9. This explains the greenhouse effect. All surfaces in a space with glazing absorb the short-wave, high temperature solar radiation transmitted by that glazing and re-emits it as low temperature IR radiation, which the glass absorbs. This turns conduction through the glass into the only way to temper warming indoors. Also, part of the absorbed IR radiation by the glass is released to the indoors, so, both may turn the indoors uncomfortably warm. Transparent synthetics behave the same way, be it that some also transmit IR.

1.3.3.4 Radiant Surfaces

1.3.3.4.1 *Types*

A distinction is made between black surfaces absorbing all incident radiation ($\alpha = 1$, $\rho = 0$, $\tau = 0$, $\alpha \neq f(\lambda, \phi)$), grey surfaces showing a constant absorptivity <1, white surfaces with absorptivity 0 and coloured surfaces, whose absorptivity depends on temperature and the direction of incidence. Although fictitious, real surfaces are considered grey. In order to falsify reality not too much, short wave solar radiation, subscript S, and long wave radiation, subscript L, with their absorptivity and reflectivity, are kept separate.

1.3.3.4.2 *Black Surfaces*

Independent of temperature, black surfaces have an emissivity 1. In fact, according to the second law of thermodynamics, if at different temperature in a closed system, they must evolve towards temperature equilibrium. Once there, all emit and absorb the same amount of radiation, meaning absorptivity equals emissivity, thus 1. What concerns the direction radiation is emitted, black bodies obey Lambert's law: luminosity constant. Or, the radiant intensity is (Figure 1.29):

$$I_{b\phi} = L_b \cos(\phi) \tag{1.72}$$

This equation is known as the cosine law. It offers a simple relation between emittance M_b and luminosity. From the definitions in Table 1.2, a full hemisphere integration gives:

$$M_b = L_b \int_\omega \cos(\varphi) \, d\omega$$

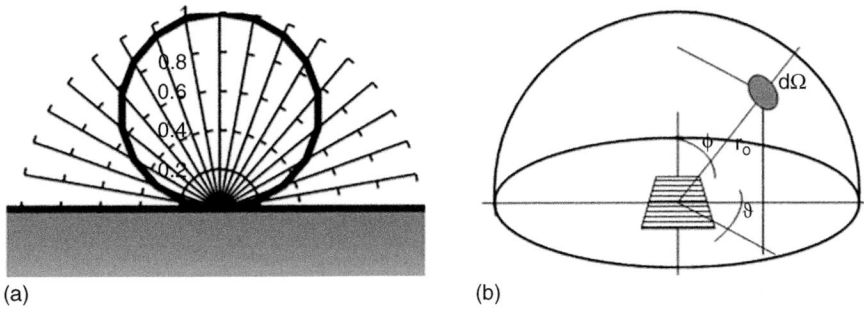

Figure 1.29 (a) Thick line showing the effect of the cosine law on radiation intensity, (b) proving the law.

The angle dω is calculated assuming the hemisphere with radius r_0, surrounds the infinitesimal surface dA with luminosity L_b completely (Figure 1.29). Or, dω equals:

$$d\omega = r_0^2 \sin(\varphi) d\varphi\, d\vartheta$$

On that hemisphere, the intensity drops to $I_{b\varphi}/r_0^2$, or:

$$M_b = L_b \int_0^{2\pi} \int_0^{\pi/2} \frac{\cos(\varphi)}{r_0^2} r_0^2 \sin(\varphi) d\varphi\, d\vartheta = [-\pi L_b \cos^2(\varphi)]_0^{\pi/2} = \pi L_b \qquad (1.73)$$

According to Planck's law, the spectral density of the emittance is:

$$M_{b\lambda} = \frac{2\pi c^2 h \lambda^{-5}}{\exp\left(\dfrac{ch}{k\lambda T}\right) - 1} \qquad (1.74)$$

with c the speed of the electromagnetic waves in m/s, h Planck's constant (6.624 · 10^{-34} J·s) and k the Boltzmann constant (1.38047 · 10^{-23} J/K). The products $2\pi c^2 h$ and ch/k figure as radiation constants for black bodies, symbols C_1 (=3.7415 · 10^{-16} W·m²) and C_2 (=1.4388 · 10^{-2} m·K). Figure 1.30 shows the spectral density of the emittance for a few absolute temperatures.

The emittance, given by the surface bounded by the curves, increases quickly with temperature, while the maximum occurs at ever shorter wavelengths. In the

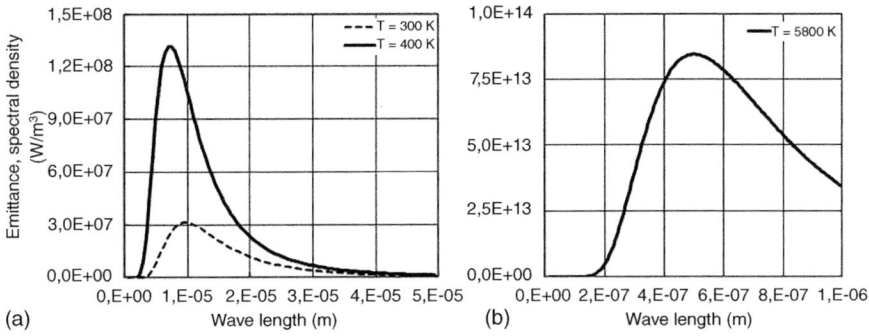

Figure 1.30 Black surface emittance, (a) spectral density at ambient and higher temperature, (b) the same for the sun as black surface.

[λ, $M_{B\lambda}$]-plane their geometric locus is a hyperbole of the fifth order, while the wavelengths obey Wien's law:

$$\lambda_M T = 2898 \ (\lambda_M \text{ in } \mu m) \tag{1.75}$$

At 20 °C, $\lambda_M = 2898/293.15 = 9.9\,\mu m$, a maximum in the IR part. For the sun, with a radiant temperature of 5800 K, $\lambda_M = 2898/5800 = 0.5\,\mu m$, a maximum in the midst of the visible light.

The emittance M_b now follows from integrating Planck's law over the wavelength:

$$M_b = \int_0^\infty M_{b\lambda}\, d\lambda = \frac{2\pi^5 k^4}{15 c^2 h^3} T^4 = \sigma T^4 \tag{1.76}$$

a result, known as the Stefan–Boltzmann law with σ Stefan's constant, $5.67 \cdot 10^{-8}\,W/(m^2 \cdot K^4)$. With Wien, Stefan–Boltzmann preceded Planck's law, which needed quantum mechanics to get formulated. Stefan–Boltzmann is mostly written as:

$$M_b = C_b \left(\frac{T}{100}\right)^4 \tag{1.77}$$

with C_b the black surface constant, $5.67\,W/(m^2 \cdot K^4)$, and $T/100$ the reduced radiant temperature in K. The luminosity and radiation intensity so become:

$$L_b = \frac{M_b}{\pi} = \frac{C_b}{\pi}\left(\frac{T}{100}\right)^4 \quad I_{b\phi} = \frac{d^2 \Phi_{Rb}}{dA\, d\omega} = L_b \cos(\phi) = \frac{C_b}{\pi}\left(\frac{T}{100}\right)^4 \cos(\phi)$$

When two infinitely small black surfaces dA_1 and dA_2, separated by a transparent gas, see each other as in Figure 1.31, the elementary radiant heat flow from dA_1 to dA_2 becomes:

$$d^2 \Phi_{R,1\to 2} = I_{b1}\, dA_1\, d\omega_1 = \frac{M_{b1}}{\pi} \cos(\phi_1) dA_1\, d\omega_1$$

with $d\omega_1$ the angle under which dA_1 sees dA_2:

$$d\omega_1 = dA_2 \cos(\varphi_2)/r^2$$

This allows rewriting the flow exchanged:

$$d^2 \Phi_{R,1\to 2} = \frac{M_{b1}}{\pi} \cos(\phi_1) \cos(\phi_2) dA_1 \frac{dA_2}{r^2}$$

Vice versa, the elementary heat flow from dA_2 to dA_1 is:

$$d^2 \Phi_{R,2\to 1} = \frac{M_{b2}}{\pi} \cos(\phi_1) \cos(\phi_2) dA_1 \frac{dA_2}{r^2}$$

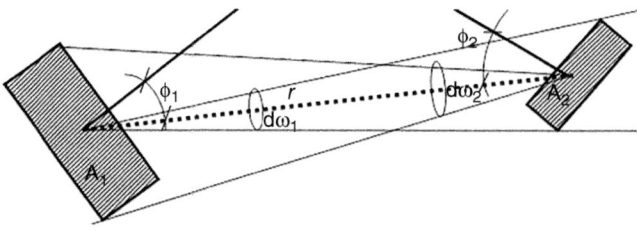

Figure 1.31 View factor between the surfaces A_1 and A_2.

The resulting radiant heat flow between the two so becomes:

From dA_1 to dA_2:
$$d^2\Phi_{R,12} = d^2\Phi_{R,1\to 2} - d^2\Phi_{R,2\to 1}$$
$$= \frac{(M_{b1} - M_{b2})\cos(\phi_1)\cos(\phi_2)dA_1\,dA_2}{\pi r^2}$$

From dA_2 to dA_1:
$$d^2\Phi_{R,21} = d^2\Phi_{R,2\to 1} - d^2\Phi_{R,1\to 2}$$
$$= \frac{(M_{b2} - M_{b1})\cos(\varphi_1)\cos(\varphi_2)dA_1\,dA_2}{\pi r^2}$$

If both surfaces have finite dimensions, the exchange looks:

From A_1 to A_2:
$$\Phi_{R,12} = (M_{b1} - M_{b2})A_1 \left[\frac{1}{\pi A_1}\int_{A_1}\int_{A_2}\frac{\cos(\varphi_1)\cos(\varphi_2)dA_2\,dA_1}{r^2}\right] \quad (1.78)$$

From A_2 to A_1:
$$\Phi_{R,21} = (M_{b2} - M_{b1})A_2 \left[\frac{1}{\pi A_2}\int_{A_2}\int_{A_1}\frac{\cos(\varphi_1)\cos(\varphi_2)dA_1\,dA_2}{r^2}\right] \quad (1.79)$$

The terms between brackets in both are called the view factor, symbol F. Other names are angle factor, shape factor or configuration factor. If surface A_1 is the emitting, the view factor writes as F_{12}, if it's A_2, the view factor writes as F_{21}. View factors indicate which fraction of the radiant flow emitted by one surface reaches a seen other. Their size, form, distance, and viewing angle define the value, 1 if all emitted radiation reaches the other.

Looking to the properties of view factors, first comes their reciprocity: $A_1 F_1 = A_2 F_2$. This follows from the definition. If a surface A_2 surrounds A_1, the view factor from A_1 to A_2 turns 1 (Figure 1.32), a result that holds for any surface surrounded by $n-1$ other surfaces, so forming a closed volume:

$$\sum_{j=2}^{n} F_{1j} = 1$$

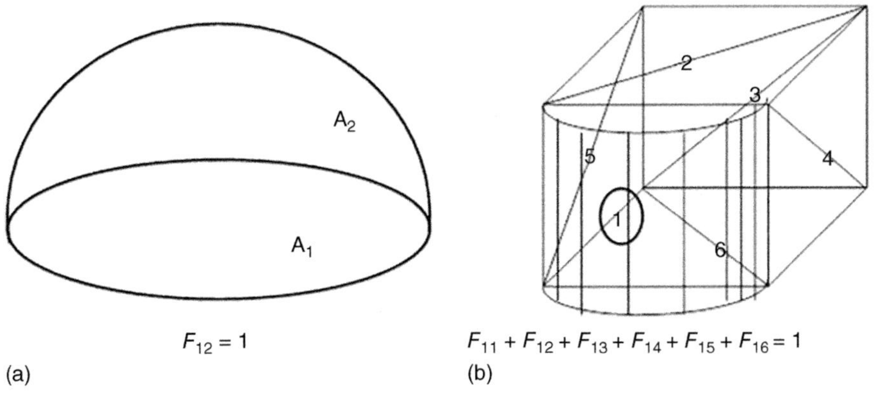

(a) $F_{12} = 1$

(b) $F_{11} + F_{12} + F_{13} + F_{14} + F_{15} + F_{16} = 1$

Figure 1.32 View factor between surface 1 (a) completely surrounded by surface 2 or (b) by the surfaces 2–6. In (b) surface 1 also radiates to itself.

A view factor 1 also holds for two infinitely mutually parallel surfaces, take both end faces of a cavity. For a few other simple surface configurations, the view factor can be calculated analytically. A point situated at a distance D from a rectangle with sides L_1 and L_2, surface $L_1 \cdot L_2 = A_2$, has a view factor with that rectangle equal to the ratio between the angle, under which it sees the rectangle, and the 4π of the sphere surrounding the point:

$$F_{12} = \frac{1}{4\pi} \int_{A_2} \frac{\cos(\phi)}{r^2} dA_2$$

For a point above a corner of that rectangle, the formula turns into ($\cos(\varphi) = D/r$, $r^2 = D^2 + x^2 + y^2$):

$$F_{12} = \frac{1}{4\pi} \int_0^{L_1} \int_0^{L_2} \frac{D}{(D^2 + x^2 + y^2)^{3/2}} dy\, dx \text{ giving } F_{12} = \frac{1}{8} - \frac{1}{4\pi} a\tan\left(\frac{D\sqrt{L_1^2 + L_2^2}}{L_1 L_2}\right)$$

Other positions of the point can be converted to the corner case by using the approach shown in Figure 1.33, giving as view factor:

$$F_{12} = F_{1a} + F_{1b} + F_{1c} + F_{1d}$$

Radiation between the human head and a ceiling is an example of such point to surface situation. Another case concerns an infinitesimal small surface dA_1 parallel to but at a distance D of a rectangle with sides L_1 and L_2 and surface $L_1 \cdot L_2 = A_2$. The formula for the view factor then is:

$$F_{12} = \frac{1}{dA_1} \int_{dA_1} \int_{A_2} \frac{\cos(\phi_1)\cos(\phi_2)dA_2\, dA_1}{\pi r^2}$$

The way dA_1 sees each infinitesimal surface dA_2 in A_2 is independent of dA_1's position, or:

$$F_{12} = \frac{1}{\pi} \int_{A_2 \text{ seen by } dA_1} \frac{\cos(\phi_1)\cos(\phi_2)dA_2}{r^2}$$

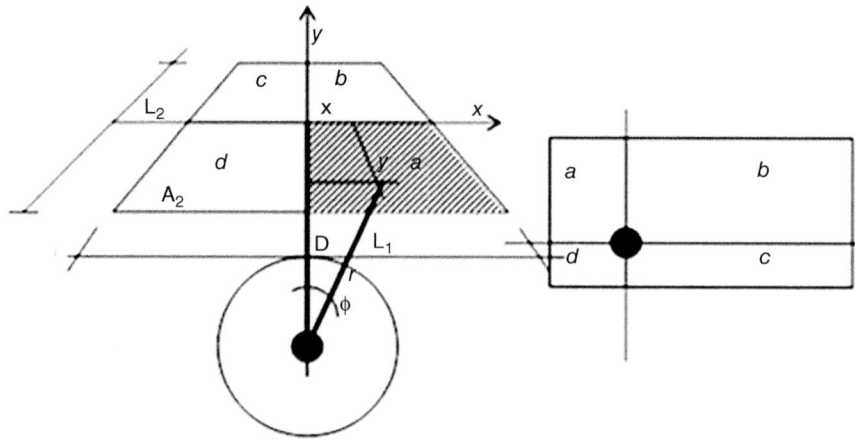

Figure 1.33 Left view factor between a point at distance D of surface A_2.

Figure 1.34 View factor between an infinitesimal small surface dA_1 parallel to and at a distance D from the corner (0,0) of surface A_2.

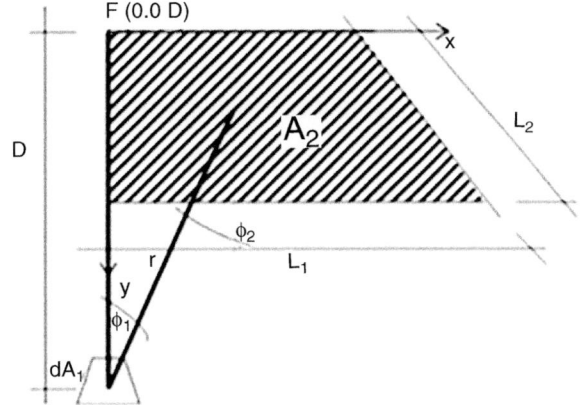

With dA_1 situated above the corner (0, 0) of the rectangle A_2 (Figure 1.34), the view factor becomes:

$$F_{12} = \frac{1}{\pi} \int_0^{L_1} \int_0^{L_2} \frac{\cos(\phi_1)\cos(\phi_2)}{r^2} dx\, dy$$

With $\cos\phi_1 = \cos\phi_2 = D/r$ and $r^2 = D^2 + x^2 + y^2$, this equation simplifies to:

$$F_{12} = \frac{1}{\pi} \int_0^{L_1} \int_0^{L_2} \frac{D^2 dy\, dx}{(D^2 + x^2 + y^2)^2}$$

with as solution:

$$F_{12} = \frac{1}{2\pi}\left[\frac{L_1}{\sqrt{D^2+L_1^2}}a\tan\left(\frac{L_2}{\sqrt{D^2+L_1^2}}\right) + \frac{L_2}{\sqrt{D^2+L_2^2}}a\tan\left(\frac{L_1}{\sqrt{D^2+L_2^2}}\right)\right]$$

Other configurations can be solved numerically, an example being an infinitely small surface dA_1 standing at a distance D perpendicular to a rectangle with sides L_1 and L_2 and surface $L_1 \cdot L_2 = A_2$. Then, dA_1 does not see the part of A_2, which extends behind the intersection with the plane to which it belongs. The numerical formula used is:

$$F_{12} = \frac{D\Delta x \Delta y}{2\pi}\left[\sum_{x=\Delta x/2 \text{ to } L_1-\Delta x/2 \text{ step } \Delta x}\sum_{y=\Delta y/2 \text{ to } L_2-\Delta y/2 \text{ step } \Delta y} \frac{\sqrt{x^2+y^2}}{(x^2+y^2+D^2)^2}\right]$$

Common in beam-shaped rooms are three pairs of equally large parallel surfaces, two for the walls and one for the floor and ceiling. The three are perpendicular to each other and have common edges and corners. An analytical calculation of their view factors is not doable but a numerical solution is easily programmed on a spreadsheet.

Included the view factor, the radiant heat flows and fluxes write as:

$$\begin{aligned}\Phi_{R,12} &= A_1 F_{12}(M_{b1} - M_{b2}) & q_{R,12} &= F_{12}(M_{b1} - M_{b2}) \\ \Phi_{R,21} &= A_2 F_{21}(M_{b2} - M_{b1}) & q_{R,21} &= F_{21}(M_{b2} - M_{b1})\end{aligned} \quad (1.80)$$

For a number of black surfaces radiating to each other, the flow and flux per surface so equal:

$$\phi_{R1n} = A_1 \sum_{j=2}^{n} [F_{1j}(M_{b1} - M_{bj})] \quad q_{R1n} = \sum_{j=2}^{n} [F_{1j}(M_{b1} - M_{bj})] \quad (1.81)$$

with $j = 1$ the surface considered and $j = 2$ to $=n$ the $n-1$ other.

1.3.3.4.3 Grey Surfaces

The laws governing radiation between grey surfaces are similar to the black surface ones, except for the radiant exchange. Per wavelength and direction, grey surfaces emit a constant fraction of the black. Related ratio is called the emissivity (e). As conservation of energy imposes that absorptivity (α) and emissivity must be equal, the reflectivity (ρ) becomes:

$$\rho = 1 - \alpha = 1 - e$$

Grey surfaces with reflectivity 1 are called white. As the radiant heat flux must follow the cosine law, their emittance equals:

$$M = \pi L$$

The spectral density follows from Planck's law multiplied with the emissivity e, giving as total emittance:

$$M = e\, C_b \left(\frac{T}{100}\right)^4 \quad (1.82)$$

Grey surfaces reflect part of the incident radiation. Is eM_b the one's emittance and E the irradiation by the other, then the radiosity of that one is:

$$M' = eM_b + \rho E \quad (1.83)$$

The difference between radiosity and irradiation now fixes the flux emitted:

$$q_R = M' - E \quad (1.84)$$

Eliminating the irradiation E from both equations so gives:

$$q_R = M' - \frac{M' - eM_b}{\rho} = -\frac{e}{\rho}(M' - M_b) \quad (1.85)$$

The radiant flux received thus is:

$$q_R = \frac{e}{\rho}(M' - M_b) \quad (1.86)$$

Otherwise said, a grey surface behaves as a black, be it with a grey filter in front and a radiant resistance given by the ratio of the grey reflectivity to its emissivity (ρ/e) in between that filter with a radiosity M' and a fictitious black surface with emittance M_b. For two grey surfaces A_1 and A_2, separated by a transparent gas-like air, the radiant flow exchanged so is:

From fictitious black surface 1 to grey filter 1: $\Phi_{R,11} = \frac{e_1}{\rho_1}(M_{b1} - M'_1)A_1$

From grey filter 1 to grey filter 2: $\Phi_{R,12} = F_{12}(M'_1 - M'_2)A_1$

From grey filter 2 to fictitious black surface 2: $\Phi_{R,22} = \frac{e_2}{\rho_2}(M'_2 - M_{b2})A_2$

From grey filter 1 to grey filter 2 the radiant heat flow looks like the one between two black surfaces. Indeed, the emittance of a black and the radiosity of a grey both obey Lambert's law for diffuse radiation. As the three flows now must be equal, elimination of the unknown radiosities M_1' and M_2' gives as flow from grey surface A_1 to grey surface A_2 and vice versa:

$$\Phi_{R,12} = \left[\frac{1}{\dfrac{\rho_1}{e_1} + \dfrac{1}{F_{12}} + \dfrac{\rho_2 A_1}{e_2 A_2}} \right] (M_{b1} - M_{b2}) A_1$$

$$\Phi_{R,21} = \left[\frac{1}{\dfrac{\rho_2}{e_2} + \dfrac{1}{F_{21}} + \dfrac{\rho_1 A_2}{e_1 A_1}} \right] (M_{b2} - M_{b1}) A_2 \qquad (1.87)$$

The term between brackets is called the radiation factor F_R, written as $F_{R,12}$ if A_1 is the emitter and $F_{R,21}$ if it's A_2. Dividing both equations by the surface of the emitter gives the fluxes.

A simple configuration is an infinite cavity having two isothermal end faces at different temperature. Then, $F_{12} = F_{21} = 1$, $A_1 = A_2$, and the flux exchanged becomes ($\rho = 1 - e$):

$$q_{R,12} = \frac{\Phi_{R,12}}{A_1} = \left[\frac{1}{\dfrac{1}{e_1} + \dfrac{1}{e_2} - 1} \right] (M_{b1} - M_{b2}) \qquad (1.88)$$

The term between brackets represents the radiation factor between both faces. Is one white, face 1, then $F_{R,12} = 1/(1/0 + 1/e_2 - 1) = 0$. Is one black, face 2, then $F_{R,12} = 1/(1/e_1 + 1/1 - 1) = e_1$

Another configuration is an isothermal surface A_1 surrounded by another isothermal surface A_2 at different temperature. F_{12} then is 1 and the radiant flow becomes:

$$\Phi_{R,12} = \frac{e_1 e_2}{e_2 \rho_1 + e_1 e_2 + \dfrac{e_1 \rho_2 A_1}{A_2}} (M_{b1} - M_{b2}) A_1$$

Are both nearly black ($e > 0.9$), then the denominator nears 1 and:

$$\Phi_{R,12} = e_1 e_2 (M_{b1} - M_{b2}) A_1.$$

More, if A_1 is very small compared to A_2, giving a ratio ≈ 0, the equation further simplifies to:

$$\Phi_{R,12} = e_1 (M_{b1} - M_{b2}) A_1 \qquad (1.89)$$

Or, the radiant heat flow exchanged then only depends on the emissivity of the surrounded surface A_1. When multiple isothermal grey surfaces, all at different but uniform temperature, see each other, the radiant flow between one (A_1) and the $n-1$ other writes as:

From black 1 to grey filter 1: $\Phi_{R,11} = \frac{e_1}{\rho_1}\left(M_{b1} - M_1'\right)A_1$

From grey filter 1 to the $n-1$ other grey filters: $\Phi_{R,1\,\text{to}\,j} = \sum_{j=2}^{n} F_{1j}\left(M_1' - M_j'\right)A_1$

As both are equal, the black emissivity becomes: $M_{b1} = \left(1 + \frac{\rho_1}{e_1}\sum_{j=2}^{n} F_{1j}\right)M_1'$
$- \frac{\rho_1}{e_1}\sum_{j=2}^{n}\left(F_{1j}M_j'\right)$

For the 2 to n surfaces surrounding, this equation converts to:

$$M_{b1} = \frac{M_1'}{e_1} - \frac{\rho_1}{e_1}\sum_{j=2}^{n}\left(F_{1j}M_j'\right) \tag{1.90}$$

With the radiosities M_j' of the grey surfaces unknown and the black emittances M_{b1} known, the result is a system of n equations with: n unknowns:

$$[M_{bj}]_n = [F]_{n.n}\left[M_j'\right]_n \tag{1.91}$$

In it, $[F]_{n.n}$ is the radiation matrix of the n isothermal grey surfaces. Solving the system gives the radiosities M_j' as function of the black surface emittances M_{bj}. The radiant flows then follow from inserting the M_j''s in the equations given for the heat exchange between these fictitious black surfaces and the grey filter in front.

1.3.3.4.4 Coloured Surfaces

For coloured surfaces, emissivity, absorptivity and reflectivity change with wavelength, so with temperature, sometimes even with the direction of the incident or emitted radiation. Kirchhoff's law ($e = \alpha$) still applies but Lambert's law does not as it requires a direction-independent emission. The spectral emittance per wavelength differs from black, though the ratio between coloured and black at a same temperature still fixes the emissivity of the coloured surface at that temperature.

To simplify things, coloured surfaces are considered grey, be it with a temperature-dependent emissivity. For the emittance and irradiance at strongly differing temperatures, take the terrestrial versus the solar temperature, Kirchoff's law no longer applies since the solar short-wave absorptivity (α_S) can be very different from the terrestrial long wave emissivity (e_L), an example being polished aluminium with $\alpha_S = 0.2\text{--}0.4$ versus $e_L = 0.05$.

1.3.3.5 Simple Formulae

At first sight, thermal radiation looks accessible. However, calculating all angle factors is cumbersome, while the system of equations, when multiple grey surfaces interfere, can be very large. A simpler approach so is welcomed. First, reality is reduced to two radiant surfaces: the considered surface 1 and the $n-1$ other, seen as one, the environment. Then, the environment is assumed black with as uniform radiant temperature θ_r the one of a black surface exchanging the same radiant flow with surface 1 as in reality. Solving the system of equations for all surfaces present then gives the radiosity of surface 1 (M_1') as linear combination of the black body emittances of all n surfaces present:

$$M'_1 = \sum_{i=1}^{n} a_{ri} M_{bi}$$

Insertion in the equation for the grey surface radiant flow received and equating with the radiant flow found for the surrounded surface 1, which is usually small compared to the environment, fixes that radiant temperature θ_r:

$$\theta_r = \sqrt[4]{\frac{1}{\rho_1}\left(\sum_{i=1}^{n} a_{ri} T_i^4 - e_1 T_1^4\right)} - 273.15$$

In analogy with convection, the radiant flux and flow to surface 1 now are written as:

$$q_r = h_r(\theta_{s1} - \theta_r) \qquad \Phi_r = h_r(\theta_{s1} - \theta_r)A_1 \tag{1.92}$$

In both, h_r is the surface film coefficient for radiation (W/(m²·K)), which varies with the configuration considered, and θ_{s1} is the temperature of the receiving surface 1. If the environment surrounds surface 1, then the surface film coefficient for radiation follows from equating this simple flux equation to the one derived for surface 1 being small compared to the surrounding environment:

$$h_r = eC_b \left[\frac{\left(\frac{T_{s1}}{100}\right)^4 - \left(\frac{T_r}{100}\right)^4}{\theta_{s1} - \theta_r}\right] \tag{1.93}$$

The term between brackets is called the temperature ratio for radiation F_T:

$$F_T = \frac{T_m}{5000}\left[\left(\frac{T_{s1}}{100}\right)^2 + \left(\frac{T_r}{100}\right)^2\right] \approx \frac{4}{100}\left(\frac{T_m}{100}\right)^3 \tag{1.94}$$

As this ratio hardly varies for temperatures between −10 and 50 °C, the expression at the right side, making the flux quasi linear, suffices. This gives for the surface film coefficient:

$$h_r = e_1 C_b F_T \tag{1.95}$$

For two isothermal parallel surfaces, one at temperature θ_{s1}, and the other at temperature θ_{s2}, no detour via the radiant temperature is needed. The surface film coefficient is directly written as:

$$q_r = h_r(\theta_{s1} - \theta_{s2}) \quad \text{with} \quad h_r = \frac{5.67 F_T}{\frac{1}{e_1} + \frac{1}{e_2} - 1}$$

If part of the surrounding surfaces has the same temperature as surface 1, the radiant temperature (θ'_r) then only includes those at different temperature, giving as surface film coefficient:

$$h_r = \frac{e_1 C_b F_{12} F_T}{e_1 + \rho_1 F_{12}}$$

with F_{12} the view factor between surface 1 and that part of the surroundings at different temperature. If surface 1 is almost black, the denominator turns 1 and:

$$h_r = e_1 \, C_b \, F_{12} \, F_T = 5.67 e_1 \, F_{12} \, F_T \tag{1.96}$$

In case two identical outer walls, equally warm, form a corner, the radiant exchange is restricted to half the environment with surfaces at different temperature. The view factor then is 0.5 and the surface film coefficient for radiation becomes:

$$h_r = e_1 \, C_b \, F_T / 2 = 2.84 \; e_1 \, F_T \tag{1.97}$$

Of course, the one equally warm can be included in the radiant temperature. The view factor then remains 1 but the radiant temperature changes.

1.4 Building-related Applications

1.4.1 Surface Film Coefficients and Reference Temperatures

1.4.1.1 Methodology

In most of the applications, conduction, convection and radiation mix up. Through any assembly without cavity, heat conduction is the mover, while between indoors and the inner face and outdoors and the outer face convection and radiation take over. Related heat fluxes write as the difference between the surface temperature and a reference temperature for the mode and ambient considered, multiplied by a surface film coefficient (h_c, h_r). Yet, because convection and radiation are coupled, a common ambient temperature replacing the mode linked ones is used, while the surface film coefficients in- and outdoors are both combined to one, h_i and h_e. Of course, if needed, the two modes can remain split, which for complex cases can be necessary. Then, as well an air as a radiant temperature intervenes in- and outdoors, while both modes keep their own surface film coefficient.

1.4.1.2 Indoors

For a surface at temperature θ_{si}, the convective heat flux with the air is:

$$[q_{ci}] = h_{ci,s}(\theta_{i,ob} - \theta_{si})$$

In it, $h_{ci,s}$ is the area averaged convective surface film coefficient and $\theta_{I,ob}$ the average air temperature backing the boundary layer. If not this, but the air temperature in the space's centre, 1.7 m above floor level (θ_i), is taken as reference, that flux equation changes to:

$$q_{ci} = h_{ci}(\theta_i - \theta_{si})$$

with h_{ci} the average convective surface film coefficient linked to the new reference temperature:

$$h_{ci} = h_{ci,s} \left(\frac{\theta_{i,ob} - \theta_{si}}{\theta_i - \theta_{si}} \right) \tag{1.98}$$

The radiant heat flux touching the surface in turn equals:

$$q_{ri} = h_{ri}(\theta_{ri} - \theta_{si})$$

with θ_{ri} the radiant temperature indoors. The total heat flux at the surface so becomes:

$$q_i = q_{ci} + q_{ri} = h_{ci}(\theta_i - \theta_{si}) + h_{ri}(\theta_{ri} - \theta_{si}) = (h_{ci} + h_{ri}) \left[\underbrace{\left(\frac{h_{ci}\theta_i + h_{ri}\theta_{ri}}{h_{ci} + h_{ri}} \right)}_{\theta_{ref,i}} - \theta_{si} \right] \quad (1.99)$$

The sum $h_{ci} + h_{ri}$ is called the inside surface film coefficient, symbol h_i, units W/(m²·K) for the weighted average between central air and radiant temperature, named $\theta_{ref,i}$, as reference. Values for the convective part h_{ci} were given when discussing convection. For a surface facing the indoors, the radiant part is:

$$h_{ri} = 5.67 e_L F_T$$

with e_L the long wave emissivity of the surface and F_T the temperature ratio for radiation in the interval $\theta_{si} - \theta_{ri}$, mostly set equal to 0.95. Since most finishes have a long wave emissivity 0.8–0.9, that radiant part becomes:

$$4.3 \leq h_{ri} \leq 4.95 \text{ W/(m}^2 \cdot \text{K)}$$

As standard values for the (combined) inside surface film coefficient (W/(m²·K)) are so advanced:

Vertical surfaces, slope > 45°	Horizontal surfaces, slope ≤ 45°
≈7.7	Heat flow up ($q\uparrow$) 10
	Heat flow down ($q\downarrow$) 6

The 2021 ASHRAE Handbook of Fundamentals gives a more complete set that takes into account the long wave emissivity of the surface:

Position	Heat flow direction	h_i (W/(m² · K) for a surface emissivity		
		0.9	0.2	0.05
Horizontal	Upward	9.26	5.17	4.32
Sloping 45°	Upward	9.09	5.00	4.15
Vertical	Horizontal	8.29	4.20	3.35
Sloping 45°	Downward	7.50	3.41	2.56
Horizontal	Downward	6.13	2.10	1.25

None can be used for cases deviating substantially from the assumptions made. Then, the theory should help in fixing case-specific values.

How to value the reference temperature? Calculating the radiant temperature is quite complex, which is why, provided the room is beam shaped and all surfaces are grey with long wave emissivity ≈ 0.9, an area-weighted average of all surface temperatures figures as quite a good estimate:

$$\theta_{ri} = \sum_{k=1}^{n}(A_k \theta_{sk}) \Big/ \sum_{k=1}^{n} A_k \tag{1.100}$$

In case of grey vertical, sloped and horizontal envelope assemblies with the heat flowing up, the reference so becomes $\theta_{ref,i} = 0.44\,\theta_i + 0.56\,\theta_{ri}$, a result close to the average between the central air and the radiant temperature. That average is called the operative temperature θ_o:

$$\theta_{ref,i} = \theta_o = (\theta_i + \theta_{ri})/2 \tag{1.101}$$

For reflective surfaces, the operative temperature nears the central air temperature as convection then dominates, or $\theta_o \approx \theta_i$. For grey horizontal partitions and envelope assemblies with the heat flowing down, the reference temperature turns into:

$$\theta_{ref,i} = 0.2\,\theta_i + 0.8\,\theta_{ri} \tag{1.102}$$

However, to note is that the larger the impact of envelope assemblies with really cold inside surfaces on the radiant temperature seen by vertical, sloped and horizontal inside partitions is, the less evident the use of the reference temperatures just defined becomes.

1.4.1.3 Outdoors

Three heat fluxes touch the outer surface of envelope assemblies. A first is convection between surface and the outside air:

$$q_{ce} = h_{ce,j}(\overline{\theta}_{e,j} - \theta_{se})$$

In it, $h_{ce,j}$ is the average convective surface film coefficient and $\overline{\theta}_{e,j}$ the average temperature of the air outside the boundary layer, usually assumed equal to the temperature measured by the nearest weather station under thermometer hut 1.7 m above grade (θ_e), or:

$$q_{ce} = h_{ce}(\theta_e - \theta_{se})$$

with h_{ce} the weather station linked convective surface film coefficient:

$$h_{ce} = h_{ce,j} \frac{\overline{\theta}_{e,j} - \theta_{se}}{\theta_e - \theta_{se}}$$

A second is long wave radiation between the surface, the terrestrial environment (e) and the sky (sk) seen as a black. The black emittance from the surface considered (s) to these two equals:

$$M_{bs} = \left[1 + \frac{\rho_{Ls}}{e_{Ls}}(F_{se} + F_{ssk})\right] M'_s - \frac{\rho_{Ls}}{e_{Ls}}(F_{se}M'_e + F_{ssk}M_{bsk})$$

1.4 Building-related Applications

From the terrestrial environment (e) to the surface and to the sky the emittance in turn is:

$$M_{be} = \left[1 + \frac{\rho_{Le}}{e_{Le}}(F_{es} + F_{esk})\right] M'_e - \frac{\rho_{Le}}{e_{Le}}\left(F_{es}M'_s + F_{esk}M_{bsk}\right)$$

In both equations, e_{Ls} and ρ_{Ls} are the long wave emissivity and reflectivity of the surface, e_{Le} and ρ_{Le} the average long wave emissivity and reflectivity of the terrestrial environment, F_{se} the view factor between the surface and the terrestrial environment, F_{es} the view factor for the inverse, F_{ssk} the view factor between the surface and the sky, and F_{esk} the view factor between the terrestrial environment and the sky. M_{bsk} is the black body emittance of the sky while M'_s and M'_e are the radiosities of the surface and the terrestrial environment. As the other two surround the surface, $F_{se} + F_{ssk} = 1$ and due to the surface being infinitely small compared to the terrestrial environment, $F_{es} \approx 0$, while the view factor F_{esk} between the terrestrial environment and the sky ≈ 1 as nearly all its radiation goes to the sky. Both balances so simplify to:

$$M_{bs} = \frac{1}{e_{Ls}}M'_s - \frac{\rho_{Ls}}{e_{Ls}}\left(F_{se}M'_e + F_{ssk}M_{bsk}\right) \quad M_{be} = \frac{1}{e_{Le}}M'_e - \frac{\rho_{Le}}{e_{Le}}M_{bsk}$$

Solving both for M'_s and inserting the result in $q_{rse} + q_{rssk} = e_{Ls}(M_{bs} - M'_s)/r_{Ls}$ knowing that $e_{Ls}(F_{se} + F_{ssk})M_{bs} = e_{Ls}M_{bs}$ gives:

$$q_{rse} + q_{rssk} = q_{re} = e_{Ls}F_{se}(M_{bs} - e_{Le}M_{be}) + e_{Ls}F_{ssk}\left[M_{bs} - \left(\rho_{Le}\frac{F_{se}}{F_{ssk}} + 1\right)M_{bsk}\right]$$

Assumed now is that the terrestrial environment is black and at outside air temperature. The fact that during clear nights the sky temperature falls $\approx 21\,°C$ below the air temperature in the atmosphere, simplifies the radiant flux between surface and outside environment to:

$$q_{rs} = e_{Ls}C_b[(F_{se}F_{Tse} + F_{ssk}F_{Tssk})(\theta_e - \theta_{se}) - 21F_{ssk}F_{Tssk}(1 - 0.87c)]$$

In it, F_{Tse} is the temperature ratio for radiation between surface and terrestrial environment and F_{Tws} the temperature ratio between surface and sky. The factor c represents the cloudiness, 0 for a clear, 1 for an overcast sky.

A third is insolation. Although a daytime reality, the average impact on the heat received is large. Each m^2 of outside face absorbs the beam, diffuse and reflected radiation (E_{ST}) proportionally to its short-wave absorptivity (α_S):

$$q_{se} = \alpha_S E_{ST}$$

Summing up the convective, radiant and solar flux gives:

$$q_e = h_{ce}(\theta_e - \theta_{se}) + 5.67\,e_{Ls}(F_{se}F_{Tse} + F_{ssk}F_{Tssk})(\theta_e - \theta_{se})$$
$$- 120\,e_{Ls}F_{ssk}F_{Tssk}(1 - f_c) + \alpha_K E_{ST}$$

With $5.76\,e_{Ls}(F_{se}F_{Tse} + F_{ssk}F_{Tssk})$ the outside surface film coefficient $h_{e,r}$ for radiation, the equation rewrites as:

$$q_e = (h_{ce} + h_{re})\left\{\left[\theta_e + \frac{\alpha_K E_{ST} - e_{Ls}120\,F_{ssk}F_{Tssk}(1-f_c)}{h_{ce} + h_{re}}\right] - \theta_{se}\right\} \quad (1.103)$$

The term between [ ] has as units °C and is called the average sol-air temperature θ_e^* over a given period (one hour, one day, one week, one month). Its value depends on the radiant properties, the inclination and orientation of the surface, the weather and the period, all factors that differ between applications and ensure that the three fluxes discussed drive the heat exchange between outdoors and outer faces. With the sum $h_{ce} + h_{re}$ representing the outside surface film coefficient h_e, related flux becomes:

$$q_e = h_e \left(\theta_e^* - \theta_{se} \right) \tag{1.104}$$

As stated above, the convective part in h_e is 19 W/(m²·K). For the radiant part, the temperature ratios for radiation F_{Tse} and F_{Tssk} equal to F_T and $F_{se} + F_{ssk} = 1$, gives:

$$h_{re} = 5.67 e_L F_T \tag{1.105}$$

As the temperature factor F_T could scatter between 0.8 and 0.9, the value is:

$$4.4 \, e_L \leq h_{re} \leq 5.1 \, e_L \text{ W/(m}^2\cdot\text{K)}$$

Provided the outside surfaces have a long wave emissivity 0.9, 4–4.6 W/(m²·K) looks likely. Or, the outside surface film coefficient for heat transfer will near 23 W/(m²·K). The EN standard takes 25 W/(m²·K), while the 2021 ASHRAE Handbook of Fundamentals differs between winter and summer:

	Direction of heat flow	h_e (W/(m²·K))
Winter (wind speed 6.7 m/s)	Any	34.0
Summer (wind speed 3.4 m/s)	Any	22.7

A more accurate calculation should consider the whole heat balance, included a better guess of the average wind speed. Long-term measurements at the leeward side of an existing building for example gave on average a much lower outside surface film coefficient than 25 W/(m²·K).

1.4.2 Steady-state, Flat Assemblies

1.4.2.1 Thermal Transmittance of Envelope Assemblies, Partitions and Party Walls

The use of surface film coefficients largely simplified the calculation of the steady-state heat flux ambient to ambient through flat assemblies. Take the outer wall of Figure 1.35.

Indoors, heating keeps the average reference temperature equal to θ_o °C. Outdoors, cold weather gives on average θ_e^* °C. From indoors to the wall's inner face, the heat flux now is:

$$q_1 = h_i (\theta_o - \theta_{si})$$

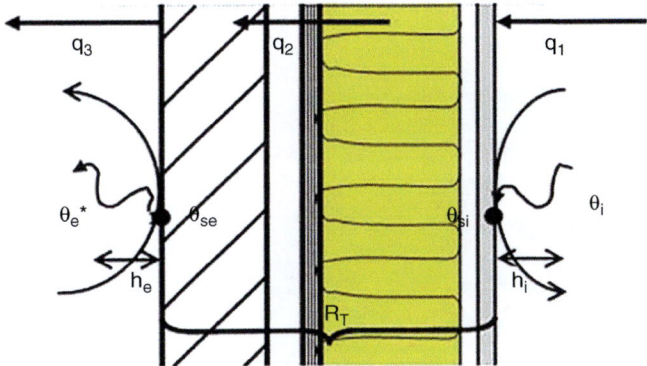

Figure 1.35 Outer wall, thermal transmittance.

with θ_{si} the inside surface temperature. Through the wall, the flux equals:

$$q_2 = (\theta_{si} - \theta_{se})/R_T$$

with θ_{se} the outside surface temperature and R_T the total thermal resistance of the wall. From the outer face to outdoors, the flux becomes:

$$q_3 = h_e \left(\theta_{se} - \theta_e^*\right)$$

In steady state, the three must be equal, common value q. Rearrangement and addition gives:

$$q/h_i = \theta_o - \theta_{si}$$
$$+qR_T = \theta_{si} - \theta_{se}$$
$$+q/h_e = \theta_{se} - \theta_e^*$$

Sum : $\quad q\left(1/h_i + R_T + 1/h_e\right) = \theta_o - \theta_e^*$

This result can be rewritten as $q = U\left(\theta_o - \theta_e^*\right)$ with:

$$U = \frac{1}{1/h_e + R + 1/h_i} \tag{1.106}$$

U is called the thermal transmittance of the outer wall, units W/(m²·K). The lower its value, the less heat flows through, or, the property reflects the insulating quality of outer assemblies. The inverse is called the thermal resistance ambient to ambient, symbol R_a, units m²·K/W. Calculating it looks simple: add the two surface resistances to the total resistance: indoors $1/h_i$, marked R_i, = 0.13 m²·K/W for vertical surfaces, = 0.1 m²·K/W for sloped and horizontal surfaces with the heat flowing up and = 0.17 m²·K/W for horizontal surfaces with the heat flowing down, outdoors $1/h_e$, marked R_e, = 0.04 m²·K/W. The thermal transmittance so defined is called 'clear wall', since two and three dimensional effects are not considered. The property is nonetheless used to characterize heterogeneous walls, take masonry, where the mortar joints and the vertically perforated bricks, see Figure 1.36, both with different λ-value, make the heat flow three-dimensionally.

Figure 1.36 Masonry wall, heat transfer developing three-dimensionally.

For partitions and party walls, the surface film coefficients at both sides are those inside (h_i), giving as thermal transmittance:

$$U = 1/(R + 2/h_i) \qquad (1.107)$$

The reference temperatures are the operative temperatures in the spaces the walls separate.

One remark anyhow. The U-value of multilayer assemblies as advanced does not consider contact resistances between layers. A main reason is these are so low that their impact is negligible. There is one exception anyhow: metal layers contacting each other. In such cases, possible contact resistances may have impact. The reason is clear: metals have such high thermal conductivity that the unintended enclosure of even thin air layers matters. This was the case with walls, consisting of cellular metal boxes filled with mineral wool and finished at the outside with a metal cladding, see Figure 1.37 for a wall with a thermal break between boxes and cladding.

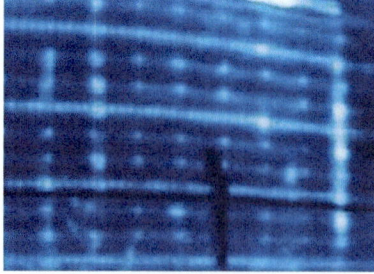

Figure 1.37 Sheet-metal assembly made of cellular metal boxes filled with mineral wool and finished with a metal cladding.

In theory, the clear wall thermal admittance should have been 0.36 W/(m²·K). The value measured without neither a thermal break nor an air layer left between the boxes and the cladding was 1.1 W/(m²·K), a direct consequence of thermal bridging where the box lips are fixed to and touch the cladding, see the figure right. Leaving an air layer between box and cladding with successive thicknesses of 0.5, 1.0 or 1.5 mm lowered the thermal transmittance to respectively 0.85, 0.78 and 0.74 W/(m²/K), still quite higher than 0.36 W/(m²·K).

1.4.2.2 Average Thermal Transmittance of Envelope Parts in Parallel

A facade with surface A_T consists of n different parts in parallel with surfaces A_i (Figure 1.38).

All face the outdoors. If lateral heat exchanges are negligible and if all see the same operative temperature indoors, the heat flow through each of them is:

$$\Phi_i = U_i A_i \Delta\theta$$

If facing different operative temperatures indoors ($\theta_{o,j}$) this equation converts to:

$$\Phi_i = a_i U_i A_i \Delta\theta$$

with 'a_i' a reduction factor equal to:

$$a_i = \frac{\theta_{o,j} - \theta_e}{\theta_{o,ref} - \theta_e}$$

In it, $\theta_{o,ref}$ is the operative temperature indoors taken as reference. The heat flow through the whole now must equal the sum of the heat flows through each part separately, or:

$$\Phi_T = \sum_{i=1}^{n} \Phi_i = \Delta\theta \sum_{i=1}^{n}(a_i U_i A_i) = \Delta\theta U_m \sum_{i=1}^{n} A_i \qquad (1.108)$$

with U_m the average clear wall thermal transmittance of the *n* parts connected in parallel, with a value equal to the sum of the products of all their weighted clear

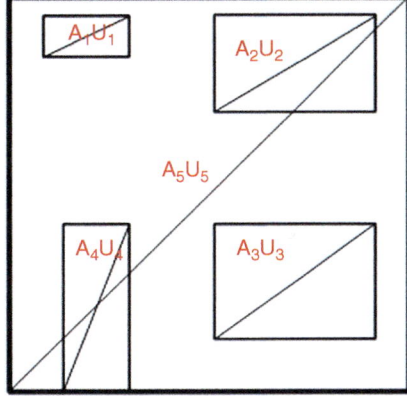

Figure 1.38 Assembly composed of parallel parts with hardly any lateral heat exchange.

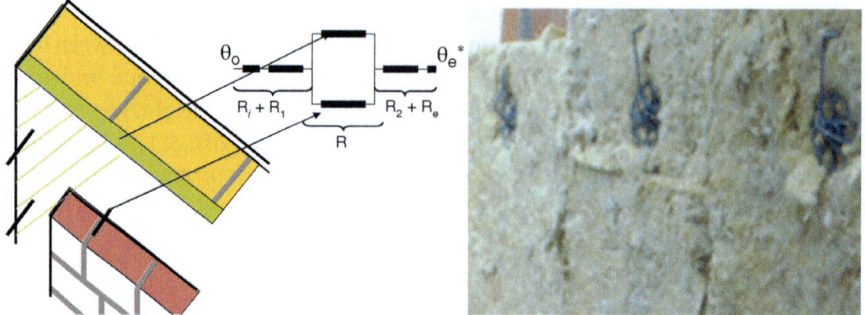

Figure 1.39 Cavity wall, electrical analogy accounting for the ties that perforate the fill.

wall thermal transmittances with their surface, divided by the whole surface:

$$U_m = \sum_{i=1}^{n}(a_i U_i A_i) \bigg/ \sum_{i=1}^{n} A_i = \sum_{i=1}^{n}(a_i U_i A_i) \bigg/ A_T \qquad (1.109)$$

Conversion to an overall thermal resistance gives:

$$R_{am} = A_T \bigg/ \sum_{i=1}^{n} \left(\frac{a_i A_i}{R_{ai}}\right) \qquad (1.110)$$

1.4.2.3 Electrical Analogy

As long as lateral conduction between the composing parts is negligible, electric analogies allow solving quite complex cases. Take a cavity wall where the ties linking the veneer to the inside leaf perforate the cavity fill. Is A_t the whole tie section, R_t related thermal resistance, A_{is} the wall's surface and R_{is} the thermal resistance of the insulation, then the overall thermal resistance (R) of the insulation as perforated becomes (Figure 1.39):

$$R = \frac{A_{is}}{\frac{A_{is} - A_t}{R_{is}} + \frac{A_t}{R_t}}$$

With $R_1 + R_i$ the thermal resistance from the insulation to the indoors and $R_2 + R_e$ the thermal resistance from the insulation to the outdoors, the overall value in- to outdoors so becomes:

$$R_T = (R_i + R_1) + R + (R_2 + R_e) = (R_i + R_1) + \frac{A_i}{\frac{A_i - A_t}{R_{is}} + \frac{A_t}{R_t}} + (R_2 + R_e) \qquad (1.111)$$

1.4.2.4 Thermal Resistance of Non-ventilated Cavities

In an infinitely extending non-ventilated cavity, the distance and temperature difference between both end faces, their radiant properties, its slope, the heat flow direction and the mean temperature of the gas fill, all affect conduction, convection and radiation across, giving as heat flux:

$$q_T = \left(\frac{\lambda_g \, \text{Nu}}{d} + \frac{C_b F_T}{1/e_{L1} + 1/e_{L2} - 1}\right)(\theta_{c1} - \theta_{c2})$$

with λ_g the thermal conductivity of the gas, Nu the case-specific Nusselt number, e_{L1} and e_{L2} the long wave emissivity and θ_{c1} and θ_{c2} the temperatures of both end faces. The thermal resistance so is:

$$R_c = \left(\frac{\lambda_g \, \text{Nu}}{d} + \frac{C_b \, F_T}{1/e_{L1} + 1/e_{L2} - 1} \right)^{-1} \tag{1.112}$$

Figure 1.40 gives values for a vertical cavity filled with 10 °C warm air as function of thickness, temperature difference between and long wave emissivity of both end faces. That the thermal resistance increases when this long wave emissivity is lower and that, when low, the temperature difference between both end faces gains impact, underlines the dominance of radiation in the thermal resistance. The absence of any additional gain once the thickness passes 20–30 mm, in turn, underlines that radiation does not depend on the thickness of the cavity, while increased convection gradually compensates the drop in conduction through the gas.

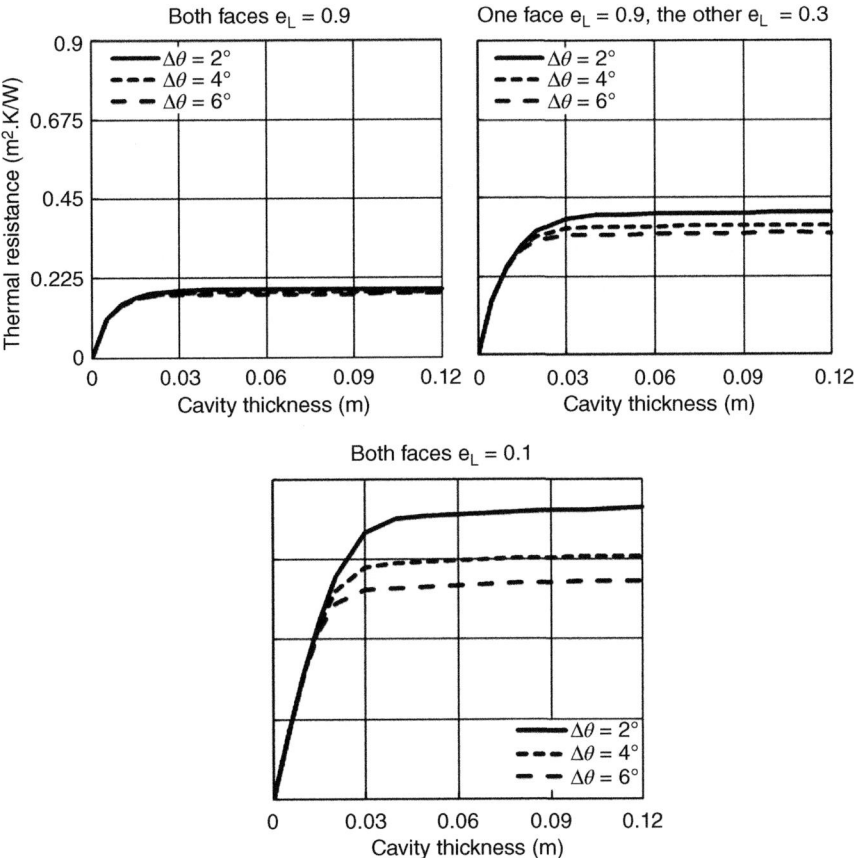

Figure 1.40 Thermal resistance of an infinite vertical cavity.

Table 1.3 Cavity, thermal resistance.

Thickness mm	Vertical cavity		Horizontal cavity	
	R_c (m² K/W), both surfaces grey	R_c (m² K/W), one surface reflecting	R_c (m² K/W) Heat flow up	R_c (m² K/W) Heat flow down
$0 < d < 5$	0.00		0.00	0.00
$5 \leq d < 7$	0.11		0.11	0.11
$7 \leq d < 10$	0.13		0.13	0.13
$10 \leq d < 15$	0.15		0.15	0.15
$15 \leq d < 25$	0.16	0.35	0.17	0.17
$25 \leq d < 50$	0.16	0.35	0.17	0.19
$50 \leq d < 100$	0.16	0.35	0.18	0.21
$100 \leq d < 300$	0.16	0.35	0.18	0.22
$d > 300$	0.16	0.35	0.18	0.23

For finite non-ventilated cavities, the values in Table 1.3 allow a first-order calculation.

1.4.2.5 Interface Temperatures

Considered is an envelope assembly. The surface resistances R_i and R_e are assumed representing the thermal resistance of a 1 m thick air layer with thermal conductivity h_i or h_e, added to the inner and outer face with as temperature differences over them the one between the reference in- ($R = R_a$, $\theta = \theta_i$) and outside ($R = 0$, $\theta = \theta_e^*$) and the temperature on these faces.

For multi-layer envelope assemblies, the temperature course in a $[R, \theta]$ axis system then is a straight line linking $[0, \theta_e^*]$ to $[R_a, \theta_i]$ with as slope the heat flux. Idem so for party walls and partitions, then linking $[0, \theta_{i,1}]$ to $[R_a, \theta_{i,2}]$. Of course, when tracing this line, the layer sequence must be respected, see Figure 1.41. The temperature on the inner face so is:

$$\text{Envelope assembly: } \theta_{si} = \theta_o - R_i \frac{\theta_o - \theta_e^*}{R_a} = \theta_o - \frac{U_{h_i}}{h_i}(\theta_o - \theta_e^*) \quad (1.113)$$

$$\text{Partitions and party walls: } \theta_{si} = \theta_{i.1} - R_i \frac{\theta_{o,1} - \theta_{o,2}}{R_a} = \theta_{o,1} - \frac{U_{h_i}}{h_i}(\theta_{o,1} - \theta_{o,2}) \quad (1.114)$$

The suffix h_i underlines that the clear wall thermal transmittance in the ratio with the surface film coefficient as denominator must be calculated using the same value as used for that surface film coefficient. The temperatures in the interfaces in turn equal:

$$\theta_x = \theta_i - q\left(R_i + R_{si}^x\right)$$

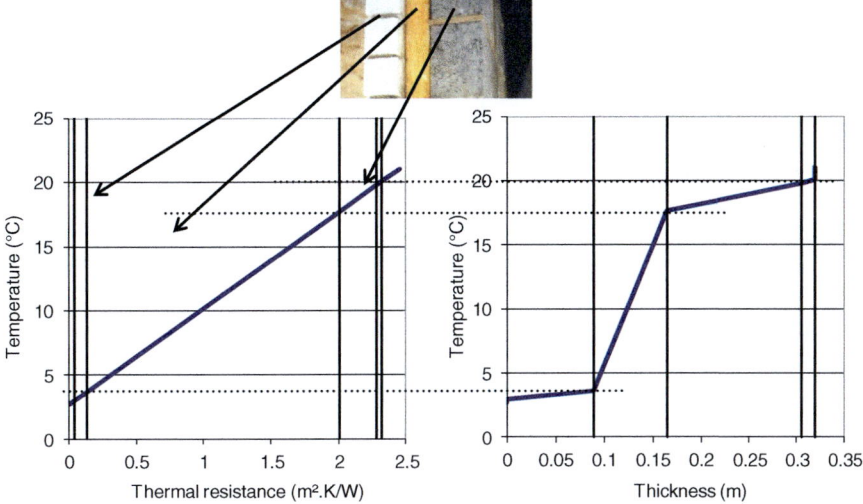

Figure 1.41 Composite envelope assembly (filled cavity wall): temperature line.

Transposing the straight line from this [R, θ] to the [d, θ] axis system goes as explained when discussing conduction through multi-layer flat assemblies, see Figure 1.41.

A same approach applies to party walls and partitions, be it with a surface resistance R_i at both faces and as reference the operative temperatures θ_i in both spaces.

1.4.2.6 Effect of Ever Thicker Insulation Layers on the Thermal Transmittance

The formula for the thermal transmittance of a multi-layer envelope assembly learns that a thicker thermal insulation pushes its value down, which, if done for all envelope parts, should result in less energy used for heating, as needed looking to the pursuit of zero carbon buildings by 2050. Question however is what happens with that less when an ever-thicker insulation layer is built-in? The answer is shown by the U-value as function of the insulation thickness and by its derivative to the insulation thickness:

$$\frac{dU}{dd_{ins}} = \frac{d}{d_{ins}}\left(\frac{1}{R_o + d_{ins}/\lambda_{ins}}\right) = \frac{-1/\lambda_{ins}}{\left(R_o + d_{ins}/\lambda_{ins}\right)^2}$$

In it, R_o is the thermal resistance of all layers included the surface film resistances but without the insulation layer, while d_{ins} is the thickness in m and λ_{ins} the thermal conductivity in W/(m·K) of that insulation layer.

To clarify the result, considered again is the cavity wall of Figure 1.41. The thermal resistance ambient to ambient excluded the cavity fill is 0.34 m²·K/W. The glass wool insulation used as fill has as thermal conductivity 0.04 W/(m·K). Increasing its thickness stepwise from 4 to 40 cm gives a hyperbolic decrease of the U-value, while the derivative to this thickness underlines how rapidly the fall (−) goes, see Figure 1.42, which shows that the extra gains by lowering the U-value diminish quickly with increasing insulation thickness, meaning that a stepwise lowering of the values mandated brings ever less additional gain, so, has its limits.

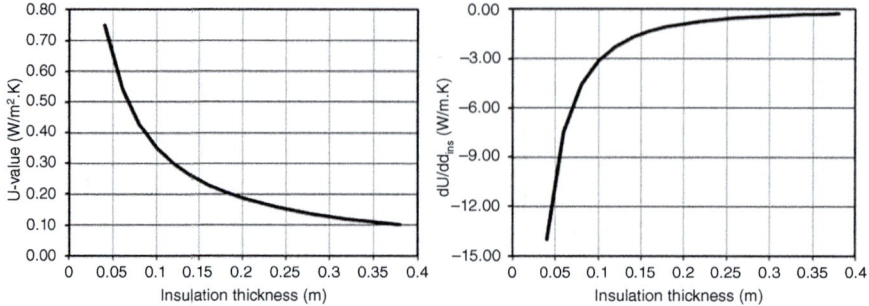

Figure 1.42 Impact of the insulation thickness on the thermal transmittance of the cavity wall of Figure 1.40.

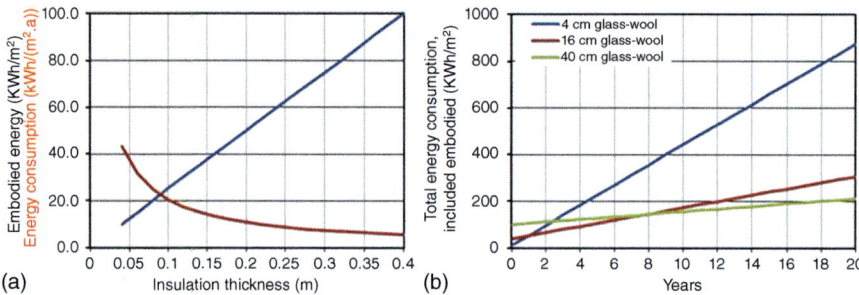

Figure 1.43 (a) Embodied energy in 1 m² of the glass wool, compared to the annual energy use for heating, (b) the total over a period of 20 years, included the embodied part.

Producing glass wool of course also requires energy, called the embodied energy. The thicker the insulation layer, the higher the embodied, see Figure 1.43. The figure shows that the total heating energy used per 1 m² of wall over a period of 20 years, included the embodied in the insulation is represented by the jump in energy used at year zero. While with 4 cm glass wool, the embodied part hardly matters, with 16 cm, giving a U-value 0.23 W/(m²·K), it takes one year to see the energy needed dropping below the 4 cm case. For 40 cm, giving a U-value 0.10 W/(m².K), less energy needed compared to 4 cm demands some 3 years to drop below but compared to 16 cm up to 8 years are needed to drop below. There is even more. With 40 cm insulation in the cavity, the wall thickness turns ≈65 cm instead of the 30 cm with 4 cm in the cavity. In case an equal usable floor surface is requested, this 40 cm demands more built-up floor surface, or, in case built-up must remain the same, gives less usable floor surface. All this does not counteract the importance of an excellent thermal insulation, though without exaggerating how low the thermal transmittances mandated must be.

1.4.2.7 Solar Transmittance

The solar heat flux through an envelope part can be written as:

$$q_S = gE_{ST} \tag{1.115}$$

with E_{ST} the incident solar radiation on the outer face and q_S the heat flux transmitted, both in W/m². The quantity g, called the solar transmittance, encompasses the direct and indirect solar gains. The direct are:

$$q_{Sd} = \tau_S E_{ST}$$

with τ_S the total shortwave transmissivity of the part. For opaque parts, τ_S and the direct gains are zero. Not so for transparent parts. Indirect gains occur because opaque and transparent parts absorb a fraction of the incident solar heat. That warming conducts part of it to the inside face, where convection and long wave radiation dissipates it into the indoor ambient.

For single glass with shortwave absorptivity α_S, this resembles a surface-linked heat exchange with the indoors, equal to:

$$q_{Si} = h_i(\theta_{si} - \theta_o) \tag{1.116}$$

where θ_{si} is the unknown solar induced inside surface temperature. As such single pane warms up nearly homogeneously, its temperature remains close to constant over the glazing's thickness, giving as heat balance per m²

$$\alpha_S E_{ST} + h_e(\theta_e - \theta_x) + h_i(\theta_o - \theta_x) = 0$$

In it, θ_e is the air temperature outdoors, θ_o the operative temperature indoors and θ_x the glass temperature with as value:

$$\theta_x = \theta_{si} = \frac{\alpha_S E_{ST}}{h_i + h_e} + \frac{h_e \theta_e + h_i \theta_i}{h_i + h_e}$$

The second term on the right side in this equation stands for the temperature the glass should have without sun, and the first term for the increase due to the absorbed solar heat. Combination with the equation for the heat flux to the indoors gives:

$$q_{Si} = \frac{h_i \alpha_S E_{ST}}{h_i + h_e} + \frac{h_i h_e(\theta_e - \theta_i)}{h_i + h_e} \tag{1.117}$$

Only the first term on the right side is sun-linked, so represents the indirect gains:

$$q_{Si} = \frac{h_i \alpha_S E_{ST}}{h_i + h_e}$$

The solar transmittance for single glass so becomes:

$$g = \frac{q_{Sd} + q_{Si}}{E_{ST}} = \tau_S + \frac{\alpha_S}{1 + h_e/h_i} \tag{1.118}$$

Or, the gains not only depend on the short-wave transmissivity of the glass but also on its short-wave absorptivity and the ease with which the absorbed heat is dissipated to the indoors.

For double glass, calculating the solar transmittance is more demanding. Let τ_{S1}, ρ_{S1}, α_{S1} and τ_{S2}, ρ_{S2}, α_{S1} be the transmissivity, reflectivity and absorptivity, all shortwave, of the one and the other pane. Reflection in the cavity breaks the solar radiation transmitted into a geometric series of added transmissions with ratio $\rho_{S1}\rho_{S2}$, having as sum (Figure 1.44):

$$q_{Sd} = \frac{\tau_{S1}\tau_{S2}}{1 - \rho_{S1}\rho_{S2}} E_{ST} \tag{1.119}$$

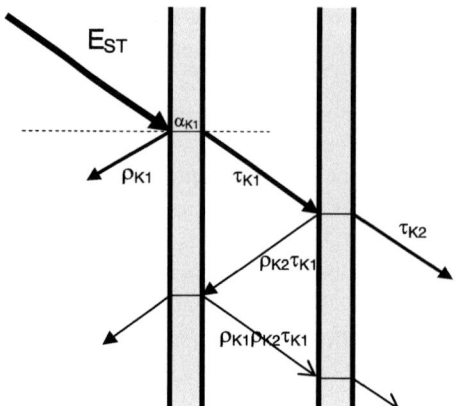

Figure 1.44 Double glass, solar transmittance.

As the denominator $1-\rho_{s1}\rho_{s2}$ is ≈ 1, it's the product of the transmissivities of both panes that mainly fixes the transmission through double glass.

For $\theta_e = \theta_o = 0\,°C$ and both panes isothermal, the indirect gains become $q_{Si} = h_i\theta_{x2}$ with θ_{x2} the temperature of the sun shone inner pane, a value ensuing from the heat balances per pane (1 is the outer, 2 the inner):

$$\text{Pane 1: } \alpha_{S1}\frac{1-\rho_{s1}\rho_{s2}+\tau_{s1}\rho_{s2}}{1-\rho_{s1}\rho_{s2}}E_{ST} - h_e\theta_{s1} + \frac{\theta_{x2}-\theta_{x1}}{R_c} = 0$$

$$\text{Pane 2: } \alpha_{S2}\frac{\tau_{s1}}{1-\rho_{s1}\rho_{s2}}E_{ST} + \frac{\theta_{x1}-\theta_{x2}}{R_c} - h_i\theta_{x2} = 0$$

Solving gives:

$$\theta_{x2} = \frac{\dfrac{\alpha_{S1}f_1}{R_c}+\alpha_{S2}f_2\left(h_e+\dfrac{1}{R_c}\right)}{\left(h_i+\dfrac{1}{R_c}\right)\left(h_e+\dfrac{1}{R_c}\right)-\dfrac{1}{R_c^2}}E_{ST} \qquad (1.120)$$

Inserting this value in the equation for the indirect gains and adding the direct gives as solar transmissivity of double glazing:

$$g = \tau_{S1}\tau_{S2} + h_i\frac{\dfrac{\alpha_{S1}f_1}{R_c}+\alpha_{S2}f_2\left(h_e+\dfrac{1}{R_c}\right)}{\left(h_i+\dfrac{1}{R_c}\right)\left(h_e+\dfrac{1}{R_c}\right)-\dfrac{1}{R_c^2}} \qquad (1.121)$$

This result shows how to decrease it: limit the direct transmission and lower either the inside surface film coefficient or the shortwave absorptivity of the panes.

For multiple glazing and the combination of glass and solar shading, the same approach applies: the transmittance for the direct gains equal to the product of the shortwave transmissivities of the panes and the shading, for the indirect gains equal to the result of solving the system of heat balances per pane plus the shading.

1.4.3 Local Inside Surface Film Coefficients

The surface film coefficients as given are of help in quantifying the area averaged heat losses and gains. To get the surface temperatures or heat fluxes at specific spots, local values should be used. As start holds:

$$h_{ix}(\theta_{ref,i} - \theta_{six}) = h_{cix}(\theta_{ix} - \theta_{six}) + h_{rix}(\theta_{rix} - \theta_{six}) \quad (1.122)$$

with h_{ix} the local surface film coefficient linked to a reference temperature $\theta_{ref,i}$, h_{cix} the local convective surface film coefficient linked to the local air temperature θ_{ix} outside the boundary layer, h_{rix} the local radiant surface film coefficient linked to the radiant temperature θ_{rix} seen by the spot considered and θ_{six} the local surface temperature. If R' is the equivalent thermal resistance linking the inside face at that spot to the ambient at the other side, then:

$$h_{ix}(\theta_{ref,i} - \theta_{six}) = (\theta_{six} - \theta_j)/R' \quad (1.123)$$

This equation is an approximation because in reality, such equivalent thermal resistance depends on how the local inside surface film coefficients (h_{ix}) are distributed over the inner face. For envelopes, θ_j is the temperature outdoors ($j = e$), for partitions and party walls the reference temperature in the adjacent space. Eliminating the local surface temperature θ_{six} from the two equations above and solving for the local surface film coefficient gives:

$$h_{ix} = \frac{h_{cix} + h_{rix} - p_T}{1 + R'p_T} \quad \text{with} \quad p_T = \frac{h_{cix}(\theta_{ref,i} - \theta_{ix}) + h_{rix}(\theta_{ref,i} - \theta_{rix})}{\theta_{ref,i} - \theta_j} \quad (1.124)$$

In case the reference temperature indoors ($\theta_{ref,i}$), its relation with the air temperature just outside the boundary layer (θ_{ix}), its relation with the radiant temperature seen by the spot (θ_{rix}) are known and the local inside surface film coefficients h_{cix} and h_{rix} are quantified, then, on condition the equivalent thermal resistance is known, the equations above allow quantifying the heat flux at the spot considered.

Questions anyhow left are: how to link the local reference temperature indoors to the overall reference indoors ($\theta_{ref,i}$), values of the local surface film coefficients? Taken as overall reference indoors is the air temperature θ_i in the room's centre, 1.7 m above floor level. Assuming the air temperature there (θ_{ix}) increases linearly along the vertical, be it less when the enclosure is better insulated and heating less convective, the relation with the overall reference could be:

$$\frac{\theta_{ix} - \theta_j}{\theta_i - \theta_j} = 1 + 0.2\, p_c U_m (y - 1.7) \quad (1.125)$$

with y a spot on that vertical, θ_j the reference temperature in the adjacent space or outdoors, p_c a convection factor (1 for air heating, 0.9 for convector heating, 0.4–0.8 for radiator heating, 0.4 for floor heating), and U_m the average thermal transmittance of all walls enclosing the space considered. The relation comes from data gathered in a test room with varying enclosure insulation and warmed with several heating systems. Computer simulations of the radiant heat exchange in spaces with different shape on the other hand showed that the local radiant temperature (θ_{rix}) could be given a value proportional to the reference, with gradient depending on the local convective surface film coefficient, the convection factor and the average U_m-value:

$$\frac{\theta_{rix} - \theta_i}{\theta_i - \theta_j} = \frac{h_{cix}}{h_{cix} + \frac{(p_c - 0.4) U_m}{0.6}} \qquad (1.126)$$

Further on, the local convective surface film coefficient (h_{cix}) could be assumed being 2.5 W/(m²·K), while following values can be advanced for the radiant surface film coefficient (h_{rix}):

Corner between three envelope assemblies or two and a partition	
Spot more than 0.5 m from an edge	5.5 e_L
Spot less than 0.5 m from an edge but more than 0.5 m from a corner	3.4 e_L
Spot less than 0.5 m from a corner	2.2 e_L
Edge between two envelope assemblies or one and a partition	
Spot more than 0.5 m from an edge	5.5 e_L
Spot less than 0.5 m from an edge	3.4 e_L
Envelope assembly or partition	
Spot anywhere	5.5 e_L

Where furniture hides a wall, a combined inside surface film coefficient of 2 W/(m² · K) is used.

1.4.4 Steady-state: Two and Three Dimensions

1.4.4.1 Pipes

The heat flow between the pipe's outer face and the ambient all-around is (Figure 1.45):

$$\Phi_{n+1} = 2\pi R_{n+1} h_2 (\theta_{s,2} - \theta_{ref,2}) \qquad (1.127)$$

with h_2 the surface film coefficient, $\theta_{ref,2}$ the reference temperature in the ambient, $\theta_{s,2}$ temperature on the pipe's outside face and R_{n+1} the radius between the pipe's centre and that face.

At the pipe's inner face, the heat flow is:

$$\Phi_1 = 2\pi R_1 h_1 (\theta_{ref,1} - \theta_{s,1}) \qquad (1.128)$$

with h_1 the surface film coefficient and $\theta_{ref,1}$ the temperature of the fluid in the pipe, $\theta_{s,1}$ related surface temperature and R_1 the pipe's inner radius.

Through the pipe, the flow becomes:

$$\Phi_{1,n+1} = \frac{\theta_{s,1} - \theta_{s,2}}{\sum_{i=1}^{n} \left[\frac{\ln(R_{i+1}/R_i)}{2\pi \lambda_i} \right]} \qquad (1.129)$$

Figure 1.45 The pipe problem.

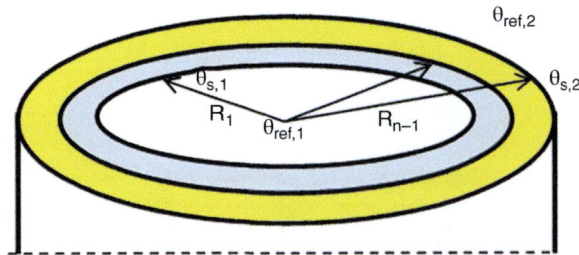

Σ because the pipe may be composed of several layers. In steady state, the three heat flows must be equal. Reshuffling and adding so gives:

$$\frac{\Phi_{n+1}}{2\pi R_{n+1} h_2} = \theta_{s,2} - \theta_{ref,2}$$

$$+ \Phi_{1,n+1} \sum_{i=1}^{n} \left[\frac{\ln(R_{i+1}/R_i)}{2\pi \lambda_i} \right] = \theta_{s,1} - \theta_{s,2}$$

$$+ \frac{\Phi_1}{\theta_{ref,1} - \theta_{s,1}} = 2\pi R_1 h_1$$

$$\overline{\Phi_{1,n+1} \left\{ \frac{1}{2\pi R_{n+1} h_2} + \sum_{i=1}^{n} \left[\frac{\ln(R_{i+1}/R_i)}{2\pi \lambda_i} \right] + \frac{1}{2\pi R_1 h_1} \right\} = \theta_{ref,1} - \theta_{ref,2}}$$

As for flat assemblies, this sum can be rewritten as:

$$\Phi_{1,n+1} = U_{pipe}(\theta_{ref,1} - \theta_{ref,2}) \tag{1.130}$$

where U_{pipe} is the thermal transmittance per meter run of the pipe:

$$U_{pipe} = \frac{1}{\frac{1}{2\pi R_{n+1} h_2} + \sum_{i=1}^{n} \left[\frac{\ln(R_{i+1}/R_i)}{2\pi \lambda_i} \right] + \frac{1}{2\pi R_1 h_1}} \quad (W/(m \cdot K)) \tag{1.131}$$

Insulating a pipe lowers its heat loss or gain, depending on whether the fluid transported is warm or cold. The benefit of thicker insulation however drops faster than for flat assemblies.

1.4.4.2 Floors on Grade

The heat loss of a floor on grade to the soil is a three-dimensional reality. Although the concept 'thermal transmittance' does not fit in such case, a simplified flat surface approach, using an equivalent thermal transmittance, is used. For that, the thermal transmittance is rewritten as:

$$U = a U_{o,floor} \tag{1.132}$$

with 'a' a reduction factor and $U_{o,floor}$ the thermal transmittance of the floor as if flat and facing the outdoors. Calculating 'a' starts with fixing the characteristic floor dimension:

$$B' = 2A_{fl}/P \quad (m) \tag{1.133}$$

Figure 1.46 Arrow showing a fraction of the free perimeter.

In it, A_{fl} is the floor surface and P that part of the floor perimeter contacting the outdoors and called the free perimeter, see Figure 1.46.

Next, the equivalent soil thickness (d_t) of the floor is calculated as:

$$d_t = d_{fw} + \lambda_{gr}\left(\frac{1}{h_e} + R_{T,fl} + \frac{1}{h_i}\right) \text{ (m)} \tag{1.134}$$

with d_{fw} the average thickness of the foundation walls along the free perimeter in m, λ_{gr} the thermal conductivity of the soil, $R_{T,fl}$ the thermal resistance of the floor, h_i the surface film coefficient indoors, 6 W/(m²·K)), and h_e the surface film coefficient outdoors, 25 W/(m²·K). The reduction factor a finally depends on how the equivalent soil thickness compares to B':

If $d_t < B'$, then $a = \frac{1}{U_{o,floor}}\left(\frac{2\lambda_{gr}}{\pi B' + d_t}\right)\ln\left(\frac{\pi B'}{d_t} + 1\right)$ If $d_t \geq B'$, then $a = \frac{1}{U_{o,floor}}\left(\frac{\lambda_{gr}}{0.457 B' + d_t}\right)$

1.4.4.3 Thermal Bridges

A thermal bridge means not only more heat loss or gain than through the surrounding flat parts, except if single or double glazing, but also the inside face there colder during the heating season. To calculate the isoflux lines and the isotherms, CVM is used, taking into account that the heat always enters a thermal bridge perpendicular to the inner face and leaves it perpendicular to the outer face. Calculating surface temperatures require the use of local, the heat flows the use of the standard surface film resistances. The energy balance for a control volume with its centre on an in- or on an outer face so combines six heat flows: four from the four adjacent centres on that face, one from the adjacent control volume in the material layer at that face and one perpendicular to the face from the ambient at reference temperature through the surface film resistance, see Figure 1.47.

The control volumes in the material touching an end face extending parallel to the [y, z]-plane have a surface a^2 and a depth $a/2$. The heat coming from the ambient at temperature $\theta_{i,m,n}$ and flowing to one of these, having as surface temperature $\theta_{s,m,n}$, equals:

$$\Phi^{i,m,n}_{s,m,n} = h_i(\theta_{i,m,n} - \theta_{s,m,n})a^2$$

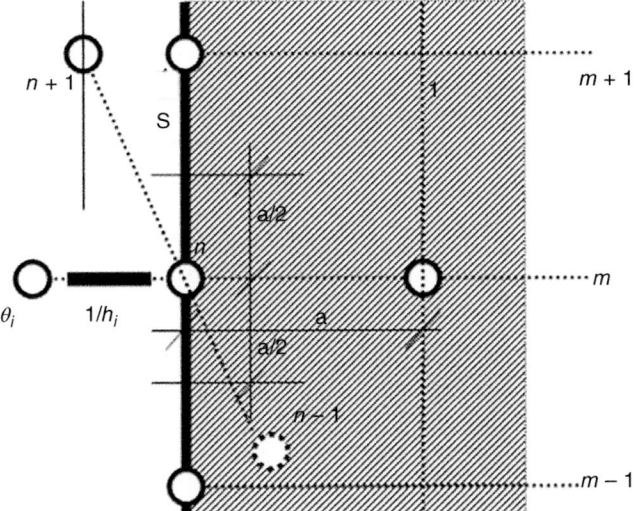

Figure 1.47 CVM-method, control volumes at the inside or outside surface.

The heat flows from the adjacent four at the end face crossing the distance a over a surface $a^2/2$ are

$$\Phi^{s,m,n}_{s,m+1,n} = \lambda_1(\theta_{s,m+1,n} - \theta_{s,m,n})\frac{a}{2} \quad \Phi^{s,m,n}_{s,m-1,n} = \lambda_1(\theta_{s,m-1,n} - \theta_{s,m,n})\frac{a}{2}$$

$$\Phi^{s,m,n}_{s,m,n+1} = \lambda_1(\theta_{s,m,n+1} - \theta_{s,m,n})\frac{a}{2} \quad \Phi^{s,m,n}_{s,m,n-1} = \lambda_1(\theta_{s,m,n-1} - \theta_{s,m,n})\frac{a}{2}$$

The heat flow from the adjacent control volume in the material with centre at distance a from the end face to the central on the end face in turn writes:

$$\Phi^{s,m,n}_{l,m,n} = \lambda_1(\theta_{l,m,n} - \theta_{s,m,n})a$$

Sum zero gives:

$$ah_i\theta_{i,m,n} + \lambda_1 \frac{\theta_{s,m+1,n} + \theta_{s,m-1,n} + \theta_{s,m,n+1} + \theta_{s,m,n-1}}{2}$$
$$+ \lambda_1\theta_{l,m,n} - (ah_i + 3\lambda_1)\theta_{s,l,m,n} = 0$$

with $ah_i\theta_{i,m,n}$ known. Central control volumes at corners give analogous equations.

A difference now is made between geometrical and structural thermal bridges (Figure 1.48).

The geometrical are linked to the form building enclosures have. Structural ones follow from structural decisions: concrete beams and columns contacting the outdoors, discontinuities in the thermal insulation, etc. Often the need for structural integrity is a cause. Take a balcony. The cantilever moment requires continuity with the floor slab, thus a thermal bridge. Excluding requires continuity of the thermal insulation. But complete continuity cannot, although the higher heat flows and lower surface temperatures the continuity needed induces should remain acceptable.

To facilitate calculating the heat losses and gains due to thermal bridging, the concept of linear and local thermal transmittances has been introduced. The first,

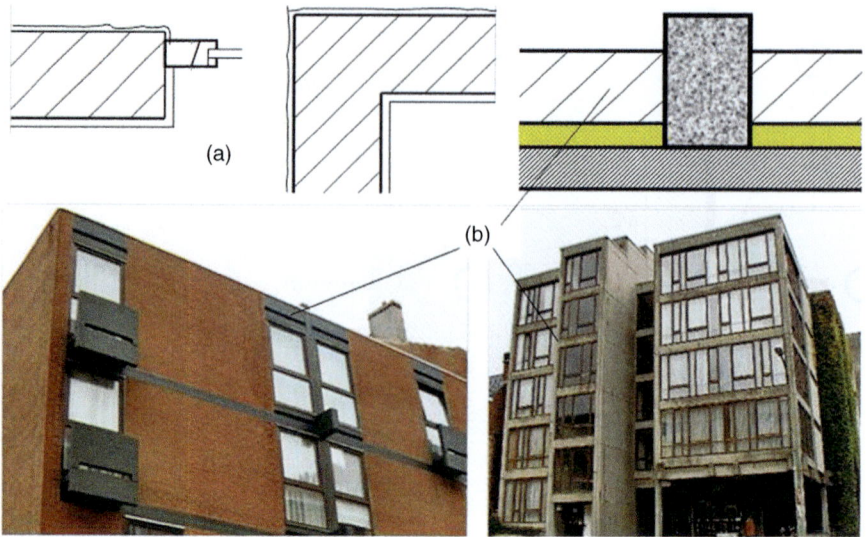

Figure 1.48 (a) Geometric thermal bridges, (b) structural thermal bridges.

symbol ψ, units W/(m·K), quantify the extra heat flow a two-dimensional thermal bridge induces per running metre and 1 K temperature difference between the ambient at both sides compared to no thermal bridge being there. The second, symbol χ, units W/K, does the same for a three-dimensional thermal bridge.

Calculating linear thermal transmittances demands a well-defined clear wall reference and agreement on which end face to consider, in- or outside. 'Outside' is preferred as it allows using facade drawings. First, the detail forming the thermal bridge is omitted and the one-dimensional clear wall heat flow calculated. Then, the linear thermal bridge using the correct drawings is added, after which CVM-meshing allows quantifying the two-dimensional heat flow and the inside surface temperatures (Figure 1.49). Are Φ_{2D} and Φ_o the heat flows through the wall with and without thermal bridge, then the linear thermal transmittance with L its length equals:

$$\psi = \frac{\Phi_{2D} - \Phi_o}{L \Delta\theta} \quad (1.135)$$

If an assembly contains a local thermal bridge, the local thermal transmittance (χ) so becomes:

$$\psi = \frac{\Phi_{3D} - \Phi_o}{L \Delta\theta} \quad (1.136)$$

Often local thermal bridges emerge where linear cross each other. If so, three cases must be calculated: first the clear wall, then the wall with consecutively each of the linear thermal bridges and finally the wall three-dimensionally. After, the local thermal transmittance is extracted as:

$$\chi = \frac{\Phi_{3D} - \Phi_{2D}}{\Delta\theta} \quad (1.137)$$

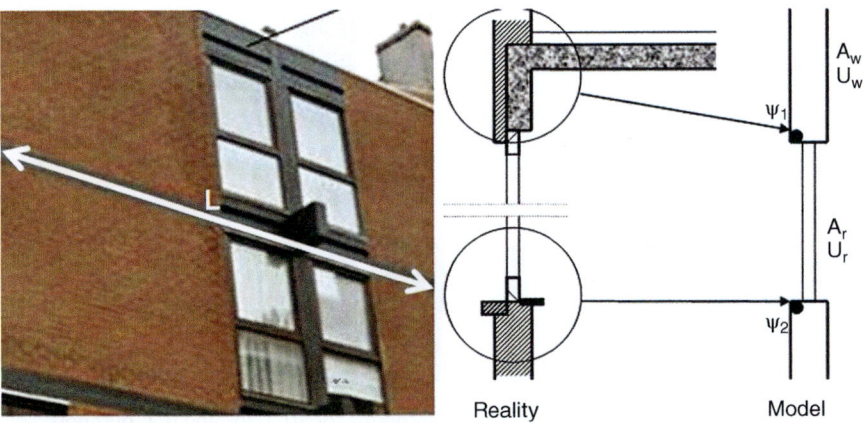

Figure 1.49 Linear thermal transmittances. The dummy consists of flat parts with lines, in cross-section points, that replace the linear thermal bridges.

Once all linear and local thermal transmittances are quantified, the whole wall thermal transmittance of a flat assembly containing thermal bridges follows from:

$$U = U_o + \frac{\sum_{i=1}^{n}(\psi_i L_i) + \sum_{j=1}^{m}\chi_j}{A} \quad (1.138)$$

with U_o the clear wall thermal transmittance, A the surface considered, n the number of linear thermal bridges over that surface, L_i their length and m the number of local thermal bridges over it.

A two or three dimensional calculation with the local inside surface film coefficient (see above) in turn will give the lowest inside surface temperature $\theta_{s,\min}$, which allows characterizing the impact of thermal bridges on the inside surface temperature by using a non-dimensional temperature factor:

$$f_{h_i} = \frac{\theta_{s,\min} - \theta_e}{\theta_o - \theta_e} \quad (1.139)$$

where θ_o and θ_e are the reference temperatures in- and outdoors. The suffix h_i reminds that the local surface film coefficient must be used. A CVM calculation with 1 K temperature difference between the ambient at both sides directly gives that temperature factor.

The higher the linear or local thermal transmittance and the lower the temperature factor, the more problematic a thermal bridge. While often taking a disproportionate share in the heat losses or gains by degrading insulation efficiency, their inside surfaces may collect more dirt, may see mould growth risk increase and may turn into preferred spots for surface condensation and crack formation.

Available is software that allows calculating the two and three dimensional heat flows through, the isotherms plus isoflux lines across and related linear or local thermal transmittances and temperature factors for any thermal bridge, requiring as input the geometry, the meshing and the material properties of the composing layers, see Figure 1.50 for a result.

Figure 1.50 Roof edge as thermal bridge, the isotherms for 20 °C in- and 0 °C outdoors, calculated using appropriate software.

1.4.4.4 Windows

Windows transfer heat three-dimensionally as the IR image of Figure 1.51 shows. Quantifying their thermal transmittance (U_{window}) so requires appropriate software. However, as using this software for any case encountered could be far too time-consuming, frames are characterized by an equivalent thermal transmittance ($U_{eq,frame}$), multi-pane glasses by their central thermal transmittance ($U_{o,glass}$) and the spacer/frame combinations, see the white rectangle in Figure 1.51, by a linear thermal transmittance (ψ_{spacer}). This allows writing U_{window} as:

$$U_{window} = \frac{A_{glass} U_{o,glass} + A_{frame} U_{eq,frame} + \psi_{spacer} L_{spacer}}{A_{window}} \quad (1.140)$$

The frame surface (A_{frame}) is assumed equal to its normal projection on a plane parallel to the window's outer face. The visible glass surface (A_{glass}) is fixed the same way, while the length of the spacers (L_{spacer}) follows from the sum of the perimeters of

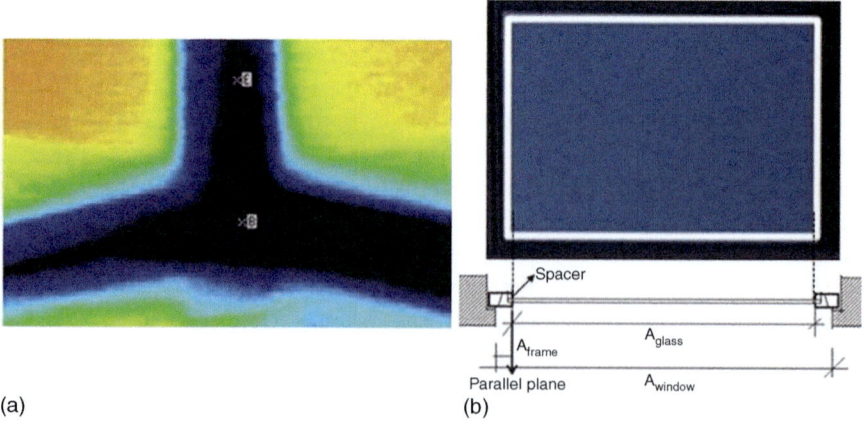

(a)   (b)

Figure 1.51 Window, (a) an IR-image of the frame containing double glazing, (b) calculating related thermal transmittance.

Table 1.4 Frames, glass and spacers, thermal transmittances and linear thermal transmittances.

Frames	$U_{eq,frame}$ (W/(m²·K))	Glazing	$U_{o,glass}$ (W/(m²·K))
Hardwood, $d = 70$ mm	2.08	Double	2.8
Aluminium, 20 mm thermal cut	2.75	Double, low-e, argon-filled	1.1
PVC, three air holes frame	2.00	Triple, low-e, argon filled	0.6

		Spacers		
Metal		ψ_{spacer} W/(m·K)	Insulating	ψ_{spacer} W/(m·K)
$U_{eq,frame} < 5.9$ W/(m²·K) $U_{o,glass} > 2.0$ W/(m²·K)		0.06	$U_{eq,frame} < 5.9$ W/(m²·K) $U_{o,glass} > 2.0$ W/(m²·K)	0.05
$U_{eq,frame} < 5.9$ W/(m²·K) $U_{o,glass} < 2.0$ W/(m²·K)		0.11	$U_{eq,frame} < 5.9$ W/(m²·K) $U_{o,glass} < 2.0$ W/(m²·K)	0.07

all separate glass parts. Table 1.4 lists practical values for the thermal transmittances and linear thermal transmittances of different frames, glazing and spacers.

1.4.4.5 Building Envelopes

Building envelopes, also called enclosures, have as functions protecting the indoors from the weather and shielding it from the soil, from crawlspaces, unheated basements, sometimes channels. They consist of low-slope roofs, sloped roofs, outer walls, glazed surfaces, party walls (often not counted, the other side assumed as warm as the indoors considered) and floors, see Figure 1.52.

They are by definition three-dimensional. Quantifying the time-averaged heat flow for a 1 °C difference between in- and outdoors is done by decomposing envelopes in flat and curved parts, coupled in parallel and having as surfaces A_j and as clear wall thermal transmittances $U_{o,j}$. The contact lines between parts, the structural system applied and the details applied may add linear thermal transmittances over lengths L_k and local thermal transmittances. The result is a mean thermal transmittance, equal to:

$$U_m = \frac{\sum_{j=1}^{n}(a_j A_j U_{o,j}) + \sum_{k=1}^{m}(a_k L_k \psi_k) + \sum_{l=1}^{p}(a_l \chi_l)}{A_T} \quad (1.141)$$

with the multipliers a_j, a_k and a_l reduction factors with value 1 for parts separating the indoors from outdoors, a value <1 for parts separating the indoors from unheated adjacent spaces, from floors on grade, floors above unheated basements or crawlspaces and vertical walls contacting the soil and, if mandated, a value 0 for party walls. The a for parts contacting water channels is:

$$a = \frac{1}{1 - 0.04 U_o}$$

Figure 1.52 The envelope, also called enclosure.

with U_o their clear wall thermal transmittance, calculated with a surface film coefficient h_e zero at the waterside.

Of course, how to measure related surfaces and lengths has to be decided upon. Handy is using the outside dimensions because these can be read from the architectural plans. When for simplicity reasons thermal bridging is not included, using the outside dimensions also limits the error. Bad workmanship of course can result in an average thermal transmittance larger than calculated. A formula reflecting this could be:

$$U_m = \frac{\sum_{j=1}^{n}(a_j A_j U_{o,j}/\eta_{\text{ins},j}) + \sum_{k=1}^{m}(a_k L_k \psi_k/\eta_{\text{ins},k}) + \sum_{l=1}^{p}(a_l \chi_l/\eta_{\text{ins},l})}{A_T} \quad (1.142)$$

with $\eta_{\text{ins},j}$, $\eta_{\text{ins},k}$ and $\eta_{\text{ins},l}$ the insulation efficiencies, 1 for perfect and <1 for lazy workmanship giving gaping joints between, air looping around, wind washing behind and indoor air washing in front of the insulation boards.

1.4.5 Heat Balances

Surface film coefficients do not exactly reflect reality. In cases, where the concept does not work, a return to solving the separate heat balances is preferred. For that, first, the surfaces or interfaces with unknown temperature and heat flow across are replaced by calculation dots, whose number fixes the totality of heat balances needed. Then, per dot, conservation of energy states that the sum of all heat flows coming from or going to the ambient or neighbour dots is zero. This way, each dot gives an equation with its temperature and the temperature of some or all of its adjacent dots as unknown and those fixed as known. Solving that system gives the requested temperatures and heat flows or fluxes. The challenge resides in not overlooking heat flows.

1.4.6 Non-steady-state

1.4.6.1 Periodic Boundary Conditions: Flat Assemblies

Assuming that surface film resistances can be seen as the resistance of a 1 m thick air layer with thermal conductivity h_i or h_e and volumetric specific heat capacity zero turns the reference temperatures into a kind of fictitious surface temperatures, for which applies:

$$\omega_n = \sqrt{\frac{2 \text{ in } \pi\rho c\lambda}{T}} = 0 \quad \cosh(\omega_n R) = 1$$

$$\omega_n \sinh(\omega_n R) = 0 \quad \frac{\sinh(\omega_n R)}{\omega_n} = \frac{0}{0} = \lim_{n \to \infty}\left(\frac{\sinh(\omega_n R)}{\omega_n}\right) = R$$

The complex surface matrixes so become:

$$W_i = \begin{bmatrix} 1 & 1/h_i \\ 0 & 1 \end{bmatrix} \quad W_e = \begin{bmatrix} 1 & 1/h_e \\ 0 & 1 \end{bmatrix}$$

Transposing these into real ones gives:

$$W_i = \begin{bmatrix} 1 & 0 & 1/h_i & 0 \\ 0 & 1 & 0 & 1/h_i \\ 0 & 0 & 1 & 0 \\ 0 & 0 & 0 & 1 \end{bmatrix} \quad W_e = \begin{bmatrix} 1 & 0 & 1/h_e & 0 \\ 0 & 1 & 0 & 1/h_e \\ 0 & 0 & 1 & 0 \\ 0 & 0 & 0 & 1 \end{bmatrix} \quad (1.143)$$

For multi-layered envelope assemblies, the system matrix ambient to ambient so becomes:

$$W_{na} = W_i W_{n1} W_{n2} W_{n3} \cdots W_{nn} W_e$$

For a multi-layer partition, it is:

$$W_{na} = W_i W_{n1} W_{n2} W_{n3} \cdots W_{nn} W_i$$

For single-layer assemblies, the product is $W_{na} = W_i W_n W_e$ for a part belonging to the enclosure and $W_{na} = W_i W_n W_i$ for a partition or party wall.

Handling face-related transient heat exchanges this way of course is a simplification. Radiation engages all other surfaces seen, the volumetric specific heat capacity of air is not 0 but ≈ 1200 J/(kg·K), the air velocity plays and the interactions with other surfaces and furniture may cause some convective inertia.

1.4.6.2 Periodic Boundary Conditions: Spaces

Assumed is that the envelope and partitions enclosing a space are decomposable in flat parts coupled in parallel, while windows do not show thermal inertia. For simplicity reasons, ventilation by outside air is kept constant, air exchanges with neighbour spaces do not exist and all gains, solar and internal, are injected in the space's centre. The operative temperature θ_o in that centre is coupled to all parts by surface film coefficients h_i, combining convection and radiation (Figure 1.53).

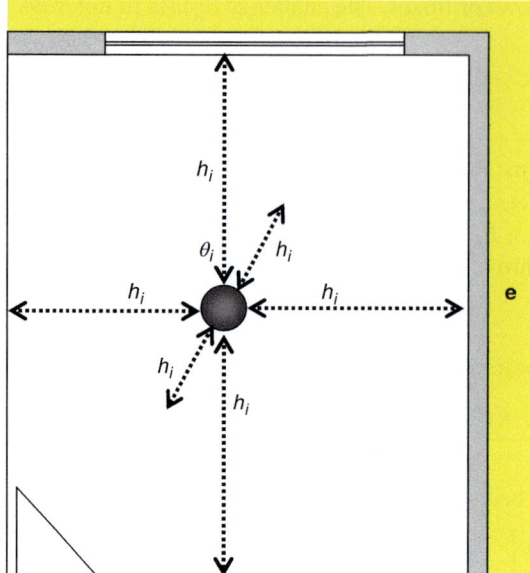

Figure 1.53 Replacing a space by its centre.

The response to a periodic heat input consists of a steady state average, a first harmonic with period T, equal to the time interval considered, for example 1 day, and higher harmonics with periods $T/2$, $T/3$. The steady state average with the operative temperature (θ_o) unknown equals:

$$\sum_{j=1}^{n}\left[a_{e,j}U_{e,j}A_{e,j}\left(\theta_{e,j}^{*}-\theta_{o}\right)\right]+\sum_{k=1}^{m}\left[U_{w,k}A_{w,k}\left(\theta'_{e,k}-\theta_{o}\right)\right]$$

$$+\sum_{l=1}^{p}\left[U_{w,l}A_{w,l}\left(\theta_{e}-\theta_{o}\right)\right]$$

$$+\underbrace{0.34nV(\theta_{e}-\theta_{i})}_{(1)}+\underbrace{\sum_{k=1}^{m}(g_{w,k}f_{w,k}r_{w,k}E_{\text{sun},w,k})+\overline{\Phi}_{\text{intern}}}_{(2)}=0 \quad (1.144)$$

The suffix e stands for opaque envelope assemblies, w for windows, and i for partitions. $\theta_{e,j}^{*}$ is the sol–air temperature per opaque envelope assembly j. The temperature θ_i in the ventilation term (1) is the air temperature in the space, assumed equal to the operative temperature θ_o. Term (2) gives the solar gains through the windows. The θ_i's represent the operative temperatures in the adjacent spaces, the A's all surfaces and the U's the related clear wall thermal transmittances. V is the air volume in the space, n the ventilation rate (ach), g the solar transmittance of the windows included their shading devices, $E_{\text{sun},w,k}$ the solar radiation touching the glass, f the ratio between glass and total window area and 'r' the shadow factor. The product $g_{w,k}f_{w,k}r_{w,k}E_{\text{sun},w,k}$ gives the average solar and the value Φ_{intern} the average internal gain over the period considered. $\theta'_{e,k}$ finally, is the specific sol air temperature per window:

$$\theta'_{e,k} = \theta_e - \frac{120 e_L F_{w,sk}(1-f_c)}{h_e}$$

with θ_e the air temperature outside, e_L the long wave emissivity of the glass, $F_{w,sk}$ the view factor between glass and sky, f_c the cloudiness factor and h_e the outside surface film coefficient.

The harmonics in turn write as:

$$\sum_{j=1}^{n}\Phi_{e,j}^n + \sum_{k=1}^{m}\Phi_{w,k}^n + \sum_{l=1}^{p}\Phi_{i,l}^n + \Phi_{vent}^n + \sum_{k=1}^{m}\Phi_{sun,w,k}^n$$

$$+ \Phi_{intern}^n = \underbrace{\left(\rho_a c_a + \frac{c_f M_f}{V}\right) V_j \frac{d\theta_o^n}{dt}}_{(3)} \quad (1.145)$$

with $\Phi_{e,j}^n$ the nth harmonic of the heat flows through the opaque envelope assemblies, $\Phi_{i,k}^n$ the nth harmonic of the heat flows through the partitions, $\Phi_{w,l}^n$ the nth harmonic of the heat flows through the windows, Φ_{vent}^n the nth harmonic of the enthalpy flow by ventilation, $\Phi_{sun,w,k}^n$ the nth harmonic of the solar gains, Φ_{intern}^n the nth harmonic of the internal gains, θ_o^n the nth harmonic of the operative temperature, c_f the specific heat capacity and M_f the weight of all furniture and furnishings in the space.

The operative temperature and heat flows, written as complex quantities, look:

Operative temperature	$\theta_o^n = \alpha_o^n \exp(2in\pi t/T)$
Transmission	$\Phi^n = \hat{\Phi}^n \exp(2in\pi t/T)$
Ventilation	$\Phi_{vent}^n = \hat{\Phi}_{vent}^n \exp(2in\pi t/T)$
Solar gains	$\Phi_{sun}^n = \hat{\Phi}_{sun}^n \exp(2in\pi t/T)$
Internal gains	$\Phi_{internal}^n = \hat{\Phi}_{internal}^n \exp(2in\pi t/T)$

In these formulas, α_o^n is the complex operative temperature, the $\hat{\Phi}_x^n$'s all complex heat flows, T the base period, n the order of the harmonic and i the imaginary unit. Rewriting the harmonic heat balance so gives:

$$\sum_{j=1}^{n}\hat{\Phi}_{e,j}^n + \sum_{k=1}^{m}\hat{\Phi}_{w,k}^n + \sum_{l=1}^{p}\hat{\Phi}_{i,l}^n + \hat{\Phi}_{v,j}^n + \sum_{k=1}^{m}\hat{\Phi}_{sun,w,k}^n + \hat{\Phi}_{intern}^n = i(\omega_n \rho_a cV)\alpha_o^n \quad (1.146)$$

with ω_n the pulsation of the nth harmonic. The value c, the equivalent specific heat capacity in the space, now is set equal to five times the specific heat capacity of air, so replacing the capacitive term (3, see equation 1.145):

$$c = c_a + c_f M_f/(\rho_a V) \approx 5 c_a \approx 5000 \quad (1.147)$$

If necessary, the value of term (3) of course can be used.

Applying the concepts 'temperature damping, dynamic thermal resistance and admittance', discussed for flat assemblies, to the space, allows rewriting the separate complex heat flows. The results are given for the first harmonic. Higher harmonics generate identical expressions, be it with the transient properties, complex temperatures and complex heat fluxes of the harmonic considered. Assuming all heat flows go from out to in, the ones through the opaque envelope assemblies write as:

$$\hat{\Phi}^n_{e,j} = \alpha'_{e,j} A_{e,j} = \left(\frac{1}{D_{q,e,j}} \alpha^*_{e,j} - \frac{D_{\theta,e,j}}{D_{q,e,j}} \alpha_o \right) A_{e,j} = \left(\frac{1}{D_{q,e,j}} \alpha^*_{e,j} - Ad_{e,j} \alpha_o \right) A$$

For windows, the thermal transmittance remains the property intervening, giving as heat flow:

$$\hat{\Phi}^n_{w,k} = \alpha'_{w,k} A_{i,l} = \left[U_{w,k} \left(\alpha'_{e,k} - \alpha_o \right) \right] A_{w,k}$$

The heat flows crossing the opaque partitions to adjacent spaces look:

$$\hat{\Phi}^n_{i,l} = \alpha'_{i,l} A_{i,l} = (\alpha_l / D_{q,i,l} - Ad_{i,l} \alpha_o) A_{i,l}$$

The constant ventilation rate gives as equation for the related heat flow:

$$\Phi_{vent} = 0.34 \ nV(\alpha_e - \alpha_o)$$

If besides the solar irradiation also the solar transmittance of a window, included its shading, varies, then the complex component of the solar gains turns into:

$$\hat{\Phi}_{sun,w,k} = f_{w,k} A_{w,k} \left(\alpha'_{sun,w,k} \right) \quad \text{with} \quad \alpha'_{sun,w,k} = \text{Harm} \left(g_{w,k} f_{w,k} r_{w,k} q_{sun,w,k} \right)$$

with $\alpha'_{sun,w,k}$ the flux touching the outside face of the shading. Harm(…) indicates that the product between brackets forms a Fourier series.

The complex components of the internal gains finally follow from a Fourier analysis:

$$\hat{\Phi}^n_{intern} = \text{Harm}(\Phi_{intern})$$

Transposing all these flow equations into the balance equation and solving it for the complex operative temperature gives:

$$\alpha_o = \frac{\left[\sum_{j=1}^{n} \left(\frac{A_{e,j}}{D_{q,e,j}} \alpha^*_{e,j} \right) + \sum_{k=1}^{m} \left(U_{w,k} A_{w,k} \alpha'_{e,k} \right) + \sum_{l=1}^{p} \left(\frac{A_{i,l}}{D_{q,i,l}} \alpha_l \right) + 0.34 nV \alpha_e \right]}{\sum_{j=11}^{n} (A_{e,j} Ad_{e,j}) + \sum_{k=1}^{m} (U_{w,k} A_{w,k}) + \sum_{l=1}^{p} (A_{i,l} Ad_{i,l}) + 0.34 nV + i(6000 \omega V)}$$

Using this equation in practice demands a turn to real numbers. If the sol–air and specific sol–air temperatures are presumed equal to the air temperature outdoors

($\theta'_e = \theta^*_e = \theta_e$ and $\alpha'_e = \alpha^*_e = \alpha_e$), which excludes solar radiation and under-cooling, then, for a ventilation rate zero and all adjacent spaces at the same operative temperature as the space considered, the formula simplifies to:

$$\alpha_o = \left\{ \frac{\sum_{j=1}^{n}\left(\frac{A_{e,j}}{D_{q,e,j}}\right) + \sum_{k=1}^{m}(U_{w,k}A_{w,k})}{\sum_{j=1}^{n}(A_{e,j}Ad_{e,j}) + \sum_{k=1}^{m}(U_{w,k}A_{w,k}) + \sum_{l=1}^{q}\left[A_{i,l}\left(Ad_{i,l} - \frac{1}{D_{i,l}}\right)\right] + i(6000\omega V)} \right\} \alpha_e$$
(1.148)

The term between large brackets contains only construction-related characteristics: surfaces, the inverses of the dynamic thermal resistance and admittance of the opaque envelope parts representing the heat storage capacity, the surface and thermal transmittance of all windows, the surface and heat storage capacity of all partitions and the heat storage capacity of the air, the furniture and the furnishings in the space, as indicated set equal to $\rho_a c \approx 6000\,\text{J}/(\text{m}^3\cdot\text{K})$. The inverse, called the room damping for the harmonic considered, so stands for the ratio between the complex outdoor air temperature and the complex indoor operative temperature:

$$D_{\theta,\text{space}} = \left\{ \frac{\sum_{j=1}^{n}(A_{e,j}Ad_{e,j}) + \sum_{k=1}^{m}(U_{w,k}A_{w,k}) + \sum_{l=1}^{q}\left[A_{i,l}\left(Ad_{i,l} - \frac{1}{D_{i,l}}\right)\right] + i(6000\omega V)}{\sum_{j=1}^{n}\left(\frac{A_{e,j}}{D_{q,e,j}}\right) + \sum_{k=1}^{m}(U_{w,k}A_{w,k})} \right\}$$
(1.149)

The property reflects how well the daily temperature fluctuation outdoors will be dampened after entering a space. The first harmonic usually suffices to classify spaces as showing a high or a low dampening.

1.4.6.3 Any Boundary Conditions: Thermal Bridges

For that, 'conduction, non-steady state, any boundary condition, flat assemblies' must be combined with what is discussed about the surface resistance approach under 'construction-related applications, steady state, thermal bridges'. The fictitious transposition of the surface resistance into a 1-m-thick air layer with the surface film coefficient as thermal conductivity, a capacitance 0 and heat transferred perpendicular to the end faces again applies here.

Problems and Solutions

Problem 1 Calculate the thermal transmittance of an outer wall, composed of from the inside out:

	d (cm)	λ (W/(m · K))	R (m² · K/W)
$h_i = 7.7\,\text{W}/(\text{m}^2 \cdot \text{K})$			
Plaster	1	0.3	
Inside leaf	14	0.5	
Cavity fill	8	0.04	
Unvented air cavity	4		0.17
Brick veneer	9	0.9	
$h_e = 25\,\text{W}/(\text{m}^2\cdot\text{K})$			

Solution 1 All quantities must be in SI units. So, metres (m), not centimetres (cm):

$$U_o = \frac{1}{1/h_i + \sum R_j + 1/h_e}$$
$$= \frac{1}{1/8 + 0.01/0.3 + 0.14/0.5 + 0.08/0.04 + 0.17 + 0.09/0.9 + 1/25}$$
$$= 0.36\,\text{W}/(\text{m}^2\cdot\text{K})$$

As all property values are to some extent loaded with uncertainty, limit the result to two digits.

Problem 2 Calculate the thermal transmittance of a low-slope roof, composed of from the inside out:

	d (cm)	λ (W/(m · K))
$h_i = 10\,\text{W}/(\text{m}^2 \cdot \text{K})$		
Plaster	1	0.3
Concrete floor	14	2.5
Screed	10	0.6
Vapour barrier	1	0.2
Thermal insulation	12	0.028
Membrane	1	0.2
$h_e = 25\,\text{W}/(\text{m}^2\cdot\text{K})$		

Solution 2 $U_o = 0.21\,\text{W}/(\text{m}^2\cdot\text{K})$

Problem 3 Calculate the clear wall thermal transmittance of a timber frame wall, composed of from the inside out (the studs concealed):

Layer	d (cm)	λ (W/(m·K))	R (m²·K/W)
$h_i = 7.7\,W/(m^2 \cdot K)$			
Gypsum board	1.2	0.2	
Air space	2		0.17
Airflow retarder	0.02	0.2	
Thermal insulation	20	0.04	
Outside sheathing	2	0.14	
Unvented air cavity	2		0.17
Brick veneer	9	0.9	
$h_e = 25\,W/(m^2 \cdot K)$			

Solution 3 $U_o = 0.17\,W/(m^2 \cdot K)$

Problem 4 Calculate the sol–air temperature for a horizontal surface receiving 750 W/m² solar irradiation. The long wave losses to the clear sky touch 100 W/m². The outdoor air temperature is 30 °C, the outside surface film coefficient 12 W/(m²·K), the shortwave absorptivity of the outer face 0.9 and its long wave emissivity 0.8. Indoors, the surface film coefficient is 7.8 W/(m²·K). Assume steady state.

Repeat the calculation for 24 °C as daily mean outdoor air temperature, 169 W/m² as daily mean solar irradiation, 50 W/m² as daily mean long wave losses to the clear sky and the surface film coefficients just given. The numbers are representative for a south-oriented vertical wall during a hot summer day in a temperate climate.

Redo the exercise for a cold winter day with −15 °C as daily mean outdoor air temperature, 109 W/m² as daily mean solar irradiation, 50 W/m² as daily mean long wave losses if the sky were clear with 0.8 as cloudiness factor. The shortwave absorptivity and long wave emissivity of the wall's outside surface equal 0.5, respectively 0.8, the outside surface film coefficient is 16 W/(m²·K) and the one inside shows no change with the value in the first step of this exercise.

Solution 4 The sol–air temperature in the first situation is:
$$\theta_e^* = \theta_e^* + \frac{a_K E_S - e_L q_L}{h_e} = 30 + \frac{0.9 \cdot 750 - 0.8 \cdot 100}{12} = 79.6\,°C$$

which is high. The mean sol–air for the south-oriented wall during the summer day touches:
$$\theta_e^* = \theta_e^* + \frac{a_K E_S - e_L q_L}{h_e} = 24 + \frac{0.5 \cdot 169 - 0.8 \cdot 50}{16} = 26.8\,°C$$

During the cold winter day, it drops to:
$$\theta_e^* = \theta_e^* + \frac{a_K E_S - e_L q_L}{h_e} = -15 + \frac{0.5 \cdot 109 - 0.8 \cdot 0.8 \cdot 50}{16} = -13.6\,°C$$

Problem 5 Return to Problem (1). Calculate the highest and lowest daily mean temperatures in all interfaces, knowing that the sol-air temperature has the same value as in the repeat part of Problem (4). The operative temperature indoors is 21 °C in winter and 25 °C in summer. The surface film coefficient outside is 16 W/(m²·K), the surface film resistance inside 0.13 m²·K/W. Draw the result.

Solution 5 The temperatures follow from $\theta_j = \theta_i - (\theta_i - \theta_e^*) \sum_{i=1}^{j} R/R_a$. The formula gives as table and figure:

Layer	ΣR m²·K/W	Temp = cold winter day °C	Temp = warm summer day °C
	0	21.0	25.0
	0.13	19.5	25.3
1	0.16	19.1	25.4
2	0.44	15.9	26.0
3	2.44	-7.1	30.3
4	2.61	-9.0	30.6
5	2.71	-10.2	30.9
	2.78	-10.9	31.0

The insulation buffers most of the temperature difference, splitting the wall in an inner part with quite stable temperature and an outer part, suffering the temperature change.

Problem 6 Repeat Problem (5) for the timber frame wall of Problem (3). The outdoor sol–air and air temperature in winter and summer, the operative temperature indoors in winter and summer and the in- and outside surface film coefficients are the same as in Problem (5). Draw the result.

Solution 6

Interface	ΣRj m²·K/W	Temperature (°C) Winter	Temperature (°C) Summer
1/hi	0.00	21	24
Gypsum board	0.13	20.3	24,2
Air space	0.19	20.0	24,2
AFVR	0.36	19.0	24,4
Thermal insulation	0.36	19.0	24,4
Outside sheathing	5.36	-8.3	30,4
Air cavity	5.50	-9.1	30,6
Brick veneer	5.67	-10.0	30,8
1/he	5.77	-10.6	30,9

where AFVR: airflow and vapour retarder.

Problem 7 Take the low-sloped roof of Problem (2). Calculate the highest and lowest daily mean temperature in all interfaces, knowing that the daily mean sol–air temperature outdoors in summer reaches 40 °C for a daily mean outdoor air temperature of 24 °C, while in winter these values drop to −19.5 and −15 °C. The average outside surface film coefficient during windless days is 12 W/(m²·K). Inside, the operative temperature in winter is 21 °C, in summer 25 °C. The inside surface film coefficient is 6 W/(m²·K) in summer and 10 W/(m²·K) in winter. Draw the result.

Solution 7

Interface	Winter		Summer	
	ΣR_j m²·K/W	Temp. (°C)	ΣR_j m²·K/W	Temp. (°C)
$1/h_i$	0	21	0.00	25.0
Render	0.17	19.6	0.10	25.3
Concrete floor	0.20	19.3	0.13	25.4
Screed	0.26	18.9	0.19	25.6
Vapor barrier	0.42	17.5	0.36	26.1
Thermal insulation	0.47	17.1	0.41	26.3
Membrane	4.76	−18.4	4.69	39.6
$1/h_e$	4.81	−18.8	4.74	39.7

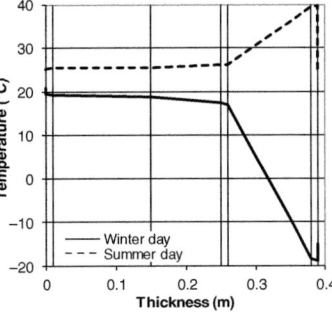

Problem 8 A manufacturer introduces a new sandwich panel composed of from the inside out:

	Thickness (d) (cm)	λ (W/(m·K))	R (m²·K/W)
Aluminium	0.2	230	
VIP (vacuum insulation)	2	0.006	
Air cavity	2		0.15
Glass pane	1	Assume ∞	

The panel is used in a curtain wall. Outdoors, it's 35 °C, indoors 24 °C. Solar irradiation on the glass pane reaches 500 W/m². No long wave radiation intervenes. The outside surface film coefficient is 15 W/(m²·K), the inside 7.7 W/(m²·K). Short wave radiant properties of the glass pane are $a_S = 0.05$, $\rho_S = 0.20$, $\tau_S = 0.75$. The VIP facing the cavity behind has a shortwave absorptivity 1. Which temperature will be noted in the glass pane? How large is the heat flux entering the building through the panel?

Solution 8 The problem is solved by writing two heat balances: one for the glass pane and one for the surface of the VIP face, side of the cavity:

Glass (temperature θ_1) $\quad h_e(\theta_e - \theta_1) + a_S E_S + \dfrac{\theta_{s2} - \theta_1}{R_{cav}} = 0$

VIP (temperature θ_{s2}) $\quad \dfrac{\theta_1 - \theta_{s2}}{R_{cav}} + \tau_S E_S + \dfrac{\theta_i - \theta_{s2}}{R_{VIP} + R_{alu} + 1/h_i} = 0$

or:

$$\begin{cases} -(15 + 1/0.15)\theta_1 + \dfrac{\theta_{s2}}{0.15} = -0.05 \cdot 500 - 15 \cdot 35 \\[1em] \dfrac{\theta_1}{0.15} - \theta_{s2}\left(\dfrac{1}{0.15} + \dfrac{1}{0.02/0.005 + 0.002/230 + 1/7.7}\right) \\[1em] = -0.75 \cdot 500 - \dfrac{1}{0.02/0.005 + 0.002/230 + 1/7.7} \cdot 24 \end{cases}$$

Solving this system gives $\theta_1 = 60.2\,°C$, $\theta_{s2} = 113.2\,°C$. The heat flux to the inside so is 21.6 W/m². The high temperatures show the panel acts as solar collector. Heat flux to the inside equals the one an assembly with clear wall U-value 1.96 W/(m²·K) should transfer in the absence of solar radiation, while the clear wall U-value of the manufactured panel is only 0.23 W/(m²·K). Which measures be taken to could lower the temperatures in the panel and the heat flux to the inside?

Problem 9 Solve Problem (8) for the case heat absorbing glass, $a_S = 0.3$, $r_S = 0.19$, $\tau_S = 0.51$, is used and the shortwave absorptivity and reflectivity of the VIP's face side cavity is 0.5.

Solution 9 The temperature of the glass is 52.8 °C, and the temperature at the cavity side of the VIP 70.2 °C. The heat flux to the inside equals 11.2 W/m², which corresponds to a U-value 1.02 W/(m²·K). The real U-value remains 0.23 W/(m²·K).

Problem 10 On the roof of a mountain chalet lays 40 cm snow ($\lambda = 0.07$ W/(m·K), $a_S = 0.15$). The outdoor temperature is −15 °C, the indoor 22 °C. Solar irradiation touches 600 W/m². The outside surface film coefficient is 15 W/(m²·K), the inside 10 W/(m²·K). Which insulation thickness is needed to restrain the snow from melting in the contact with the membrane, if the insulation material has as apparent thermal conductivity 0.023 W/(m·K)? What heat flux will cross the roof? Without insulation, the thermal resistance face to face of the roof is 0.5 m²·K/W.

Solution 10 Thickness needed is 21 cm. The heat flux across equals 2.24 W/m².

Problem 11 An intensely ventilated attic receives an insulated ceiling, composed of metal girders, mounted 60 cm centre-to-centre with 120 mm thick thermal insulation boards in between. The girder section is given in the figure below. Suppose the insulation has as thermal conductivity 0 W/(m·K), the metal ∞ W/(m·K). The surface film coefficients are 25 W/(m²·K) at the attic side and inside 6 W/(m²·K). The attic temperature is −10 °C, and the temperature indoors is 20 °C. Does the heat loss differ between the profile mounted with the broader flange down or up? What is the metal temperature in both cases? Calculate the U-value of the ceiling?

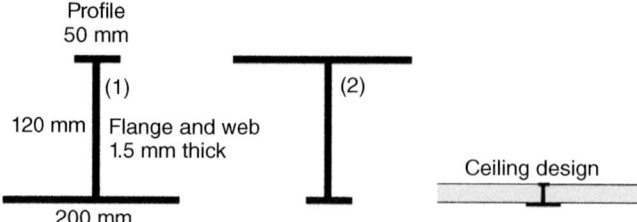

Solution 11 The heat balances are:
broader flange down (1) is: $0.2·6·(20 − \theta_x) + 0.06·25·(−10 − \theta_x) = 0$,
broader flange up (2): $0.05·6·(20 − \theta_x) + 0.2·25·(−10 − \theta_x) = 0$.

This leads to following results:

Heat loss different? Yes
Broader flange down, the metal temperature is 4.7 °C; up, it is −8.3 °C
For 1/0.6 girders per running metre broader flange down, $U = 1.02\,W/(m^2 \cdot K)$, broader flange
   o up, $0.47\,W/(m^2 \cdot K)$

Problem 12 Solve Problem (11) for a metal profile with flanges of 100 mm each.

Solution 12 The temperature of the steel profiles is 4.2 °C, the U-value of the ceiling $0.53\,W/(m^2 \cdot K)$.

Problem 13 A reinforced concrete column with sides 0.4 m stands between two glass panels in a way the glass lines with its inner face. The glass is assumed having thickness 0 m. The temperature indoors is 21 °C, and the temperature outdoors 0 °C. The inside surface film coefficient is $8\,W/(m^2 \cdot K)$ and the outside $25\,W/(m^2 \cdot K)$. Calculate the temperature field in and the heat loss through the column?

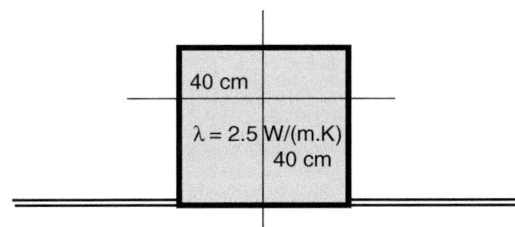

Solution 13 Assumed first is that the column reacts as a flat wall. The U-value then is:

$$\frac{1}{1/25 + 0.4/2.5 + 1/8} = 3.1\,W/(m^2 \cdot K).$$

The temperature at the inside surface so equals $21 - 3.1 \cdot (21 - 0)/8 = 12.9\,°C$, giving a temperature factor 0.65. The heat loss equals $3.1 \cdot 0.4 \cdot 21 = 25.8\,W/m$. Or, the temperature factor and thermal transmittance are close to double glass. In the column's centre, the temperature is 7.8 °C.

A first upgrade consists of applying a very simple CVM grid with one centre point, the column's centre (point 1, 2). Its heat balance is:

$$\frac{0.2(21 - \theta_x)}{1/8 + 0.2/2.5} + 3\frac{0.2(0 - \theta_x)}{1/25 + 0.2/2.5} = 0$$

giving as central temperature 3.4 °C and as inside surface temperature 10.3 °C, convening with a temperature factor 0.49, a value 24.6% lower than what the flat wall assumption gave. Heat loss now is 34.3 W/m, so 33.7% higher than with the flat wall assumption.

In a second upgrade, the grid over half the column gets up to a way 6 centre points, of which 5 on the perimeter.

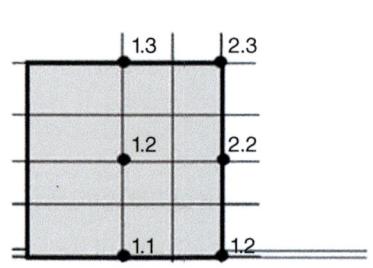

Red: 19 to 20 °C
Dark blue: 0 to 1 °C
Each colour in between is a step of 1 °C

The heat balances now become:

Point 1.1 $8 \cdot 0.1 \cdot (21 - \theta_{1.1}) + \dfrac{2.5 \cdot 0.1}{0.2}(\theta_{1.2} - \theta_{1.1})$

$+ \dfrac{2.5 \cdot 0.1}{0.2}(\theta_{2.1} - \theta_{1.1}) = 0$

Point 2.1 $8 \cdot 0.1 \cdot (21 - \theta_{2.1}) + \dfrac{2.5 \cdot 0.1}{0.2}(\theta_{1.1} - \theta_{2.1})$

$+ \dfrac{2.5 \cdot 0.1}{0.2}(\theta_{2.2} - \theta_{2.1}) + 25 \cdot 0.1 \cdot (0 - \theta_{2.1}) = 0$

Point 1.2 $\dfrac{2.5 \cdot 0.1}{0.2}(\theta_{1.1} - \theta_{1.2}) + \dfrac{2.5 \cdot 0.1}{0.2}(\theta_{1.3} - \theta_{1.2})$

$+ \dfrac{2.5 \cdot 0.2}{0.2}(\theta_{2.2} - \theta_{1.2}) = 0$

Point 2.2 $\dfrac{2.5 \cdot 0.1}{0.2}(\theta_{2.1} - \theta_{2.2}) + \dfrac{2.5 \cdot 0.2}{0.2}(\theta_{1.2} - \theta_{2.2})$

$+ \dfrac{2.5 \cdot 0.1}{0.2}(\theta_{2.3} - \theta_{2.2}) + 25 \cdot 0.2 \cdot (0 - \theta_{2.2}) = 0$

Point 1.3 $\dfrac{2.5 \cdot 0.1}{0.2}(\theta_{1.2} - \theta_{1.3}) + 25 \cdot 0.1 \cdot (0 - \theta_{1.3})$

$+ \dfrac{2.5 \cdot 0.1}{0.2}(\theta_{2.3} - \theta_{1.3}) = 0$

Point 2.3 $2 \cdot 25 \cdot 0.1 \cdot (0 - \theta_{2.3}) + \dfrac{2.5 \cdot 0.1}{0.2}(\theta_{2.2} - \theta_{2.3})$

$+ \dfrac{2.5 \cdot 0.1}{0.2}(\theta_{1.3} - \theta_{2.3}) = 0$

Solving this system gives as temperatures in the column:

0.4 0.9 0.4
1.4 2.9 1.4
4.9 8.1 4.9

The lowest temperature factor inside now is at the corners, 0.25, as bad as single glass. The heat loss turns into:

$$\Phi = 2 \cdot 0.1 \cdot 8 \cdot (21 - 4.9) + 0.2 \cdot 8 \cdot (21 - 8.1) = 46.4 \, \text{W/m}$$

which is 80% higher than with the flat wall assumption.

Last upgrade is calculated using software for two-dimensional heat transport. The figure above at the right shows the isotherms found, nearly the correct answer in terms of temperatures.

Problem 14 Aerated concrete is chosen as material for the outer walls. The manufacturer praises its good transient thermal response, ensuring a much higher dynamic thermal resistance than the steady state value. Is this true? The material properties are:

Situation	ρ (kg/m³)	λ (W/(m·K))	c (J/(kg·K))
Just applied (humid)	450	0.30	2700
After some years (air-dry)	450	0.13	1120

The wall thickness is either 10, 20 or 30 cm. The outside surface film coefficient touches 25 W/(m²·K), the inside 8 W/(m²·K).

Solution 14 The truth relies in the wall's harmonic properties. A high dynamic thermal resistance may confirm the manufacturer's claim but a low admittance says this does not suffice to stabilize the indoor conditions in case of important solar and internal gains. To show, temperature damping, the dynamic thermal resistance and the admittance are calculated for a humid 10 cm thick wall.

Thermal diffusivity: $a = \dfrac{\lambda}{\rho c} = 2.469 \cdot 10^{-7} \, \text{m}^2/\text{s}$.

X_n-value: $X_n = d\sqrt{\dfrac{n\pi}{aT}} = 0.1\sqrt{\dfrac{3.1418}{2.469 \cdot 10^{-7} \cdot 3600 \cdot 24}} = 1.2135$

G_{n1} 0.640 G_{n4} 0.486
G_{n2} 1.437 G_{n5} −1.431
G_{n3} 0.928 G_{n6} 2.733

Functions $G_{n1}(X_n)$ to $G_{n6}(X_n)$:
Layer matrices

Inside surface (1)

$$\begin{vmatrix} 1 & 0 & 0.125 & 0 \\ 0 & 1 & 0 & 0.125 \\ 0 & 0 & 1 & 0 \\ 0 & 0 & 0 & 1 \end{vmatrix}$$

Layer (2)

$$\begin{vmatrix} 0.640429 & 1.437199 & 0.309307 & 0.161937 \\ -1.437199 & 0.640429 & -0.161937 & 0.309307 \\ -4.29249 & 8.198851 & 0.640429 & 1.437199 \\ -8.198851 & -4.29249 & -1.437199 & 0.640429 \end{vmatrix}$$

Outside surface (3)

$$\begin{vmatrix} 1 & 0 & 0.04 & 0 \\ 0 & 1 & 0 & 0.04 \\ 0 & 0 & 1 & 0 \\ 0 & 0 & 0 & 1 \end{vmatrix}$$

Matrix multiplication:

Layer (2)·layer (2) =
$$\begin{vmatrix} 0.640429 & 1.437199 & 0.38936 & 0.341587 \\ -1.437199 & 0.640429 & -0.341587 & 0.38936 \\ -4.29249 & 8.198851 & 0.103868 & 2.462055 \\ -8.198851 & -4.29249 & -2.462055 & 0.103868 \end{vmatrix}$$ (matrix 4)

Layer (3)·(matrix 4) =
$$\begin{vmatrix} 0.46873 & 1.765153 & 0.3993515 & 0.440069 \\ -1.765153 & 0.46873 & -0.440069 & 0.3993515 \\ -4.29249 & 8.198851 & 0.103868 & 2.462055 \\ -8.198851 & -4.29249 & -2.462055 & 0.103868 \end{vmatrix}$$

This gives as harmonic properties:

Temperature damping $|D_\theta| = \sqrt{0.46873^2 + 1.765133^2} = 1.83$

$$\varphi_\theta = a\tan\left(\frac{1.765133}{0.46873}\right)\frac{12}{\pi} = 5\ \text{h}$$

Dynamic thermal resistance $|D_q| = \sqrt{0.3993515^2 + 0.38936^2}$
$$= 0.59\ \text{m}^2\cdot\text{K/W}$$

$$\varphi_q = a\tan\left(\frac{0.38936}{0.3993515}\right)\frac{12}{\pi} = 3.2\ \text{h}$$

Admittance $|Ad| = \dfrac{|D_\theta|}{|D_q|} = 3.09\ \text{W/(m}^2\cdot\text{K)}$

$\varphi_{Ad} = \varphi_\theta - \varphi_q = 1.8\ \text{h}$

It's up to the reader to calculate on spreadsheet the three for a 10 cm air-dry aerated concrete outer wall and for a 20 and 30 cm thick humid and air-dry aerated concrete outer wall. The results should be:

	Humid			Air-dry		
	Case 1 $d = 10$ cm $\lambda = 0.3$ W/(m·K)	Case 2 $d = 20$ cm $\lambda = 0.3$ W/(m·K)	Case 3 $d = 30$ cm $\lambda = 0.3$ W/(m·K)	Case 4 $d = 10$ cm $\lambda = 0.13$ W/(m·K)	Case 5 $d = 20$ cm $\lambda = 0.13$ W/(m·K)	Case 6 $d = 30$ cm $\lambda = 0.13$ W/(m·K)
D_θ	1.83	6.56	23.1	1.63	5.72	19.1
ϕ_θ (h)	5 h 00'	9 h 48'	14 h 36'	4 h 33'	9 h 19'	13 h 57'
D_q (m²·K/W)	0.59	1.93	6.56	1.02	3.16	10.4
ϕ_q (h)	3 h 18'	7 h 54'	12 h 42'	2 h 24'	6 h 53'	11 h 28'
Ad (W/(m²·K))	3.09	3.39	3.53	1.60	1.81	1.84
ϕ_{Ad} (h)	1 h 48'	1 h 52'	1 h 57'	2 h 09'	2 h 26'	2 h 29'

Clearly, aerated concrete is not the wonder in terms of thermal inertia manufacturers promise. To get a sufficient temperature damping air-dry ($D_\theta > 15$), a thickness beyond 20 cm is needed. The same holds for the dynamic thermal resistance if a value ≥ 4 m².K/W is the target. Still, the admittance remains low, surely once the aerated concrete is air-dried. The material so does not function the way claimed. On the contrary, it does not help a lot in damping the effects of the daily swings in outdoor temperature and solar radiation on the temperature indoors.

Problem 15 A living room has as surface 4×6.5 m² and a ceiling height of 2.5 m. Two of the outer walls, one 4×2.5 m², the other 6.5×2.5 m² large, are completely glazed with gas-filled, low-e double glazing, U-value 1.3 W/(m²·K) for $h_{i-} = 7.7$ W/(m²·K) and $h_{e-} = 25$ W/(m²·K). The 2 partitions and the ceiling have a thermal resistance 0.505 m²·K/W between their face in the living room and the ambient in the neighbour room. The floor heating consists of a network of pipes covered by a screed having a thermal resistance 0.1 m²·K/W. Walls, floor and ceiling have a long wave emissivity 0.9, the glass 0.92. The ventilation rate touches 1 ach, while the surface film coefficient for convection is 3.5 W/(m²·K). Calculate the glass, wall and ceiling temperatures, knowing that the in- and outside air temperatures are 21 and $-8\,°C$.

Solution 15 The room is a six grey surfaces combination: two windows, one with surface A_1, temperature T_{s1} and one with surface A_2, temperature T_{s2}, two partitions, one with surface A_3, temperature T_{s3} and one with surface A_4, temperature T_{s4}, a ceiling with surface A_5, temperature T_{s5}, the floor with surface A_6, temperature T_{s6}. The water in the floor heating has a temperature T_{fl}. Seven heat balances so are needed, one convective for the room as a whole and one per wall.

$$\text{Room:} \quad Q_v + \sum_{j=1}^{6} h_c A_j (21 - \theta_{sj}) = 0,$$

or, with $Q_v = \rho_a c_a V(\theta_e - \theta_i)$, $\theta_i = 21\,°C$, $\theta_e = -8\,°C$, $V = 65\,m^3$, $A_1 = A_3 = 10\,m^2$, $A_2 = A_4 = 16.25\,m^2$, $A_5 = A_6 = 26\,m^2$, $c_a = 1008\,J/(kg·K)$, ρ_a and $h_c = 3.5\,W/(m^2·K)$:

$$-633.36 + 35T_{s1} + 56.875T_{s2} + 35T_{s3} + 56.875T_{s4}$$
$$+ 91T_{s5} + 91T_{s6} - 7680.75 = 0$$

Wall surface-related heat balances: the radiant heat flux writes as: $q_R = e_L(M' - M_b)/\rho_L$. Linearization of the black body emittance M_b in the temperature interval 10–25 °C gives:

$$M_b = 307.75 + 5.57\,\theta_s, \quad r^2 = 0.999$$

The radiosity $M'_j = \dfrac{M_{bj}}{e_j} - \dfrac{\rho_j}{e_j}\sum_{i=2}^{6} F_{ji}M'_j$ with F_{ji} view factor between each face and the other 5:

	Surface 1	Surface 2	Surface 3	Surface 4	Surface 5	Surface 6
Surface 1	—	0.187	0.070	0.187	0.278	0.278
Surface 2	0.115	—	0.115	0.210	0.280	0.280
Surface 3	0.070	0.187	—	0.187	0.278	0.278
Surface 4	0.115	0.210	0.115	—	0.280	0.280
Surface 5	0.107	0.175	0.107	0.175	—	0.436
Surface 6	0.107	0.175	0.107	0.175	0.436	—

The black body emittance of the surfaces so becomes:

s1 $M_{b1} = \dfrac{1}{0.92}M'_{s1} - \dfrac{0.08}{0.92}$
$\cdot \left(0.187M'_{s2} + 0.07M'_{s3} + 0.187M'_{s4} + 0.278M'_{s5} + 0.278M'_{s6}\right)$

s2 $M_{b2} = \dfrac{1}{0.92}M'_{s2} - \dfrac{0.08}{0.92}$
$\cdot \left(0.115M'_{s1} + 0.115M'_{s3} + 0.21M'_{s4} + 0.28M'_{s5} + 0.28M'_{s6}\right)$

s3 $M_{b3} = \dfrac{1}{0.9}M'_{s3} - \dfrac{0.1}{0.9}$
$\cdot \left(0.07M'_{s1} + 0.187M'_{s2} + 0.187M'_{s4} + 0.278M'_{s5} + 0.278M'_{s6}\right)$

s4 $M_{b4} = \dfrac{1}{0.9}M'_{s4} - \dfrac{0.1}{0.9}$
$\cdot \left(0.115M'_{s1} + 0.210M'_{s2} + 0.115M'_{s3} + 0.28M'_{s5} + 0.28M'_{s6}\right)$

s5 $M_{b5} = \dfrac{1}{0.9}M'_{s5} - \dfrac{0.1}{0.9}$
$\cdot \left(0.107M'_{s1} + 0.175M'_{s2} + 0.107M'_{s3} + 0.175M'_{s4} + 0.436M'_{s6}\right)$

s6 $M_{b6} = \dfrac{1}{0.9}M'_{s6} - \dfrac{0.1}{0.9}$
$\cdot \left(0.107M'_{s1} + 0.175M'_{s2} + 0.107M'_{s4} + 0.175M'_{s4} + 0.436M'_{s5}\right)$

Inverting the matrix of this system of six equations gives the radiosities of the six surfaces as function of their black surface emittances:

$$\text{Matrix:} \begin{vmatrix} 1.0870 & -0.0163 & -0.0061 & -0.0163 & -0.0242 & -0.0242 \\ -0.0100 & 1.0870 & -0.0100 & -0.0183 & -0.0243 & -0.0243 \\ -0.0078 & -0.0208 & 1.1111 & -0.0208 & -0.0309 & -0.0309 \\ -0.0128 & -0.0233 & -0.0128 & 1.1111 & \cdot 0.0311 & -0.0311 \\ -0.0119 & -0.0194 & -0.0119 & -0.0194 & 1.1111 & -0.0485 \\ -0.0119 & -0.0194 & -0.0119 & -0.0194 & -0.0485 & 1.1111 \end{vmatrix}$$

$$\text{Inverted} \begin{vmatrix} H'_1 \\ H'_2 \\ H'_3 \\ H'_4 \\ H'_5 \\ H'_6 \end{vmatrix} = \begin{vmatrix} 0.9208 & 0.0150 & 0.0058 & 0.0146 & 0.0219 & 0.0219 \\ 0.0092 & 0.9214 & 0.0090 & 0.0162 & 0.0221 & 0.0221 \\ 0.0074 & 0.0187 & 0.9010 & 0.0182 & 0.0273 & 0.0273 \\ 0.0115 & 0.0207 & 0.0112 & 0.9017 & 0.0275 & 0.0275 \\ 0.0108 & 0.0176 & 0.0105 & 0.0172 & 0.9032 & 0.0408 \\ 0.0108 & 0.0176 & 0.0105 & 0.0172 & 0.0408 & 0.9032 \end{vmatrix}$$

$$\cdot \begin{vmatrix} 307.75 + 5.57\theta_{s1} \\ 307.75 + 5.57\theta_{s2} \\ 307.75 + 5.57\theta_{s3} \\ 307.75 + 5.57\theta_{s4} \\ 307.75 + 5.57\theta_{s5} \\ 307.75 + 5.57\theta_{s6} \end{vmatrix}$$

Introducing this result in the radiant heat flux equation allows eliminating the constant 307.75. The radiant, convective and conductive heat balance per surface now is $q_R + q_C + q_{\text{cond}} = 0$, or:

$$-10.1367\theta_{s1} + 0.9599\theta_{s2} + 0.3726\theta_{s3} + 0.9352\theta_{s4}$$
$$+ 1.4023\theta_{s5} + 1.4023\theta_{s6} + 0\theta_{fl} = 60.985$$

$$0.5907\theta_{s1} - 10.0972\theta_{s2} + 0.5771\theta_{s3} + 1.0391\theta_{s4}$$
$$+ 1.4129\theta_{s5} + 1.4129\theta_{s6} + 0\theta_{fl} = 60.985$$

$$0.3726\theta_{s1} + 0.9378\theta_{s2} - 10.444\theta_{s3} + 0.9136\theta_{s4}$$
$$+ 1.3699\theta_{s5} + 1.3699\theta_{s6} + 0\theta_{fl} = 115.084$$

$$0.5755\theta_{s1} + 1.0391\theta_{s2} + 0.5622\theta_{s3} - 10.410\theta_{s4}$$
$$+ 1.3765\theta_{s5} + 1.3765\theta_{s6} + 0\theta_{fl} = 115.084$$

$$0.5393\theta_{s1} + 0.8831\theta_{s2} + 0.5269\theta_{s3} + 0.8603\theta_{s4}$$
$$- 10.335\theta_{s5} + 2.0454\theta_{s6} + 0\theta_{fl} = 115.084$$

$$0.5393\theta_{s1} + 0.8831\theta_{s2} + 0.5269\theta_{s3} + 0.8603\theta_{s4}$$
$$+ 2.0454\theta_{s5} - 18.355\theta_{s6} + 10\theta_{fl} = -73.5$$

The diagonal terms in these equations come from:

$$\theta_{s1}, \theta_{s2} - \left[\frac{1}{(1/1.3 - 0.13)} + 3.5 + 0.9208 \left(\frac{0.92}{0.08} \right) \right]$$

$$\theta_{s3}, \theta_{s4}, \theta_{s5} - \left[\frac{1}{0.505} + 3.5 + (0.901, \ 0.9017, \ 0.9032) \left(\frac{0.9}{0.1} \right) \right]$$

$$\theta_{s6} - \left[\frac{1}{0.1} + 3.5 + 0.9032\left(\frac{0.9}{0.1}\right)\right]$$

Solving this system of six surface and one room heat balances gives as temperatures:

Window 4.0 × 2.5 m²	$\theta_{s1} = 17.6\,°C$	Wall 6.5 × 2.5 m²	$\theta_{s4} = 21.8\,°C$
Window 6.5 × 2.5 m²	$\theta_{s2} = 17.8\,°C$	Ceiling	$\theta_{s5} = 22.3\,°C$
Wall 4.0 × 2.5 m²	$\theta_{s3} = 21.9\,°C$	Floor	$\theta_{s6} = 29.2\,°C$

The water in the floor heating system has a temperature of 36.1 °C

Problem 16 Repeat (15) assuming the two envelope walls consist of double glazing, $U = 2.9\,W/(m^2.K)$ for $h_i = 7.7\,W/(m^2·K)$ and $h_e = 25\,W/(m^2·K)$, while all other data remain the same.

Solution 16 With normal double glazing, the temperatures become:

Window 4.0 × 2.5 m²	$\theta_{s1} = 11.8\,°C$
Window 6.5 × 2.5 m²	$\theta_{s2} = 12.1\,°C$
Wall 4.0 × 2.5 m²	$\theta_{s3} = 21.9\,°C$
Wall 6.5 × 2.5 m²	$\theta_{s4} = 21.7\,°C$
Ceiling	$\theta_{s5} = 22.6\,°C$
Floor	$\theta_{s6} = 34.7\,°C$
Floor heating	$\theta_{fl} = 47.0\,°C$

The floor is much warmer now than acceptable for feet comfort (28 °C), or, heat loss is too high to only install floor heating. The room also needs a radiator or a convector.

Further Reading

ASHRAE (2021). *Handbook of Fundamentals*, SIe. Atlanta, GA: Tullie Circle.

Blomberg, T. (1996). Heat conduction in two and in three dimensions, computer modelling of building physics applications, Report TVBH-1008, Lund University of Technology.

Cammerer, J.S. (1962). *Wärme- und Kälteschutz in der Industrie.* Berlin/Heidelberg/New York: Springer Verlag (in German).

Carslaw, H.S. and Jaeger, J.C. (1986). *Conduction of Heat in Solids.* Oxford Science Publications.

CSTC (1975). Règles Th, Règles de calcul des caractéristiques thermiques utiles des parois de construction de base des bâtiments et du coefficient G des logement et autres locaux d'habitation, DTU (in French).

De Grave, A. (1957). *Bouwfysica 1.* Brussel: Uitgeverij SIC (in Dutch).

de Wit, M.H. (1995). Warmte en vocht in constructies (Heat and moisture in building constructions), cursus TU-Eindhoven, (in Dutch).

Defraeye, T., Blocken, B., and Carmeliet, J. (2011). An adjusted temperature wall function for turbulent forced convective heat transfer for bluff bodies in the atmospheric boundary layer. *Building and Environment* 46: 2130–2141.

DIN 4701 (1983). Regeln für die Berechnung des Wärmebedarfs von Gebäuden, DNA, (in German) (German Standard).

Dragan, C. and Goss, W. (1995). Two-dimensional forced convection perpendicular to the outdoor fenestration surface-FEM solution. *ASHRAE Transactions* 101: 201–209, Pt 1.

El Sherbiny, S., Raithby, G., and Hollands, K. (1982). Heat transfer by natural convection across vertical and inclined air layers. *Journal of Heat Transfer* 104: 96–102.

Emmel, M., Abadie, O. and Mendes, N. (2007). New external convective heat transfer coefficient correlations for isolated low-rise buildings. IEA-EBC Annex 41, paper A41-T3-Br-07-2.

Feynman, R., Leighton, R., and Sands, M. (1977). *Lectures on Physics*, vol. 1. Reading, Massachusetts: Addison-Wesley Publishing Company.

Fischer, D. (1995). An experimental investigation of mixed convection heat transfer in a rectangular enclosure, Ph.D. Thesis, University of Illinois, Urbana-Champaign.

Haferland, F. (1970). *Das wärmetechnische Verhalten mehrschichtiger Außenwände*. Wiesbaden-Berlin (in German): Bauverlag GmbH.

Hagentoft, C.-E. (2001). *Introduction in Building Physics*. Lund: Studentlitteratur.

Häupl, P. (2008). Bauphysik; Klima, Wärme, Feuchte, Schall, Grundlagen, Anwendungen, Beispiele, Ernst & Sohn, 550 pp (in German).

Hauser, G. and Stiegel, H. (1990). *Wärmebrücken Atlas für den Mauerwerksbau*. Wiesbaden, Berlin: Bauverlag GmbH (in German).

Hens, H. (2007). *Building Physics – Heat, Air and Moisture, Fundamentals and Engineering Methods with Examples and Exercises*, 1e. Ernst & Sohn.

Hens, H. (1978, 1981). *Bouwfysica, Warmte en Vocht, Theoretische grondslagen*, 1^e en 2^e uitgave. Leuven (in Dutch): ACCO.

Hens, H. (1991). *Bouwfysica 1, Warmte en Massatransport*. Leuven (in Dutch): ACCO.

IEA-Annex 14 (1990). *Condensation and Energy: Guidelines and Practice*. Leuven: ACCO.

Judkoff, R. and Neymark, J. (1995). Building Energy Simulation Test (BESTEST) and Diagnostic Method, NREL/TP-472-6231, Golden, Colorado NREL.

Kreith, F. (1976). *Principles of Heat Transfer*. New York: Harper & Row Publishers.

Kumaran K. and Sanders C., (2008). Boundary conditions and whole building HAM analysis, final report IEA-EBC Annex 41, p. 65–110.

Lecompte, J. (1989). De invloed van natuurlijke convectie op de thermische kwaliteit van geïsoleerde spouwconstructies (Impact of natural convection on the thermal quality of insulated cavity walls), doktoraal proefschrift, KU-Leuven (in Dutch).

Lutz, P., Jenisch, R., Klopfer, H. et al. (1989). *Lehrbuch der Bauphysik*. Stuttgart (in German): B.G. Teubner Verlag.

Mainka, G.W. and Paschen, H. (1986). *Wärmebrückenkatalog*. Stuttgart (in German): B. G. Teubner Verlag.

Murakami, S., Mochida, A., Ooka, R., and Kato, S. (1996). Numerical prediction of flow around a building with various turbulence models: comparison of k-ε, EVM, ASM, DSM and LES with wind tunnel tests. *ASHRAE Transactions* 102, Pt 1.

NBN B62-003 (1987). Berekening van de warmtedoorgangscoëfficiënt van wanden, BIN, (in Dutch) (Belgian standard).

NEN 1068 (1981). Thermische isolatie van gebouwen, NNI, (in Dutch) (Dutch standard).

Physibel, C.V. (1996). Kobra Koudebrugatlas, (edited in several languages).

PREN 31077 (1993). Windows, Doors and Shutters, Thermal Transmittance, Calculation Method, CEN, (European standard).

Rietschel, R. (1970). *Heiz- und Klimatechnik*, 15th Auflage. Berlin-Heidelberg-New York (in German): Springer Verlag.

Roots, P. (1997). Heat transfer through a well insulated external wooden frame wall, Report TVBH-1009, Lund.

Saelens, D. (2002). Energy performance assessment of single storey multiple-skin facades. Doctoral thesis. KU-Leuven.

Standaert, P. (1984). Twee- en driedimensionale warmteoverdracht: numerieke methoden, experimentele studie en bouwfysische toepassingen, (Two- and three-dimensional heat transfer: numerical models, experimental study and Building Physics related supplications), doktoraal proefschrift, KU-Leuven (in Dutch).

Taveirne, W. (1990). *Eenhedenstelsels en groothedenvergelijkingen: Overgang naar het SI*. Wageningen (in Dutch): Pudoc.

Tavernier, E. (1985). De theoretische grondslagen van het warmtetransport, cursus 'Thermische isolatie en vochtproblemen in gebouwen', TI-KVIV (in Dutch).

TU-Delft (1975–1985). Faculteit Civiele Techniek, Vakgroep Utiliteitsbouw-Bouwfysica, Tekstboek 'Bouwfysica', naar de colleges van Prof A.C.Verhoeven, (in Dutch).

Vogel, H. (1997). *Gerthsen Physik*. Berlin Heidelberg (in German): Springer Verlag.

Welty, J., Wicks, C., and Wilson, R. (1969). *Fundamentals of Momentum, Heat and Mass Transfer*. New York: John Wiley & Sons.

2

Mass Transfer

2.1 In General

2.1.1 Facts

Examples of mass transfer in and across buildings, building assemblies and materials are airflow in and through rooms, vapour transport across a roof assembly, water and solved salts migration in bricks, diffusion of blowing agents out of PUR boards and CO_2 absorption by fresh lime plaster. To develop in materials, they must have open pores with equivalent diameters larger than the migrating molecules or molecule clusters. In materials without pores, with pores smaller than the said diameter or with only closed pores, mass transfer is excluded.

Overall, the displacement of air, water vapour and water dominates. Moving air carries heat (enthalpy) and water vapour, which has positive and negative effects. On one hand, dry air moving through an assembly increases its drying potential and may remove water vapour without causing condensation, on the other hand, air looping in fibrous insulation layers and around carelessly mounted closed cell insulation boards may dramatically increase the heat loss and gain. Building assemblies turning moist is the most destructive of all climate-related loads, reason why ensuring moisture tolerance is a challenge for designers and builders.

Looking to open-porous materials, the term 'moisture' refers to water filling the open pores, be it as solid (ice), liquid, then included dissolved substances, and gas (vapour). Below $0\,°C$, ice, liquid and vapour interfere, above $0\,°C$, only liquid and vapour do. In vapour, the water molecules move separately. In liquid, they form clusters with much larger diameter than the molecule's 0.28 nm. As a consequence, pores that let vapour pass could be unaccessible for liquid. Some materials so are water but not vapour proof. The crystalline ice finally is 10% more voluminous than liquid water, which makes frost potentially destructive.

2.1.2 Definitions

In materials, the amounts of humid air and moisture, they may contain, depend on:

Density, symbol ρ, units kg/m^3	Mass per unit volume of material. Porous materials have a lower density than their specific one at porosity 0 (ρ_s)
Total porosity, symbol Ψ, units % m^3/m^3	The volume of pores per unit volume of material
Open porosity, symbol Ψ_o, units % m^3/m^3	The volume of open pores per unit volume of material. Which fraction is 'open' depends on the fluid migrating. In general, the open porosity is lower than the total ($\Psi_o \leq \Psi$)

When all pores are filled with humid air, following relation giving the porosity holds:

$$\Psi = \frac{\rho_s - \rho}{\rho_s - \rho_a} \approx 1 - \frac{\rho}{\rho_s} \tag{2.1}$$

In it, ρ_s is the specific density and ρ_a the density of humid air.

The air presence in materials is quantified using as variables:

Air content, symbol w_a, units: kg/m^3	The mass of air per unit volume of material
Air ratio, symbol X_a, units: % kg/kg	The mass of air per unit mass of dry material
Volumetric air ratio, symbol Ψ_a, units % m^3/m^3	The volume of air per unit volume of material
Air saturation degree, symbol S_a, units %	The ratio between the current air content and the maximum possible. May also be defined as the ratio between the fraction of open pores filled with and those accessible for air

Analogous quantities are used to quantify the moisture presence:

Moisture content, symbol w, units kg/m^3	The mass of moisture per unit volume of material
Moisture ratio, symbol X, units % kg/kg	The mass of moisture per unit mass of dry material
Volumetric moisture ratio, symbol Ψ, units % m^3/m^3	The volume of moisture per unit volume of material
Moisture saturation degree, symbol S, units: %	The ratio between the current moisture content and the maximum possible. May also be defined as the ratio between the fraction of open pores filled and those accessible for moisture

Doing it this way can be extended to other liquids and gases. The first three quantities are averaging the presence over the material's weight or volume, although only the pores contain air, moisture, dissolved salts, etc. Coupling to the pore volume is done by introducing the product ρS with ρ the density of the liquid or gas and S its saturation degree.

Following relations, with ρ the density of the material and ρ_a the density of air, hold between the air ratio and the air content, the volumetric air ratio and the air content, the moisture ratio and the moisture content and the volumetric moisture ratio and the moisture content:

Air:
$$X_a = 100\frac{w_a}{\rho} \qquad \Psi_a = 100\frac{w_a}{\rho_a} \tag{2.2}$$

Moisture:
$$X = 100\frac{w}{\rho} \qquad \Psi = 100\frac{w}{1000} = \frac{w}{10} \tag{2.3}$$

The air and the moisture content are also coupled:
$$w_a = \rho_a \left(\frac{\Psi_o \text{ in } \%}{100} - \frac{w}{1000}\right) \tag{2.4}$$

If air and moisture are the only substances filling the pores, then the saturation degree for air and moisture has as sum: $S + S_a = 1$. This and the equations above indicate that in pores partly filled with liquid, what's left as pore volume than is filled by air containing water vapour.

For air, no rules say which quantity to use (w_a, X_a of Ψ_a). Not so for moisture. Albeit all apply to a same wetness, the impact of the linked numbers differs. For materials havier than 1000 kg/m³, the moisture ratio gives the lowest numbers, for those lighter than 1000 kg/m³, the volumetric moisture ratio does. Since 'moisture' carries a negative connotation, manufacturers often use the lowest and apparently best scoring number without mentioning the units. Therefore, following rules apply: use moisture content for stony materials, moisture ratio for wood-based materials and volumetric moisture ratio for highly porous materials, all with the units added.

2.1.3 Saturation Degree Scale

When water fills all open pores in a material, the moisture saturation degree turns 100% and the air saturation degree 0%. Conversely, without water in the open pores, the air saturation degree is 100% and the material dry. Reality lays in between with the two saturation degrees changing in opposite direction. A rise of the one gives a drop of the other. So, a saturation scale can be constructed with intervals and values reflecting specific situations, see Table 2.1.

For capillary open-porous materials such as wood, brick, sand-lime stone, natural stone, concrete, etc., the order is $S_{m,H} < S_{m,cr} < S_{m,c} < S_{m,fr}$. For non-capillary open-porous materials as some insulation materials, no-fines concrete, etc., saturation degrees beyond hygroscopic do not play.

2.1.4 Air and Moisture Transfer

An accurate description of the air and moisture transfer would be possible if all related physical laws were known and the pore system could be quantified as hydraulic network, which is not the case. The transport laws used are

Table 2.1 Air and moisture saturation scale in materials.

Saturation degree (%)		Meaning
Moisture	Air	
$S_m = 0$	$S_a = 100$	*Dry material.* All open pores air-filled. Never the case, though a moisture saturation degree close to 0 is possible
$0 < S_m \leq S_{mH}$	—	*Hygroscopic interval.* Relative humidity in the pores determines the moisture saturation degree with the limit value ($S_{m,H}$) fixed at 98% RH
$S_{m,cr}$	—	*Critical moisture saturation degree.* Without impacting air displacement, below liquid water present hardly moves, above it starts doing it
$S_{m,c}$	$S_{a,cr}$	*Capillary moisture saturation degree.* Above, air movement across a material stops, below it's hardly hindered. If moistening and drying are in balance, then under atmospheric conditions, water presence only occasionally exceeds this saturation degree
$S_{mc} < S_m \leq 100$ $S_{m,fr}$		To enter this interval, an open porous material must contact water for a long time. Then, dissolution of the air left in the pores will push the moisture saturation degree above capillary, possibly up to the *moisture saturation degree for frost* $S_{m,fr}$, the value above which frost damage looms, S_m lays close to 90% for materials having good tensile strength but may be as low as capillary for those barely having tensile strength
$S_m = 100$	$S_a = 0$	*Moisture saturation degree.* All air left the material. To get there, water sorption under vacuum or boiling the material is needed

approximations and constructing the pore distribution applying mercury porosimetry, electronic microscope images and X-ray tomography does not deliver an accurate three-dimensional picture of the network. Another complication resides in the changes of state between water, water vapour and ice, the more that water vapour mixes with the air in the pores. Left is a phenomenological approach with potentials and transport coefficients.

For air, the driving forces are wind, water migrating through the porous system, differences in temperature and, sometimes the way an assembly is composed. Moisture transport in turn combines water vapour and liquid water, the last included all dissolved substances, with as driving forces equivalent diffusion, air mitigation, capillary suction, gravity and external pressures. So:

Mechanism	Driving force
Water vapour	
Equivalent diffusion	Vapour pressure differences in the pores and between the ambient at both sides
Ait transported	Gradients in total air pressure. Vapour then migrates together with the air containing it in and through assemblies

Mechanism	Driving force
Water	
Capillarity	Differences in capillary suction. The wider a pore, the less it sucks but the larger the flow generated
Gravity	Linked to the difference in weight between water heads. Gravity may also cause flows in porous materials that hardly show capillary suction
Pressure	Differences in air or in water pressure. Air pressure gradients are small compared to water pressure gradients but both induce water flow through larger pores, cracks, open joints, etc.

Which of these driving forces intervenes, depends on the kind of material:

Material	Vapour		Water		
	Eq. diff.	Airflow	Capill	Gravity	Pressure
Non-capillary, non-hygroscopic	X	X		X	X
Non-capillary, hygroscopic	X	X		X	(X)
Capillary, non-hygroscopic	X		X	(X)	
Capillary, hygroscopic	X		X		

Non-capillary, non-hygroscopic materials only contain macro-pores. Non-capillary, hygroscopic materials mix macro pores too large to suck with sorption-active micro-pores that exert too much friction to allow capillary transport. Airflow is possible in the macropores, hardly in the micro-pores. In capillary non-hygroscopic materials, the pores are small enough to be suction-active but too large to make sorption important. Capillary, hygroscopic materials finally have pores narrow enough to allow suction and sorption.

2.1.5 Moisture Sources

Omnipresent as boundary condition is humid air, except for parts contacting water. Humid air is also what fills the open pores in dry material layers. At the same time, the water vapour it contains entrains hygroscopic moisture, although open-porous materials remain what's called air-dry as long as the RH in the ambient air remains low enough. Quoted as accidental water sources are leaking water pipes and sewage pipes, leaky joints around shower basins a.o. Quoted as non-accidental are construction moisture entrained in the materials during production and the building process, rising damp standing for moisture sucked bottom-up by walls and, external water heads moistening walls and floors. Though, leading in many climates is rain (Figure 2.1).

Water as source

Rising damp

Rain

External water heads

Building moisture

Vapour as source

Surface condensation

Interstitial condensation

Figure 2.1 Moisture sources.

Looking to water vapour, unavoidable is sorption/desorption, called hygroscopic behaviour, which is activated by the RH in the ambient air. Another is surface condensation on both faces of envelope assemblies, seen as harmful on the inner and unavoidable if happening on the outer, although the amounts deposited there can be comparable to what wind-driven rain delivers. Really troubling is condensation in assemblies, called interstitial condensation, a phenomenon remaining unnoticed till rot and corrosion appears or, when happening in roofs, moisture starts dripping down (Figure 2.1).

Of all sources mentioned, the water vapour triplet is most studied as their appearance looks least understandable. That's why professional literature deals extensively with them, be it often in a too simplistic way. Most professionals however fear rain and rising damp the most.

2.1.6 Air and Moisture in Relation to Durability

The presence of humid air in a material or an assembly is not alarming in itself. Only when moving through, the effects may turn negative. In fact, related airflow entrains vapour displacement and heat transfer. Vapour can condense, while the heat transferred can lower the insulation efficiency. Also, liquid moisture does not always harm. Nobody complains about a brick veneer turning wet by rain. Things change when wet stains appear on the inner face of a wall. At most a few considered surface condensation on single glass to be a problem. Only when running-off and wetting the window frame, concerns surfaced.

Nonetheless, moisture is a prime cause of damage to and degradation of buildings. Up to 70% of all deficiencies are wetness-related. Reason is the strong polarity of the water molecule, turning it into an excellent solvent and efficient chemical catalyst, while being a prerequisite for biological activity. Invisible as damage is the increase in thermal conductivity, a fact when insulation materials pick up water or turn wet by interstitial condensation. The latent heat then released enlarges the consequences. Wetness also decreases the strength and rigidity, not so for stony but for resin-bound, wood-based materials that swell with increasing moisture content and shrink when drying. Even more visible as damage is hydrolysis of the binder resin in particle board, plywood and oriented strand board (OSB) with a loss in strength and stiffness possibly causing rupture far below the stress allowed in untouched boards. These materials also suffer from mould, mildew, fungal growth and bacteria once the RH on their surface passes a given threshold. Especially fungae are damaging as they digest cellulose and lignine while producing water, a combination that may disintegrate timber completely.

Even stony materials could show visible damage by wetness. While moss and algae growth on their surface may seem a purely aesthetic problem, the moss roots can disintegrate the mortar joints in masonry while the organic acids they produce may initiate metal corrosion. Salt hydration and crystallization can pulverize bricks, while carbonatation and alkali-silicate reaction could attack concrete. Frost can damage stones and bricks, while steel starts corroding at high RH, whereby the oxide so formed is more voluminous, causing rusting bars to push away their concrete cover.

For synthetics, hydration is most feared as it decreases their cohesion over time. Finally, temperature and vapour pressure fluctuations in foams due to interstitial condensation can irreversibly deform them.

2.1.7 Links with Energy Transfer

Mass displacement means energy transferred. Indeed, moving generates kinetic energy, although most mass transfer phenomena in porous materials are too slow for the kinetic energy to play a role. The differences in potential energy due to gravity, suction and pressure gradients are what induces displacement and heat transfer. In fact, each gas, liquid and solid at temperature θ contains a quantity of heat per unit mass, called the enthalpy, symbol h, units J/kg, given by:

$$Q = h = c(\theta - \theta_0) \qquad (2.5)$$

In it, c is the specific heat capacity of the gas, liquid or solid (J/(kg·K)) and θ_0 a reference temperature, mostly 0 °C. Any mass flow (G) so induces a sensible enthalpy flow, equal to:

$$\Phi = Gc(\theta - \theta_0) \qquad (2.6)$$

If phase changes occur in the moving mass, the latent heat of transformation, symbol l_0, units J/kg, adds:

$$\Phi = G[c(\theta - \theta_0) + l_0] \qquad (2.7)$$

In building enclosures most processes are ± isobaric, so c can be taken equal to the specific heat capacity at constant pressure. This enthalpy transfer lowers the insulation efficiency of any assembly, subjected to air and moisture flows.

2.1.8 Conservation of Mass

As stated, air and moisture transfer prevails. Of course, other gasses – radon, CO_2, SO_2 – may also move in and through open-porous materials, while the liquid water migrating often carries soluble substances, take salts. These crystallise where the liquid evaporates and remain dissolved where wet. But first, some definitions:

Amount of mass, symbol M, units: kg	Quantifies the mass present or migrating. As mass is a scalar, the amount also is.
Mass flow, symbol G, units: kg/s	Stands for the mass migrating per time unit. Again a scalar.
Mass flux, symbol g, units kg/(m²·s)	Quantifies the mass migrating per unit of time through a unit surface normal to the flow. The flux is a vector with a same direction as the surface vector. The components in Cartesian coordinates are g_x, g_y, g_z, in polar coordinates g_R, g_ϕ, g_Θ

A suffix characterises which mass is considered: a for air, da for dry air, v for water vapour, m for moisture and w for water. Solving a mass transfer problem now means determining the scalar field of transfer potentials (Po(x,y,z,t)), and the vectorial field of fluxes ($\mathbf{g}(x,y,z)$). Calculation so requires a scalar and a vector equation. The first follows from mass conservation, stating that a mass flow exchanged between system and environment per unit of time, added what's produced in or removed from the system per unit of time, called a source or sink (S_x in kg/(m³·s)), must equal the mass change in the system per unit of time:

$$\mathrm{div}(\mathbf{g_x}) \pm S_x = \frac{\partial w_x}{\partial t} \qquad (2.8)$$

The flux can be diffusive, depending on the gradient of a driving force, or convective, depending on that driving force.

2.2 Air

2.2.1 In General

Dry air is a mixture of 21% m³/m³ of oxygen (O_2), 78% m³/m³ of nitrogen (N_2) plus traces of other gases (CO_2, SO_2, Ar, Xe). If an ideal gas, the equation of state should look:

$$p_{da}V = m_{da}R_{da}T \qquad (2.9)$$

with p_{da} the (partial) dry air pressure in Pa, T the temperature in K, m_{da} the mass of dry air in kg filling the volume V in m³ and R_{da} the gas constant, 287.055 J/(kg·K). In reality, dry air is not an ideal gas and obeys instead following <exact> equation of state:

$$\frac{p_{da}V}{n_{da}R_oT} = 1 + \frac{B_{aa}}{(V/n_1)} + \frac{C_{aaa}}{(V/n_1)^2}$$

In it, n_{da} is the number of moles in V (m³) and R_o the general gas constant, 8314.41 J/(mol·K), while B_{aa} and C_{aaa} are two viral coefficients with value close to 0:

$$B_{aa} = 0.349568 \cdot 10^{-4} - \frac{0.668772 \cdot 10^{-2}}{T} - \frac{2.10141}{T^2} + \frac{92.4746}{T^3} \quad (m^3/mol)$$

$$C_{aaa} = 0.125975 \cdot 10^{-2} - \frac{0.190905}{T} + \frac{63.2467}{T^2} \quad (m^6/mol^2)$$

As dry air concentration, the ideal gas law gives:

$$\rho_{da} = \frac{m_{da}}{V} = p_{da}/(R_{da}T) \qquad (2.10)$$

Moist air, a mixture of dry air and water vapour, suffix a, has as ideal gas equation and density:

$$P_aV = m_aR_aT \quad \rho_a = P_a/(R_aT)$$

In both, P_a is the total air pressure in Pa, equal to the sum of the partial dry air and the partial water vapour pressure ($P_a = p_{da} + p_v$). Compared to dry air, water vapour so modifies the air mass m_a, the gas constant R_a and the density. However, for temperatures below 50 °C these effects are so small that the following holds:

$$R_a \approx R_{da} \quad P_a \approx p_{da} \quad \rho_a \approx \rho_{da}$$

or:

$$P_a/(R_aT) \approx p_{da}/(R_{da}T).$$

Airflow through assemblies now occurs when wind, stack or fans create air pressure differences between adjacent spaces or the in- and outdoors. Of course, to happen, the enclosure and fabric parts must be air permeable, which demands air-open materials, cracks, overlaps, leaks a.o.

2.2.2 Air Pressure Differentials

2.2.2.1 Wind

The pressure, wind exerts, is given by:

$$P_w = C_p \frac{\rho_a v^2}{2} \approx 0.6\, C_p v^2 \tag{2.11}$$

an equation, that follows from Bernoulli's law applied to wind blowing horizontally with a velocity v dropping to zero when hitting an obstacle. The pressure factor C_p values the obstacle's finite dimensions. Finitude in fact deflects the wind at the top and the sides of obstacles contacting the soil. At the edges, if free, vortexes develop between the windside and the leeside of the obstacle. In practice, C_p couples the real wind pressure against an obstacle to the velocity, measured 10 m high in the open field. Of course, depending on the situation, other reference velocities can be used. For buildings, wind pressures on the façade so are often linked to the velocity measured just above the ridge. A change of the reference alters C_p.

Wind direction, building location, building geometry and the façade spot considered are the factors impacting C_p. A positive value means over-, a negative value underpressure. Looking to buildings, overpressure reigns at the windside, underpressure at the leeside and along faces more or less parallel to the wind direction (Figure 2.2). As wind direction and speed vary, related pressures also do, they even may change sign. Table 2.2 gives pressure factor data.

2.2.2.2 Stack Effect

The stack effect in gases and liquids, also called buoyancy, has as cause differences in temperature and composition. In the atmosphere, air pressure drops with the height above sea level (**h**):

$$dP_a = -\rho_a \mathbf{g}\, d\mathbf{h}$$

In it, **g** is the acceleration by gravity. Inserting the ideal gas law in this formula gives:

$$\frac{dP_a}{P_a} = -\frac{\mathbf{g}\, d\mathbf{h}}{R_a T} \tag{2.12}$$

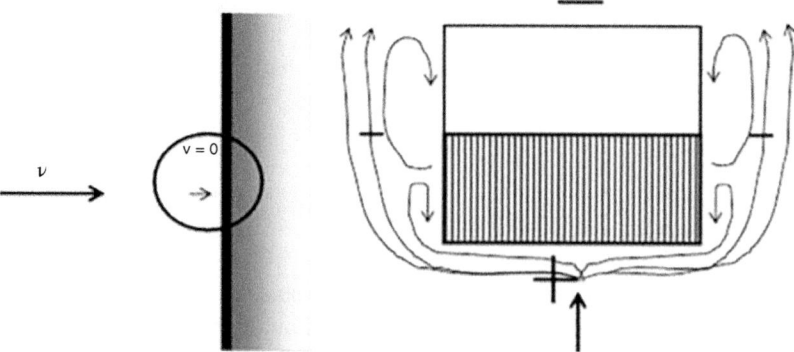

Figure 2.2 Bernoulli's law and the wind pressure field around a building.

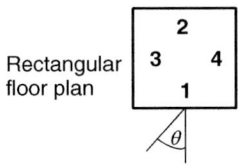

Table 2.2 Pressure factor data (from AIVC, A Guide to Energy Efficient Ventilation, 1996).

Low-rise, ≤3 storeys detached, exposed		Wind angle θ (°)							
		0	45	90	135	180	235	270	315
Face 1		0.7	0.35	−0.5	−0.4	−0.2	−0.4	−0.5	0.35
Face 2		−0.2	−0.4	−0.5	0.35	0.7	0.35	−0.5	−0.4
Face 3		−0.5	0.35	0.7	0.35	−0.5	−0.4	−0.2	−0.4
Face 4		−0.5	−0.4	−0.2	−0.4	−0.5	0.35	0.7	0.35
Roof (<10° pitch)	Front	−0.8	−0.7	−0.6	−0.7	−0.8	−0.7	−0.6	−0.5
	Rear	−0.4	−0.5	–	−0.7	−0.8	−0.7	−0.6	−0.5
Average		*−0.6*	*−0.6*	*−0.6*	*−0.6*	*−0.6*	*−0.6*	*−0.6*	*−0.6*
Roof (11–30° pitch)	Front	−0.4	−0.5	−0.6	−0.5	−0.4	−0.5	−0.6	−0.5
	Rear	−0.4	−0.5	−0.6	−0.5	−0.4	−0.5	−0.6	−0.5
Average		*−0.5*	*−0.6*	*−0.5*	*−0.4*	*−0.5*	*−0.6*	*−0.5*	*−0.5*
Roof (>30° pitch)	Front	0.3	−0.4	−0.6	−0.4	−0.5	−0.4	−0.6	−0.4
	Rear	−0.5	−0.4	−0.6	−0.4	0.3	−0.4	−0.6	−0.4
Average		*−0.1*	*−0.4*	*−0.6*	*−0.4*	*−0.1*	*−0.4*	*−0.6*	*−0.4*

Low-rise, ≤3 storeys detached, shielded[a]		Wind angle θ (°)							
		0	45	90	135	180	235	370	315
Face 1		0.2	0.05	−0.25	−0.3	−0.25	−0.3	−0.25	0.05
Face 2		−0.25	−0.3	−0.25	0.05	0.2	0.05	−0.25	−0.3
Face 3		−0.25	0.05	0.2	0.05	−0.25	−0.3	−0.25	−0.3
Face 4		−0.25	−0.3	−0.25	−0.3	−0.25	0.05	0.2	0.05
Roof (<10° pitch)	Front	−0.5	−0.5	−0.4	−0.5	−0.5	−0.5	−0.4	−0.5
	Rear	−0.5	−0.5	−0.4	−0.5	−0.5	−0.5	−0.4	−0.5
Average		*−0.5*	*−0.5*	*−0.4*	*−0.5*	*−0.5*	*−0.5*	*−0.4*	*−0.5*
Roof (11–30° pitch)	Front	−0.3	−0.4	−0.5	−0.4	−0.3	−0.4	−0.5	−0.4
	Rear	−0.3	−0.4	−0.5	−0.4	−0.3	−0.4	−0.5	−0.4

Table 2.2 (Continued)

Low-rise, ≤3 storeys detached, exposed		Wind angle θ (°)							
		0	45	90	135	180	235	270	315
Average		−0.3	−0.4	−0.5	−0.4	−0.3	−0.4	−0.5	−0.4
Roof (>30° pitch)	Front	0.25	−0.3	−0.5	−0.3	−0.4	−0.3	−0.5	−0.3
	Rear	−0.4	−0.3	−0.5	−0.3	0.25	−0.3	−0.5	−0.3
Average		−0.08	−0.3	−0.5	−0.3	−0.08	−0.3	−0.5	−0.3

a) Snielded by trees and/or other building as high as and close by the detached residence.

With the temperature (T) constant, solving that differential equation results in:

$$P_a = P_{a0} \exp\left(-\frac{gh}{R_a T}\right) \tag{2.13}$$

an outcome, known as the barometric relation. Temperature differences and changes in gas composition that make the gas constant (R_a) variable, now cause pressure gradients if at the same height. At different heights (h_1 and h_2), the pressure difference ΔP_a between two points touches:

$$\Delta P_a = g P_{a,h_o}(h_2 - h_1)/(R_a T)$$

with the suffix h_o denoting the average height. Stack at a height **h** is given now by the difference in pressure there between air at variable and air at constant temperature and composition:

$$P_{\text{Stack}} = g P_{a,h_o}\left[\left(\int_{h=0}^{h}\frac{dh}{R_a(h)T(h)}\right) - \frac{h}{(R_a T)_o}\right] = g P_{a,h_o} \mathbf{h}\left[\frac{1}{(R_a T)_m} - \frac{1}{(R_a T)_o}\right]$$

In it, $(R_a T)_m$ is the harmonic average of the product $R_a(h)T(h)$ over the height h:

$$(R_a T)_m = \mathbf{h} \Big/ \int_{h=0}^{h} \frac{dh}{R_a(h)T(h)}$$

Stack between two points so becomes:

$$P_{T,1-2} = P_{\text{Stack2}} - P_{\text{Stack1}} = g P_{a,h_o} \mathbf{h}\left[\frac{1}{(R_a T)_{m2}} - \frac{1}{(R_a T)_{m1}}\right]$$

If only the temperatures T_{m1} and T_{m2}, have a different value over the height **h**, stack simplifies to:

$$P_{T,1-2} \approx \frac{g P_{a,h_o}(R_{a1}T_1 - R_{a2}T_2)\mathbf{h}}{R_{a1}T_1 R_{a2}T_2} \approx \frac{P_{a0} g (R_{a1}\theta_1 - R_{a2}\theta_2)\mathbf{h}}{R_{am12} T_{m12}} \approx \rho_a g \beta (\theta_1 - \theta_2)\mathbf{h}$$

$$\tag{2.14}$$

with β ($=1/T_{m12}$) the compressibility of air. If, on the contrary, the vapour concentrations have a different value over that height, stack becomes:

$$P_{T,1-2} \approx \rho_a R_a g \beta z \left(\frac{1}{R_{a2}} - \frac{1}{R_{a1}}\right) \tag{2.15}$$

In case leaks at different height in a partition connect two air volumes, one warmer or more humid and the other colder or less humid, then somewhere in between the higher and lower leaks a neutral plane, stack zero, will exist. Where, depends on the leak size and leak distribution. In the absence of other pressure differentials, then through the leaks above the warmer air will pass from warm to cold or, the humid air, from higher to lower vapour concentration. Instead, through the leaks below, the cold air will pass from cold to warm, or humid air from lower to higher concentration.

Unless for high-rises, the air pressure differences by thermal stack are small, those caused by differences in vapour concentration negligible: 20 °C indoors and 0 °C outdoors in winter gives as thermal stack over a 2.5 m high room ($\rho_{a0} = 1.2 \text{ kg/m}^3$, $g = 9.81 \text{ m/s}^2$, $\beta = 1/273.16/\text{K}$, $\theta_{m12} - \theta_0 = 20\,°\text{C}$, $z = 2.5\,\text{m}$): $p_T = 1.2 \cdot 9.81 \cdot 1/273.15 \cdot 20 \cdot 2.5 = 2.15\,\text{Pa}$. Between the lowest and highest floor of a 250 m tall highrise, the result is 215 Pa.

Thermal stack gives more stable pressure differences than wind does.

2.2.2.3 Fans

Air heating, air conditioning, forced ventilation systems need fans (Figure 2.3). The pressure differentials they create are usually not as small as those stack gives and remain also quite stable.

2.2.3 Air Permeability and Air Permeances

Air permeable are (i) open porous materials, (ii) air open layers, (iii) unintended and (iv) intended leaks. Examples of (i) are no-fines concrete, mineral fibre and wood-wool cement boards. Examples of (ii) are sidings, finishes made of scaly or plate elements, take tiled, slated or corrugated roof covers, lathed ceilings and layers composed of boards, take plywood, OSB, insulation layers and underlays. Examples of (iii) are accidental joints in construction parts and boards, cracks due to too high tensile stresses, microcracks between mortar and bricks, nail holes, leaky joint fillers and too loose connections. Examples of (iv) finally are electricity boxes in walls, light spots in ceilings and trickle vents above windows (Figure 2.4). In addition, outer walls, party walls and roofs often include cavities and voids.

Figure 2.3 Fan.

Figure 2.4 Air-open layers: tiled roof covers, slated roof covers, a lathed ceiling. Spotlight and trickle vent as intended leaks.

Air displacement in open porous materials obeys Poiseuille's law: proportionality between flux and driving force, in the case being the air pressure gradient, be it with a multiplier in between:

$$\mathbf{g_a} = -k_a \mathbf{grad}\, P_a \tag{2.16}$$

The minus sign indicates that air always moves from higher to lower air pressures.

The multiplier k_a is called the air permeability, units s. It is a scalar for isotropic and a tensor with three main directions and a different value per direction for anisotropic materials. Its value increases the higher the open porosity and the number of open macro pores in a material.

Instead, for air-permeable layers, cracks, joints, cavities, leaks and openings, the equations quantifying the air passing through look:

Air open layers (per m^2) $\quad g_a = -K_a \Delta P_a, G_a$ in kg/($m^2 \cdot$s), K_a in s/m

Cracks, joints and cavities (per m) $\quad G_a = -K_a^\psi \Delta P_a, G_a$ in kg/(m·s), K_a^ψ in s

Leaks, openings (per unit) $\quad G_a = -K_a^\chi \Delta P_a, G_a$ in kg/s, K_a^χ in m·s $\tag{2.17}$

with K_a, K_a^ψ, K_a^χ the air permeances and ΔP_a the air pressure differential between the two sides of the air-permeable element. As the flow is not necessarily laminar, their value typically depends on the pressure differential:

$$K_a^\chi = a(\Delta P_a)^{b-1}$$

with a the air permeance coefficient for 1 Pa pressure difference and b the air permeance exponent. For laminar flow b is 1, for turbulent flow 0.5, for transition flow $0.5 < a < 1$. In most cases, knowing the air permeance coefficient and the air permeance exponent requires testing. For joints, leaks, cavities and openings with known geometry, hydraulics may help as the pressure losses by air passing through are frictional, so equal to:

$$\Delta P_a = f \frac{L}{d_H} \frac{\rho_a v^2}{2} \approx 0.42 f \frac{L}{d_H} \mathbf{g}_a^2 \quad (2.18)$$

with f the friction factor, d_H the section's hydraulic diameter, L the passage length and v the average air velocity. Bends, widenings, narrowings, entrances and exits add local losses:

$$\Delta P_a = \xi \frac{\rho_a v^2}{2} \approx 0.42 \xi \, \mathbf{g}_a^2 \quad (2.19)$$

with ξ the local loss factor. Table 2.3 lists values for both the friction and that local loss factor. In Re, see the table, equal to vd_H/ν, v is the average flow velocity, ν the kinematic viscosity and d_H the hydraulic diameter, for a circular section its diameter, for a rectangular section with sides a and b $2ab/(a+b)$ and for a cavity, b meter wide, wide, $2b$. Re nears $56\,000 \mathbf{g}_a \, d_H$ with $\mathbf{g}_a$ the air flux.

Calculating air permeances is based on the continuity axiom: in a serial circuit, all passages and local disturbances see a same flow passing. As the total pressure loss, the sum of the frictional and local pressure losses, thereby must equal the driving force, this allows calculating the airflow as function of the overall pressure differential.

2.2.4 Airflow in Open-porous Materials

2.2.4.1 The Conservation Law Adapted

In the absence of sources and sinks, conservation of mass simplifies to:

$$\text{div } \mathbf{g_a} = -\frac{\partial w_a}{\partial t} \quad (2.20)$$

The air content w_a in a material now depends on its open porosity Ψ_0, the air saturation degree S_a, the air pressure P_a and the temperature T, or:

$$w_a = \Psi_0 S_a \rho_a = \Psi_0 S_a P_a/(R_a T).$$

As the gas constant R_a is nearly invariable and the open porosity Ψ_0 a constant in case the open pores do not contain liquid, the variables left are the air saturation degree, the air pressure and the temperature. The partial derivative of the air content over time so can be written as:

$$\frac{\partial w_a}{\partial t} = \frac{\Psi_0}{R_a T} \left(S_a \frac{\partial P_a}{\partial t} + P_a \frac{\partial S_a}{\partial t} - \frac{S_a P_a}{T} \frac{\partial T}{\partial t} \right)$$

Since the air saturation degree is normally quite constant, this equation simplifies to:

$$\frac{\partial w_a}{\partial t} = \frac{\Psi_0 S_a}{R_a T} \left(\frac{\partial P_a}{\partial t} - \frac{P_a}{T} \frac{\partial T}{\partial t} \right)$$

When isothermal, thus no thermal stack, the derivative of the temperature versus time turns 0, or:

$$\text{div } \mathbf{g_a} = -\left(\frac{\Psi_0 S_a}{R_a T} \right) \frac{\partial P_a}{\partial t} \quad (2.21)$$

Table 2.3 Friction factor f and local loss factors ξ (Re the Reynolds number).

Reynolds number $Re = v\,d_H/v$	Flow	F
$Re \leq 2500$	Laminar	$96/Re$
$2500 \leq Re \leq 3500$	Critical	$\dfrac{0.038\,(3500 - Re) + f_{T,Re=2500}(Re - 2500)}{1000}$
$Re > 3500$	Turbulent	See (1) below
$Re \gg 3500$	Stable turbulent	$f_T = C^{te}$, single function of roughness

Local loss factor		ξ
Entering an opening		0.5
Leaving an opening		1.0
Widening	$Re \leq 1000$	$-0.036 + 9.6 \cdot 10^{-5}\,Re + \Delta\xi$
$\sigma = A_o/A_1$	$1000 < Re \leq 3000$	$1.28 \cdot 10^{-5}\,Re^{1.223} + \Delta\xi$
A_o small section	$Re > 3000$	$0.21\,Re^{0.012} + \Delta\xi$
A_1 large section	with $\sigma \leq 0.5$	$\Delta\xi = 0.78 - 1.56\sigma$
	$\sigma > 0.5$	$\Delta\xi = 0.48 - 0.96\sigma$
Narrowing		
$\sigma = A_o/A_1$	$Re \leq 1000$	$0.98\,Re^{-0.03} + A$
A_o small section	$1000 < Re \leq 3000$	$10.59\,Re^{-0.37} + A$
A_1 large section	$Re > 3000$	$0.57\,Re^{-0.01} + A$
	where $A = 0.0373\sigma^2 - 0.067\sigma$	
Leak	2.85	
Angle or curve b_0: width inlet channel b_1: width of channel after the curve $_0$: refers to the inlet channel f_0: friction factor in inlet channel	$\dfrac{k_{Re0}\xi_g k_0}{(d_H)_0}$ where $\xi_g = 0.885 \left(\dfrac{b_1}{b_0}\right)^{-0.86}$ and	

Relative roughness, ε	$3000 \leq Re < 40000$		$Re \geq 40000$	
	k_{Re0}	$k_0/(d_H)_0$	k_{Re0}	$k_0/(d_H)_0$
0	$45f_o$	1	1.1	1
0–0.001	$45f_o$	1	1.0	$1 + 0.5 \cdot 10^3 a$
>0.001	$45f_o$	1	1.1	1

(1): $f = \left[2\log\left(-4.793\log\left(\dfrac{10}{Re} + 0.2\varepsilon\right)\bigg/Re + 0.2698\varepsilon\right)\right]^{-2}$, ε: relative roughness, Figure 2.5.

Figure 2.5 Relative roughness.

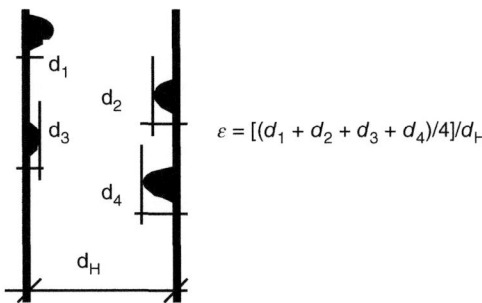

The ratio $\Psi_o S_a/(R_a T)$ is called the isothermal volumetric specific air content of the material, symbol c_a, units kg/(m³·Pa). As the open porosity and air saturation degree cannot pass 1 and $R_a = 287$ J/(kg·K), c_a obeys:

$$c_a = \Psi_o S_a/(R_a T) < 0.00348 T$$

Or, the isothermal volumetric specific air content looks negligible. Inserting the air flux equation in the isothermal mass balance gives:

$$\text{div}(k_a \, \mathbf{grad} \, P_a) = c_a \frac{\partial P_a}{\partial t}$$

If the air permeance k_a is a constant, this equation further simplifies to:

$$\nabla^2 P_a = \frac{c_a}{k_a} \frac{\partial P_a}{\partial t} \tag{2.22}$$

The inverse of the ratio c_a/k_a stands for the isothermal air diffusivity of an open-porous material, symbol D_a, units m²/s. Related value is generally so large that adapting to a changing air pressure happens within seconds. Only for pressures that fluctuate really fast, take wind gusts, the pressure response could show some damping and time shift. For mineral wool with density 40 kg/m³ the properties k_a and c_a for example have as value:

Thermal	Air
$\lambda = 0.036$ W/(m·K)	$k_a = 8 \cdot 10^{-5}$ s
$\rho c = 33\,600$ J/(m³·K)	$c_a = 1.17 \cdot 10^{-5}$ kg/(m³·Pa)

The thermal diffusivity so is $1.1 \cdot 10^{-6}$ m²/s, the isothermal air diffusivity 6.8 m²/s, a value 6 355 000 times larger! Or, in case of a transient air transfer under isothermal conditions, the air balance remains steady state, or:

$$\nabla^2 P_a = 0$$

This turns the combination with the flux equation into a copy of the steady state heat conduction.

Although in non-isothermal conditions, due to the fact the pressure differences include thermal stack, the isobaric volumetric specific air capacity is showing some variability ($\approx 3525/T^2$), the response remains nearly steady state:

$$\nabla^2 (P_{ao} - \rho_a \, \mathbf{g} \, z) = 0 \quad \mathbf{g}_a = -k_a \, \mathbf{grad} \, (P_{ao} - \rho_a \, \mathbf{g} \, z)$$

One point anyhow, because the temperature dependency of the air density ρ_a induces stack, the air and heat balances must be solved simultaneously.

2.2.4.2 One Dimension: Flat Assemblies

Consider a flat assembly composed of air-permeable layers. Given the resemblance to steady-state heat conduction, if single-layered, the air pressure profile is a straight line linking the air pressures on both faces (Figure 2.6), while the air flux is:

$$g_a = k_a \frac{\Delta P_a}{d} = \frac{\Delta P_a}{d/k_a} \tag{2.23}$$

The ratio d/k_a is called the air resistance, symbol W, units m/s. The inverse gives the air conductance, symbol K_a, units s/m.

For multi-layer assemblies, the air flux through equals:

$$g_a = \Delta P_a \bigg/ \sum_{i=1}^{n} \frac{d_i}{k_{ai}} \tag{2.24}$$

with $\sum d_i/k_{ai}$ the total air resistance, symbol W_T, units m/s. $1/W_T$ gives the total air conductance K_{aT}, while the ratio d_i/k_{ai} is the air resistance W_i and $1/W_i = k_{ai}/d_i$ the air conductance K_{ai} of layer i in the assembly. The air pressures in the interfaces become:

$$P_{aj} = P_{a1} + \frac{\sum_{i=1}^{j} W_i}{W_T} (P_{a2} - P_{a1}) = P_{a1} + g_a W_{1j} \tag{2.25}$$

In a $[W, P_a]$ axis system, the air pressure course in a multi-layer assembly looks as if single-layered (Figure 2.7). A transfer of the intersections with the successive

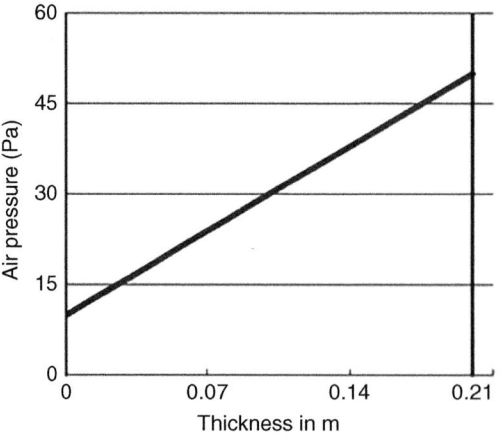

Figure 2.6 Airflow through a single-layer flat assembly, air pressure line.

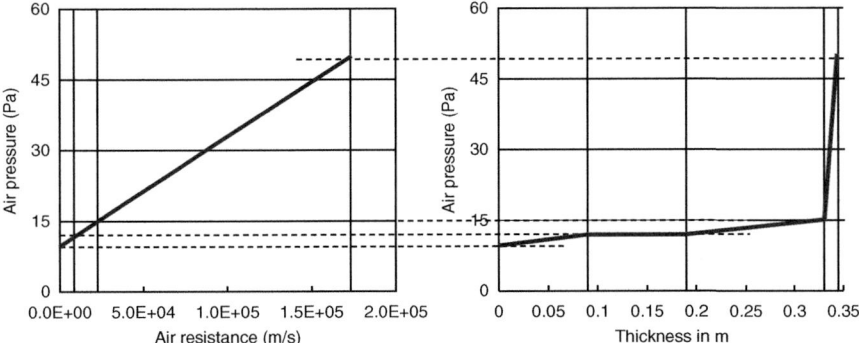

Figure 2.7 Airflow through a multi-layered flat assembly, air pressure in the $[W_a, P_a]$ and the $[d, P_a]$ axis system.

interfaces in the correct order to the assembly drawn in a $[d, P_a]$ axis system and linking them in the right order with straight line segments gives an air pressure course as shown in Figure 2.7, with as air pressure difference over a layer:

$$\Delta P_{aj} = \frac{W_j}{W_T}(P_{a2} - P_{a1})$$

Clearly, the most airtight layer takes most of the air pressure difference. Inclusion of such layer, called an air barrier, in an assembly increases its air resistance substantially, so minimizes the airflow through. Of course, air barriers must be able to resist such air pressure difference.

Is a flat assembly approach relevant? Not really. The assumption of a one-dimensional flow excludes stack along vertical and inclined surfaces as the air pressures then change along the height. Even wind is never uniform. Also, unintended cracks, leaks and other unexpected perforations will disrupt one-dimensionality.

2.2.4.3 Two and Three Dimensions

For isotropic open porous materials, the calculation is a copy of two and three dimensional heat conduction. Mass conservation replaces energy conservation, stating that the sum of the airflows from the adjacent volumes to the central volume must be zero:

$$\sum G_{a,i+j} = 0$$

The algorithm so becomes:

$$\sum_{\substack{i=l,m,n \\ j=\pm 1}} \left(K'_{a,i+j} P_{a,i+j}\right) - P_{a,l,m,n} \sum_{\substack{i=l,m,n \\ j=\pm 1}} K'_{a,i+j} = 0$$

with $K'_{a,i+j}$ the air permeance between the centres of each adjacent and the central volumes. If in a same material layer, K_a becomes:

$$K_a = k_a A/a$$

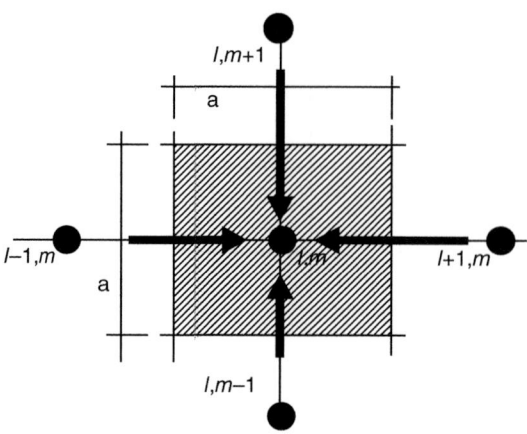

Figure 2.8 Conservation of mass: sum of airflows from the nearest surrounding to a central control volume zero.

In it, A is the contact surface between central and adjacent control volume and a the distance separating their centres (Figure 2.8).

For p control volumes, the result is a system of p equations with p unknowns:

$$[K'_a]_{p,p}[P_a]_p = [K'_{a,i,j,k} P_{a,i,j,k}]_p \tag{2.26}$$

In it, $[K'_a]_{p,p}$ is a p rows, p columns permeance matrix, $[P_a]_p$ a column matrix with the p unknown air pressures and $[K'_{a,i,j,k} P_{a,i,j,k}]_p$ a column matrix with all known air pressures. Once the unknowns are solved, the airflow exchanged between adjacent and central control volumes follows from:

$$G_{a,i,j,k} = K'_{a,i,i+j}(P_{a,i+j} - P_{a,i}) \tag{2.27}$$

For anisotropic materials, a same algorithm applies on condition each of the lines linking the centres of adjacent to central control volumes coincides with the main directions of the permeability tensor. Related $(K'_a)_{z,y,z}$-values can then be used.

Under non-isothermal conditions, per control volume thermal stack couples the air to the heat balances. Most of the time, solving requires iteration between both.

2.2.5 Airflow Through Assemblies with Air-open Layers, Leaky Joints, Leaks, Cavities, etc.

For the flow equations, see above. Most assemblies combine air-open layers, open porous materials, joints and cavities. In such case, writing the conservation law as a partial differential equation does not work. An approach exists of transforming the assembly into an equivalent hydraulic circuit composed of well-chosen points, connected by air permeances (Figure 2.9). Per point, the sum of the airflows coming from the adjacent points must be 0. As each flow can be written as $K_a^x \Delta P_a$, insertion in the conservation law gives:

$$\sum_{\substack{i=l,m,n \\ j=\pm 1}} \left(K^x_{a,i+j} P_{a,i+j} \right) - P_{a,l,m,n} \sum_{\substack{i=l,m,n \\ j=\pm 1}} K^x_{a,i+j} = 0 \tag{2.28}$$

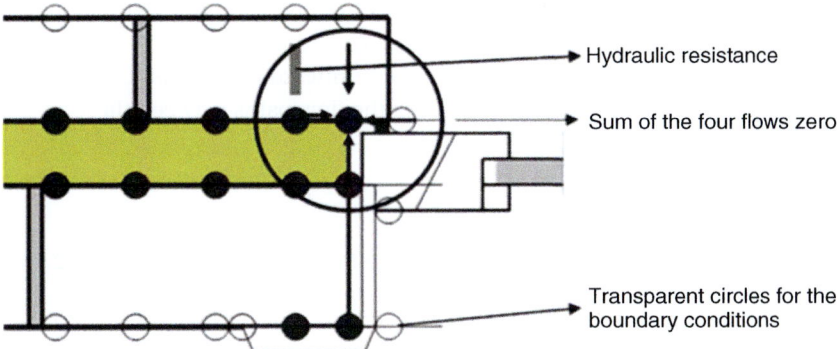

Figure 2.9 Equivalent hydraulic circuit.

A total of p points so results in a system of p non-linear equations with p unknown air pressures. If the boundary conditions are known, solving gives these unknowns. Insertion into the flow equations then ensues the airflows between the control volumes. Of course, non-linearity requires iteration, starting by assuming values for the p unknown air pressures, which allows to calculate the related air permeances as function of these and solve the system. Next, all permeances are recalculated using the p air pressures just found and the system is solved again. Iterating so should continue till the deviation between the preceding and new results dives below a predefined number.

In case the flow through a flat assembly, composed of air open layers, is by exception one-dimensional, a same air flux g_a must pass each layer with non-linear air permeance:

Layer 1 $\quad g_a = K_{a_1} \Delta P_{a_1} = a_1 \Delta P_{a_1}^{b_1} \quad$ or $\quad \left(\dfrac{g_a}{a_1}\right)^{1/b_1} = \Delta P_{a_1}$

Layer 2 $\quad g_a = K_{a_2} \Delta P_{a_2} = a_2 \Delta P_{a_2}^{b_2} \quad$ or $\quad \left(\dfrac{g_a}{a_2}\right)^{1/b_2} = \Delta P_{a_2}$

Layer n $\quad g_a = K_{a_n} \Delta P_{a_n} = a_n \Delta P_{a_n}^{b_n} \quad$ or $\quad \left(\dfrac{g_a}{a_n}\right)^{1/b_n} = \Delta P_{a_n}$

Sum: $\quad g_a \left[\sum \dfrac{g_a^{1/b_i - 1}}{a_i^{1/b_i}} \right] = \Delta P_a \quad$ (2.29)

ΔP_a is the air pressure differential between both end faces. Solving requires again iteration. Once the 'correct' air flux is known, the air pressure course in the assembly then follows from the layer equations.

2.2.6 Airflow at the Building Level

2.2.6.1 Definitions

In buildings, air moving between spaces and to and from outdoors is called interzonal flow, air movement in a space intrazonal flow. Questions which intrazonal must answer include: What with air looping? How do the zone and ventilation air mix? Which corners lack air washing? Related calculations require Computerized Fluid Dynamics (CFD), not discussed here. Only interzonal flow is. Calculating starts with replacing spaces by nodes and fixing the flow paths between these.

2.2.6.2 Thermal Stack
With the outdoors as reference, stack writes as:

$$P_a - g\rho_a z = P_a - z\frac{gP_a}{R_a}\left(\frac{1}{T_e} - \frac{1}{T_i}\right) \approx P_a - 3460z\left(\frac{1}{T_e} - \frac{1}{T_i}\right) \quad (2.30)$$

with T_i the temperature in the space in K, T_e the temperature outdoors in K and z the height considered in m. Between equally warm spaces at different height coupled by a flow path, the pressure difference with outdoors is:

$$\Delta_{12}(P_a - \rho gz) = P_{a1} - P_{a2} - 3460(z_1 - z_2)\left(\frac{1}{T_e} - \frac{1}{T_i}\right)$$

with z_1 and z_2 the height of the two zonal points above a horizontal reference plane. Both spaces at different temperature ($T_{i,1}$, $T_{i,2}$) changes this stack equation into:

$$\Delta_{12}(P_a - \rho gz) = P_{a1} - P_{a2} - 3460$$
$$\times \left[(z_o - z_1)\left(\frac{1}{T_e} - \frac{1}{T_{i,1}}\right) + (z_2 - z_o)\left(\frac{1}{T_e} - \frac{1}{T_{i,2}}\right)\right]$$

with z_o the height of the horizontal plane (o) where temperature $T_{i,1}$ becomes $T_{i,2}$.

2.2.6.3 Large Openings
The airflow through a closed window is:

$$G_a = a\, L\, \Delta P_a^{2/3}$$

with a the air permeance per running metre of joints in the casement in kg/(m·s·Pa$^{2/3}$), L their length in m, ΔP_a the pressure differential between the zone node and either outdoors or at the zone node on the other side of the window, then part of a partition. Open, doors and windows instead act as large openings with an air permeance (K_a) depending on the driving force. Where wind and fans give a uniform flow through, stack activates two equally large flows, one up from warm to cold and an equal one down from cold to warm, with the dividing line in the middle of the opening (Figure 2.10).

The airflow in case wind or a fan care for the pressure differences equals:

$$G_a = C_f BH\sqrt{2\rho_a \Delta P_a} \quad \text{(kg/s)} \quad (2.31)$$

giving as air permeance:

$$K_a = (C_f BH\sqrt{2\rho_a})\Delta P_a^{-0.5} \quad (2.32)$$

with C_f a flow factor with value 0.33–0.7, B the width and H the height of the door or window. With thermal stack, the flows in to out and out to in become:

$$G_{a1} = -G_{a2} = \frac{C'_f B}{3}\sqrt{\frac{\rho_a g P_a H^3}{R_a}\left(\frac{1}{T_1} - \frac{1}{T_2}\right)} \quad \text{(kg/s)} \quad (2.33)$$

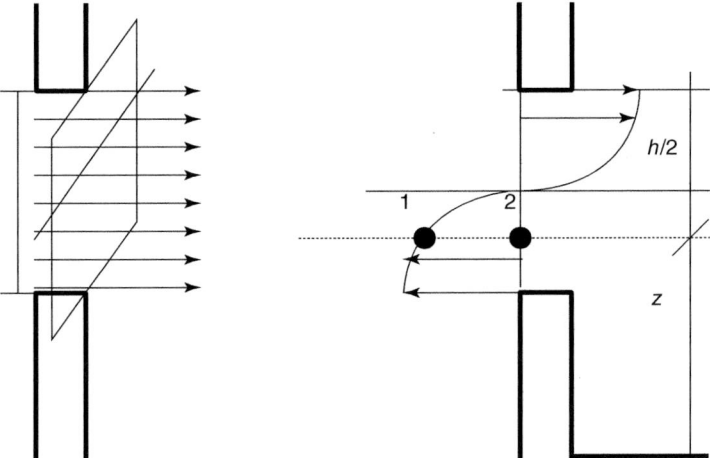

Figure 2.10 Large opening, uniform flow by wind and fans and stack-induced flow.

giving as air permeance $\left(\Delta p_{T,\max} = gP_a H/R_a \left(T_e^{-1} - T_2^{-1} - T_e^{-1} + T_1^{-1}\right) \approx 3450\,H \left(T_1^{-1} - T_2^{-1}\right)\right)$:

$$K_{a1} = -K_{a2} = \frac{C_f' B H \sqrt{\rho_a}}{3}\left[3450H\left(\frac{1}{T_1} - \frac{1}{T_2}\right)\right]^{-0.5} \quad (2.34)$$

In it, R_a is the gas constant of air and C_f' the flow factor characterising the two-way move. The velocity of the air in- and outgoing along the opening's height is parabolic, a consequence of Bernouilli's law applied to a flow path at height z above the dividing line:

$$\rho_{a1} g z = \frac{\rho_{a2} v_{a,z}^2}{2} + \rho_{a2} g z$$

This gives as velocity:

$$v_{a,z} = \sqrt{2gz\left(\frac{\rho_{a1} - \rho_{a2}}{\rho_{a2}}\right)}.$$

The air permeance equation given so follows from first integrating the air speed over half the height of the opening, then using the ideal gas law to convert air density into the related temperature and finally by multiplying the result with the flow factor.

2.2.6.4 The Conservation Law Applied

Buildings include several spaces, thus, several nodes, see Figure 2.11.

Per node, the algebraic sum of the n airflows coming in and m going out must be 0, or:

$$\sum_{i=1}^{m+n} G_{a,i} = 0 \quad (2.35)$$

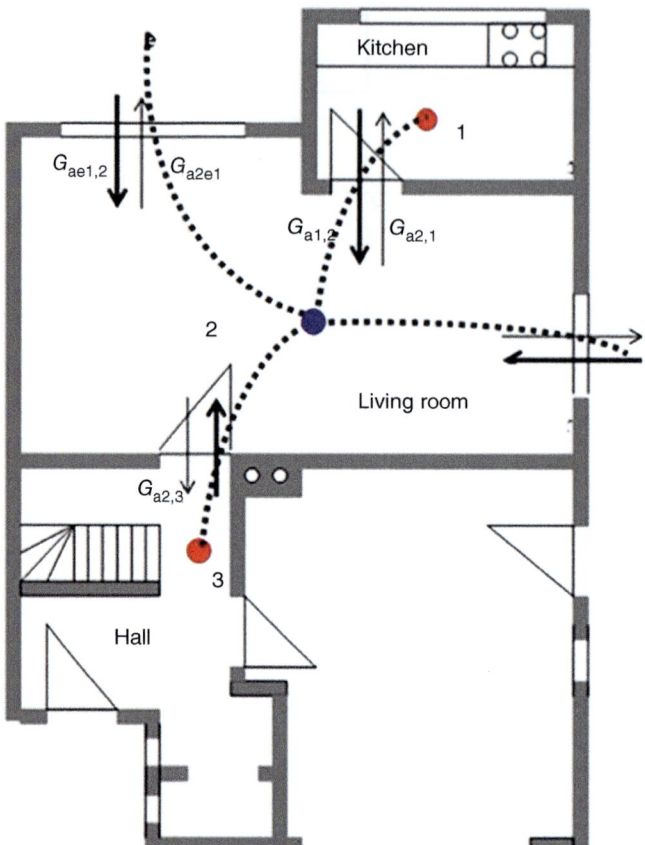

Figure 2.11 The ground floor of the building considered.

The living room, zone 2 in Figure 2.11, has two windows and two doors, one to the kitchen, zone 1, one to the hall, zone 3. All partitions and opaque outer walls are now assumed airtight, which in reality may not be the case. The building has no fan-driven ventilation and the temperatures in- and out are set equal. Wind then is the only driving force. Air will enter or leave the living room through the operable window sashes ($G_{a,2-e1}$, $G_{a,2-e2}$) and through both doors ($G_{a,2-1}$, $G_{a,2-3}$). Wind pressures outdoors are known. The four flows supposed to move inwards equal:

$$G_{a,2-e1} = K_{a,2-e1}(P_{a,e1} - P_{a,2}) \quad G_{a,2-e2} = K_{a,2-e2}(P_{a,e2} - P_{a,2})$$

$$G_{a,2-1} = K_{a,2-1}(P_{a,1} - P_{a,2}) \quad G_{a,2-3} = K_{a,2-3}(P_{a,3} - P_{a,2})$$

Summing and reshuffling give:

$$-(K_{a,2-e1} + K_{a,2-e2} + K_{a,2-1} + K_{a,2-3})P_{a,2} + K_{a,2-e1}P_{a,e1} + K_{a,2-e2}P_{a,e2} + K_{a,2-1}P_{a,2} + K_{a,2-3}P_{a,3} = 0$$

or:

$$-P_{a,2}\sum(K_{a,i-j/e}) + \sum(K_{a,i-j}P_{a,j}) = -\sum(K_{a,i-e}P_{a,e})$$

The three unknowns in this equation are the air pressure in the living room (P_{a2}), the kitchen (P_{a1}) and the hall (P_{a3}).

Besides living room, kitchen and hall, the ground floor also consists of a garage and a toilet, while the staircase links the ground floor to three bedrooms and a bathroom on the first floor. For each, the flows in and out, the last supposed inward, give sum 0. The result is as many equations as there are spaces. In matrix form:

$$[K_a]_{n,n}[P_a]_n = [K_{a,e}P_{a,e}]_n \tag{2.36}$$

If non-linear, solving that system demands iteration.

Is the building warmer or colder than outdoors, all leaks must be located. The spaces then get nodes at the height of each leak, after which thermal stack with outdoors against a reference height complements the pressure differences between adjacent nodes at different height. Between nodes one above the other in a same space, they get coupled by very large fictitious air permeances. With the temperature in all spaces and outdoors known, stack figures as a known term in the balance equations. The temperature in unheated spaces, however, depends on the transmission losses, the internal gains and the temperature of the entering air. This obliges to combine mass and energy conservation with the correct temperatures following from assuming a value and alternatively solving the heat and air balances using the temperatures and airflows given by the preceding iteration, until the differences between new and preceding drop below a preset value. Is ventilation fan-driven, these act as sources with known pressure/flow characteristics.

2.2.6.5 Applications

Airflow through a narrow horizontal opening in a façade wall of an otherwise airtight space, take a trickle vent above a window, is hardly possible. Vertically opening windows set ajar instead care for air exchange, not only by thermal stack giving a twinned flow but also by pulsating wind pushing air in and out. Openings in two opposite walls enclosing a space in turn act in series. Has the one an air permeance $K_{a,1}$, the other an air permeance $K_{a,2}$, is $P_{a,1}$ the air pressure in front of the one, $P_{a,2}$ the air pressure beyond the other, then, if isothermal, the air balance is (Figure 2.12):

$$-(K_{a,1} + K_{a,2})P_{a,x} + K_{a,1}P_{a,1} + K_{a,2}P_{a,2} = 0$$

with $P_{a,x}$ the unknown air pressure in the space.

For air to flow, $P_{a,1}$ and $P_{a,2}$ must be different, which, if windy, is likely for openings in opposite outer walls. Of course, given the wind pressure increases with height, two openings at different height in a wall also act as a series circuit. How large the airflow will be, depends on their air permeances. If constants, the mass balance gives:

$$G_{a,1} = -G_{a,2} = G_a = K_{a,1}(P_{a,1} - P_{a,x}) = \frac{1}{\frac{1}{K_{a1}} + \frac{1}{K_{a2}}}(P_{a,1} - P_{a,2})$$

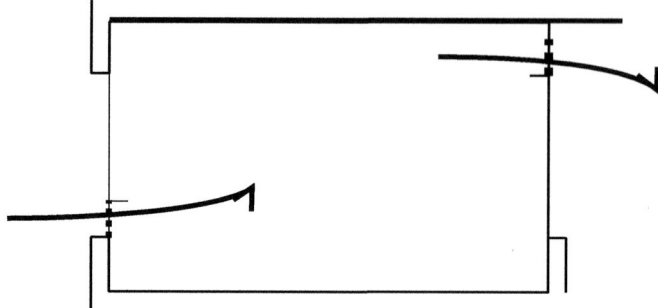

Figure 2.12 Two openings in series.

If in both permeances the pressure has a same exponent, the airflow changes to:

$$G_a = \frac{1}{\left(\dfrac{1}{a_1^{1/b}} + \dfrac{1}{a_2^{1/b}}\right)^b}(P_{a,1} - P_{a,2})^b$$

If that exponent differs, the formula becomes:

$$G_a = \frac{1}{\left[\left(\dfrac{G_a}{a_1}\right)^{\frac{1-b_1}{b_1}} + \left(\dfrac{G_a}{a_2}\right)^{\frac{1-b_2}{b_2}}\right]}(P_{a,1} - P_{a,2})$$

Solving it now requires iteration, beginning with calculating the airflow G_a, assuming both exponents b_1 and b_2 equal to 1. This gives an airflow that can be used to value the numerator, which allows a better guess of that airflow, now with the correct value of both exponents. Iterating this way must go on till the difference with the previous result drops below a preset value. Anyhow, if one of both leaks is hardly air permeable, the airflow will anyway near zero.

If several spaces are series coupled with an inlet in an outer wall of the first, an outlet in an outer wall of the last and flow through openings in the partitions in-between, then, with $P_{a,1}$ and $P_{a,n+1}$ the air pressures in front of the inlet and past the outlet, the resulting airflow through will become:

All air permeances constant

$$G_a = \frac{1}{\sum_{j=1}^{n}\dfrac{1}{K_{aj}}}(P_{a,1} - P_{a,n+1})$$

All exponents b identical

$$G_a = \frac{1}{\left(\sum_{j=1}^{n}\dfrac{1}{a_j^{1/b}}\right)^b}(P_{a,1} - P_{a,n+1})^b$$

Exponents b different

$$G_a = \frac{1}{\sum_{j=1}^{n}\left[\left(\dfrac{G_a}{a_j}\right)^{\frac{1-b_j}{b_j}}\right]}(P_{a,1} - P_{a,n+1})$$

The smallest air permeance again will define the flow's magnitude.

With openings at different heights in a same outer wall and the temperature in and out different, thermal stack then creates a pressure difference. Take two openings having a constant air permeance at ΔH meters above each other. On the horizontals across, the stack pressures are $P_{a,1}$ and $P_{a,2}$, while an infinite large, fictitious air permeance $K_{a,z}$ is coupling the two at the warm side (Figure 2.13).

The mass balances give:

$$\text{Opening 1: } -(K_{a,1} + K_{a,z})P_{a,1} + K_{a,z}P_{a,2} = -K_{a,1}P_{a,e}$$
$$+ K_{a,z}\left[3460\Delta H\left(\frac{1}{T_e} - \frac{1}{T_i}\right)\right]$$
$$\text{Opening 2: } -(K_{a,2} + K_{a,z})P_{a,2} + K_{a,z}P_{a,1} = -K_{a,2}P_{a,e}$$
$$- K_{a,z}\left[3460\Delta H\left(\frac{1}{T_e} - \frac{1}{T_i}\right)\right]$$

With $K_{a,z}$ infinite, the airflow passing through becomes:

$$G_a = \frac{1}{\frac{1}{K_{a1}} + \frac{1}{K_{a2}}}\left[3460\Delta H\left(\frac{1}{T_e} - \frac{1}{T_i}\right)\right] \approx \frac{1}{\frac{1}{K_{a1}} + \frac{1}{K_{a2}}}[0.043\Delta H(\theta_i - \theta_e)]$$

Or, the two openings act as series coupled with the smallest fixing the flow. Instead, several openings in a partition between two equally warm spaces act as coupled in parallel. When the air pressures at both sides differ, the resulting airflow will be:

All air permeances constant

$$G_a = \sum_{j=1}^{n} K_{aj}(P_{a,1} - P_{a,2})$$

Figure 2.13 Openings at different height in a same outer wall, thermal stack intervening.

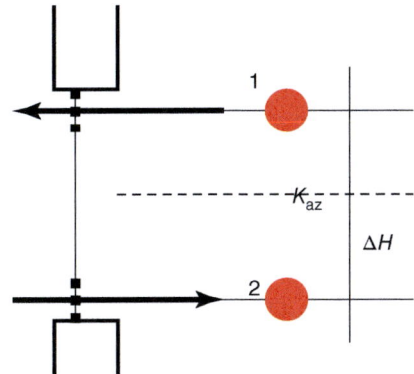

All exponents b identical

$$G_a = \sum_{j=1}^{n} a_j (P_{a,1} - P_{a,2})^b$$

Exponents b different

$$G_a = \sum_{j=1}^{n} \left(a_j \Delta P_a^{b_j - 1} \right) (P_{a,1} - P_{a,2})$$

Clearly, more air passes through when more openings or leaks perforate a partition.

2.2.7 Combined Heat and Airflow Through Assemblies Composed of Open-porous Layers

2.2.7.1 Heat Balance

When air moves through open porous layers, the enthalpy contained adds to the heat transmitted. Of course, related flow is normally so slow that along its thickness air and layer get the same temperature. Energy conservation per elementary volume dV so becomes:

$$\text{div} \, (\mathbf{q} + c_a \mathbf{g}_a \theta) = -\frac{\partial (\rho c + \rho_a c_a S_a \Psi_o) \theta}{\partial t} \pm \Phi'$$

with $\rho_a c_a$ the volumetric specific heat capacity of air, S_a the saturation degree of the layer for air and Ψ_o its open porosity. In most cases the product $\rho_a c_a S_a \Psi_o$ is negligible, simplifying that equation to:

$$\text{div} \, (\mathbf{q} + c_a \mathbf{g}_a \theta) = -\frac{\partial \rho c \theta}{\partial t} \pm \Phi'$$

Introducing Fourier's law with the thermal conductivity (λ) and the volumetric heat capacity (ρc) assumed constant and expanding the operator term, knowing that $\text{div} \, \mathbf{g}_a = 0$ gives:

$$-\lambda \nabla^2 \theta + c_a \mathbf{g}_a \nabla \theta = -\rho c \frac{\partial \theta}{\partial t} \pm \Phi' \tag{2.37}$$

2.2.7.2 Steady-state: Flat Assemblies
2.2.7.2.1 The Equations Adapted

In steady-state, the energy balance just derived simplifies to:

$$\lambda \nabla^2 \theta - c_a \mathbf{g}_a \nabla \theta = \pm \Phi' \tag{2.38}$$

In case the air and heat fluxes develop perpendicular to the assembly's faces and no heat source nor sink intervenes, the equation further simplifies to:

$$\frac{d^2 \theta}{dx^2} - \frac{c_a g_a}{\lambda} \frac{d\theta}{dx} = 0 \tag{2.39}$$

with as solution:

$$\theta = C_1 + C_2 \exp\left(\frac{c_a g_a x}{\lambda}\right) \tag{2.40}$$

C_1 and C_2 are the integration constants, ensuing from the boundary conditions. The conduction related, enthalpy related, and total heat fluxes so become:

Conduction $\quad q = -\lambda \dfrac{d\theta}{dx} = -c_a g_a C_2 \exp\left(\dfrac{c_a g_a x}{\lambda}\right)$

Enthalpy (convection) $\quad H = q_c = c_a g_a \left[C_1 + C_2 \exp\left(\dfrac{c_a g_a x}{\lambda}\right)\right]$

Total $\quad q_T = q + H = q + q_c = c_a g_a C_1$

While the conductive flux and the enthalpy flux, also called the convective flux change exponentially, their sum, remains constant over the assembly. Otherwise said, the assembly functions as heat exchanger between the conductive and enthalpy fluxes, so is warming or cooling the moving air.

2.2.7.2.2 Single-layered

With d the thickness of the single-layer wall, rewriting the solution gives:

$$\theta = C_1 + C_2 \exp\left(\dfrac{c_a g_a d}{\lambda} \dfrac{x}{d}\right) \tag{2.41}$$

The dimensionless ratio $c_a g_a d/\lambda$ is called the Péclet number, symbol Pe. It couples the enthalpy flux to the conductive flux. The more air flows through $1\,m^2$ and the higher the thermal resistance of the assembly, the larger Pe.

Type 1 boundary conditions now presume that the air flux through and the temperature at both end faces are known, or: $x = 0: \theta = \theta_{s1}; x = d: \theta = \theta_{s2}$. The integration constants then follow from: $\theta_{s1} = C_1 + C_2$ and $\theta_{s2} = C_1 + C_2 \exp(Pe)$, giving:

$$C_1 = \dfrac{\theta_{s2} - \theta_{s1}\exp(Pe)}{1 - \exp(Pe)} \quad C_2 = \dfrac{\theta_{s1} - \theta_{s2}}{1 - \exp(Pe)}$$

The temperatures in the assembly and the conductive, convective and total heat flux so become:

$$\theta = \theta_{s1} + F_1(x)(\theta_{s2} - \theta_{s1})$$

$$q = F_2(x)\lambda \dfrac{\theta_{s2} - \theta_{s1}}{d} \quad q_c = Pe \dfrac{\lambda \theta}{d} \quad q_T = \lambda \dfrac{\theta_{s2} F_2(0) - \theta_{s1} F_2(d)}{d} \tag{2.42}$$

with:

$$F_1(x) = \dfrac{1 - \exp\left(Pe\dfrac{x}{d}\right)}{1 - \exp(Pe)} \quad F_2(x) = \dfrac{Pe \exp\left(Pe\dfrac{x}{d}\right)}{1 - \exp(Pe)}$$

The functions $F_1(x)$ and $F_2(x)$ depend on the material properties and the magnitude of the air flux, which is seen as negative when moving from warm (end face d) to cold (end face 0) and as positive when the opposite. Air flowing through thus changes the temperature course from linear to convex exponential if from warm to cold, to concave exponential if from cold to warm (Figure 2.14).

The higher the Péclet number, the more air passes through and the more concave or convex the exponential. Compared to pure conduction (q_o), the total heat flux shows a change in % equal to ($x = 0, \theta_{s1} = 0\,°C; x = d, \theta_{s2} = 1\,°C$):

$$100(q_T - q_0)/q_0 = 100[abs(F_2(0)) - 1] \quad (\%)$$

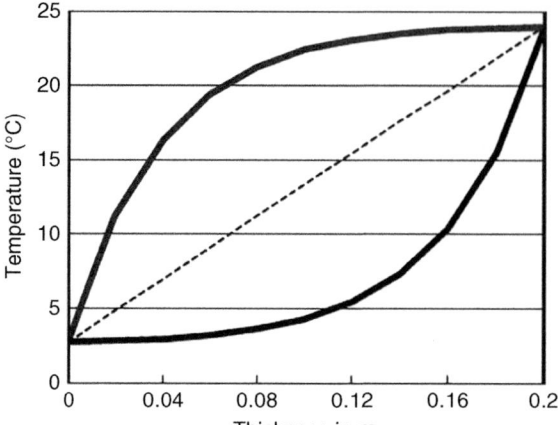

Figure 2.14 Single-layer assembly, conductive plus enthalpy fluxes. Exponential temperature course, convex if air moves from warm to cold, concave if from cold to warm.

For air moving from warm to cold, abs($F_2(0)$) passes 1 and the enthalpy flux lifts the heat flux. If from cold to warm, abs($F_2(0)$) drops below 1 and the enthalpy flux lowers the heat flux. The conductive flow changes the opposite way: lower for air moving from warm to cold, higher if moving from cold to warm.

Type 2 boundary conditions consider the air flux through and the heat transferred by convection and radiation to the end faces in both ambients as being known. Adding a fictitious surface layer with thermal resistance $1/h_1$ to end face $x = 0$ and one with thermal resistance $1/h_2$ to end face $x = d$ converts type 2 to type 1 with the temperatures θ_1 and θ_2 replacing the surface temperatures and the assembly three-layered. The thermal balance with R as independent variable then gives:

$$\frac{d^2\theta}{dR^2} - c_a g_a \frac{d\theta}{dR} = 0 \tag{2.43}$$

with as solution:

$$\theta = C_1 + C_2 \exp(c_a g_a R)$$

The boundary conditions are: $R = 0$: $\theta = \theta_1$; $R = R_T = 1/h_1 + R + 1/h_2$: $\theta = \theta_2$, giving as integration constants:

$$C_1 = \frac{\theta_2 - \theta_1 \exp(c_a g_a R_T)}{1 - \exp(c_a g_a R_T)} \qquad C_2 = \frac{\theta_1 - \theta_2}{1 - \exp(c_a g_a R_T)}$$

The temperatures in the assembly, the conduction, the convection and the total heat flux so equal:

$$\theta = \theta_1 + (\theta_2 - \theta_1) F_1(R)$$

$$q = F_2(R)(\theta_2 - \theta_1) \quad q_c = c_a g_a \theta \quad q_T = \theta_2 F_2(0) - \theta_1 F_2(R_T) \tag{2.44}$$

with:

$$F_1(R) = \frac{1 - \exp(c_a g_a R)}{1 - \exp(c_a g_a R_T)} \qquad F_2(R) = \frac{c_a g_a \exp(c_a g_a R)}{1 - \exp(c_a g_a R_T)} \tag{2.45}$$

In both, R is the thermal resistance from $R = 0$ to any parallel plane in the assembly. Again, the functions F_1 and F_2 depend on the material properties and the magnitude of the air flux.

2.2.7.2.3 Multi-layered

Usable is the solution for a single-layer assembly, subjected to type 2 boundary conditions. With $R = 0$ the outdoors or the other side and $R = R_T$ the indoors, the boundary conditions become: $R = 0$: $\theta = \theta_e$; $R = R_T = 1/h_1 + \Sigma R + 1/h_2$: $\theta = \theta_i$. With end face 1 the outside and the sequence 1, 2,... of interfaces starting at 1, the interface temperatures become ($\theta_e (R = 0) < \theta_i (R = R_T)$):

End face 1 $\theta_{s1} = \theta_e + (\theta_i - \theta_e)F_1(1/h_1)$

Interface 1 $\theta_1 = \theta_e + (\theta_i - \theta_e)F_1(1/h_1 + R_1)$

Interface 2 $\theta_2 = \theta_e + (\theta_i - \theta_e)F_1(1/h_1 + R_1 + R_2)$,

etc...

Related conductive, convective and total heat flux look:

$$q = F_2(R)(\theta_i - \theta_e) \quad q_c = c_a g_a \theta \quad q_T = \theta_i F_2(0) - \theta_e F_2(R_T)$$

with

$$F_1(R) = \frac{1 - \exp(c_a g_a R)}{1 - \exp(c_a g_a R_T)} \quad F_2(R) = \frac{c_a g_a \exp(c_a g_a R)}{1 - \exp(c_a g_a R_T)}$$

In a $[\theta, R]$-axis system, the temperature course looks as in a single-layer assembly. If the air flows to the colder outdoors, the exponential turns convex, if otherwise, concave. The move to the $[\theta, d]$-axis system goes the way explained for heat conduction, but the successive interface temperatures are linked exponentially now (Figure 2.15).

In case of a leaky envelope, infiltration in windy weather occurs at the windward side, exfiltration at the leeward side and at the sides ≈parallel to the wind. Compared to the heat losses by conduction only, the combination conduction plus enthalpy flow is moderating them. Or, an air-leaky envelope saves energy. Indeed, the windward conduction losses warm the air infiltrating, while at the leeward side and the

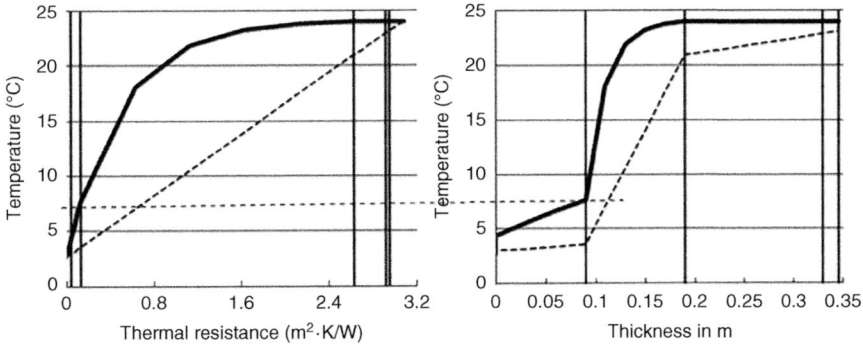

Figure 2.15 Multi-layered assembly, temperatures, indoor air moving to the cold outdoors.

sides parallel to the wind that warmed air exfiltrates, so lowers the conduction losses there. Of course, this does not compensate for the other problems a leaky envelope gives.

2.2.7.3 Steady-state, Two and Three Dimensions

Analytical solutions for the air balance hardly exist. With a numerical approach, included thermal stack, iteration waits. Once the airflows are known, the heat flow between two adjacent nodes equals:

If from node 2 to 1 $\quad \Phi_{21} = P'_{21}(\theta_2 - \theta_1) + c_a G_a \theta_2 = \theta_2 \left(P'_{21} + c_a G_a\right) - \theta_1 P'_{21}$

If from node 1 to 2 $\quad \Phi_{21} = P'_{21}(\theta_2 - \theta_1) - c_a G_a \theta_2 = \theta_2 P'_{21} - \theta_1 \left(P'_{21} + c_a G_a\right)$

$$(2.46)$$

with G_a the known airflow in between and P'_{21} the thermal permeance linking the two. An alternative is to start from the analytical solution for the heat flux through flat assemblies with as heat flow for a series connection:

$$\Phi_{21} = \theta_1 F_2(0) - \theta_2 F_2\left(R'_{21}\right)$$

where R'_{21} is the surface-linked thermal resistance between the two adjacent nodes. For a combined parallel-serial path, the heat flow per serial path is calculated first, after which combining gives:

$$\Phi_{21} = \sum_{i=1}^{2} \Phi^i_{21}$$

2.2.7.4 Non-steady-state, Flat Assemblies

For a constant air flux and periodically varying temperatures, the complex numbers approach as used for heat conduction remains applicable, giving per harmonic:

$$\theta(R) = \alpha(R)\exp\frac{2in\pi t}{T}$$

with $\alpha(R)$ the complex temperature. Without heat source or sink, the response ensues from:

$$\frac{d^2\alpha}{dR^2} - c_a g_a \frac{d\alpha}{dR} - \frac{2in\pi\rho c\lambda}{T}\alpha = 0$$

with i the imaginary unit and T the period. Solving gives:

$$\alpha(R) = C_1 \exp(r_1 R) + C_2 \exp(r_2 R) \tag{2.47}$$

with r_1 and r_2 the roots of the quadratic equation $r^2 - c_a g_a r - 2in\pi\rho c\lambda/T = 0$:

$$r_1 = \frac{1}{2}\left[c_a g_a + \sqrt{(-c_a g_a)^2 + \frac{8in\pi\rho c\lambda}{T}}\right] \quad r_2 = \frac{1}{2}\left[c_a g_a - \sqrt{(-c_a g_a)^2 + \frac{8in\pi\rho c\lambda}{T}}\right]$$

or:

$$\alpha(R) = \exp\left(\frac{1}{2}c_a g_a R\right)\left[(C_1 - C_2)\sinh\left(\frac{1}{2}aR\right) + (C_1 + C_2)\cosh\left(\frac{1}{2}aR\right)\right]$$

with:

$$a = \sqrt{(-c_a g_a)^2 + \frac{8in\pi\rho c\lambda}{T}}$$

An air flux zero gives the solution found for conduction only. The integration constants C_1 and C_2 follow from the boundary conditions, for example the complex temperature α_s and the complex heat flux α'_s at $R = 0$, the outside, known. It is up to the reader the make these and other calculations. Most important result is that air inflow from $R = 0$ decreases, air outflow at $R = 0$ increases the temperature damping and linked time shift of an assembly.

2.2.7.5 Non-steady-state, Two and Three Dimensions

The energy balance per control volume now is:

$$\sum \Phi_m \approx \rho c \, \Delta V \, \Delta\theta/\Delta t \pm \Phi' \Delta V$$

with Φ' a heat source or sink in it, Φ_m the heat flow by conduction and convection during time Δt from each of the adjacent to the central control volume with thermal capacity $\rho c \Delta V$.

2.2.7.6 Air-permeable Layers, Joints and Leaks

Calculating steady-state and transient combined heat and airflow through assemblies having air-permeable layers, joints in layers and leaks here and there demands combining the equivalent hydraulic network used for the air moving through with the control volume method (CVM) for the heat exchanged. For that the equations above for combined conduction and convection, included, if transient, the change in heat stored in the control volumes can be used.

2.2.7.7 Vented Cavities

The cavity in a flat assembly is d_{cav} cm wide and L m high. The air moves through at a velocity v (m/s) (Figure 2.16), while the thermal resistance from ambient 1 to the cavity face there is R_1 and the thermal resistance from ambient 2 to the cavity face there R_2, both in m²·K/W.

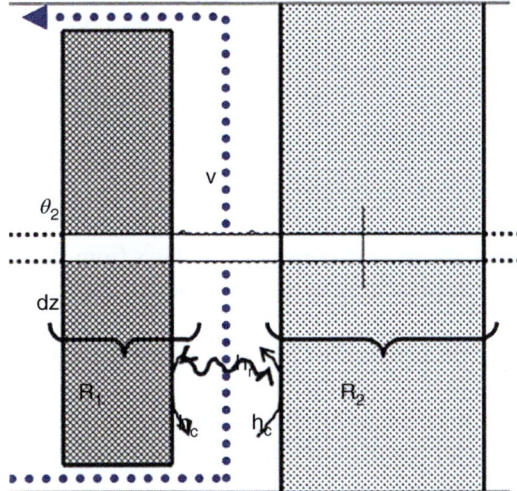

Figure 2.16 Vented cavity in a flat assembly. Combines heat conduction through both leaves with enthalpy flow in the cavity.

With z the ordinate along, the heat balances at both cavity faces and along the cavity height become:

$$\frac{\theta_1 - \theta_{s1}}{R_1} + h_c(\theta_{cav} - \theta_{s1}) + h_r(\theta_{s2} - \theta_{s1}) = 0$$

$$\frac{\theta_2 - \theta_{s2}}{R_2} + h_c(\theta_{cav} - \theta_{s2}) + h_r(\theta_{s1} - \theta_{s2}) = 0$$

$$[h_c(\theta_{s1} - \theta_{cav}) + h_c(\theta_{s2} - \theta_{cav})] \, dz = \rho_a c_a d_{sp} v \, d\theta_{cav}$$

with h_c the convective surface film coefficient at and h_r the surface film coefficient for radiation between both cavity faces, both in W/(m²·K). Unknowns are the temperature in the cavity and the temperatures at its faces ($\theta_{cav}, \theta_{s1}, \theta_{s2}$).

The solution uses the cavity's surface balances to rewrite the temperatures θ_{s1} and θ_{s2} as function of the cavity temperature θ_{cav}, after which both temperature equations are included in the cavity balance. To keep the equations simple the constants $D, A_1, A_2, B_1, B_2, C_1$ and C_2 replace following functions:

$$D = \left(h_c + h_r + \frac{1}{R_1}\right)\left(h_c + h_r + \frac{1}{R_2}\right) - h_r^2$$

$$A_1 = \frac{h_c + h_r + \frac{1}{R_2}}{DR_1} \quad A_2 = \frac{h_r}{DR_1} \quad B_1 = \frac{h_r}{DR_2} \quad B_2 = \frac{h_c + h_r + \frac{1}{R_1}}{DR_2}$$

$$C_1 = \frac{h_c\left(h_c + h_r + \frac{1}{R_2}\right) + h_r h_c}{D} \quad C_2 = \frac{h_c\left(h_c + h_r + \frac{1}{R_1}\right) + h_r h_c}{D}$$

The cavity faces temperatures so become:

$$\theta_{s1} = A_1\theta_1 + B_1\theta_2 + C_1\theta_{cav} \quad \theta_{s2} = A_2\theta_1 + B_2\theta_2 + C_2\theta_{cav} \quad (2.48)$$

while the cavity balance turns into:

$$\left[\underbrace{\frac{(A_1 + A_2)\theta_1 + (B_1 + B_2)\theta_2}{2 - C_1 - C_2}}_{=a} - \theta_{cav}\right] dz = \underbrace{\frac{\rho_a c_a d_{cav} v}{h_c(2 - C_1 - C_2)}}_{=b} d\theta_{cav}$$

with as solution:

$$\frac{z}{b} = -\ln\left(\frac{a - \theta_{cav}}{C}\right)$$

The integration constant C follows from the boundary conditions. With L being and z approaching ∞, also z/b does, giving for $z = \infty$ as solution:

$$\ln\left(\frac{a - \theta_{cav,\infty}}{C}\right) = -\infty \text{ or } a - \theta_{cav,\infty} = 0, \text{ so } a = \theta_{cav,\infty}.$$

with 'a' standing for the asymptotic temperature in an infinitely high cavity, equal to the air temperature in an unvented cavity. In fact, for an air velocity 0, z/b also turns ∞!

For $z = 0$, θ_{cav} equals $\theta_{cav,0}$, giving:

$$\ln\left(\frac{\theta_{cav,\infty} - \theta_{cav,0}}{C}\right) = 0$$

Or, $C = \theta_{cav,\infty} - \theta_{cav,0}$, meaning the air temperature in the cavity equals:

$$\theta_{cav} = \theta_{cav,\infty} - (\theta_{cav,\infty} - \theta_{cav,0})\exp(-z/b)$$

The temperature in a vented cavity so changes exponentially, from the inflow value $\theta_{cav,0}$ to a value close to if not vented. The term b has as dimension meter and is called the ventilation length. The higher the air speed in the cavity, the longer that length, the more heat exchanged, the shorter that length. The temperatures on both cavity faces (θ_{s1}, θ_{s2}) now follow from implementing the cavity temperature θ_{cav} in their respective equations:

$$\theta_{s1} = A_1\theta_1 + B_1\theta_2 + C_1\left[\theta_{cav,\infty} - (\theta_{cav,\infty} - \theta_{cav,0})\exp\left(-\frac{z}{b}\right)\right]$$

$$\theta_{s2} = A_2\theta_1 + B_2\theta_2 + C_2\left[\theta_{cav,\infty} - (\theta_{cav,\infty} - \theta_{cav,0})\exp\left(-\frac{z}{b}\right)\right]$$

With '2' the inside leaf, the heat flow through becomes:

$$\Phi = \frac{1}{R_2}\int_0^L (\theta_2 - \theta_{s2})\,dz = \frac{1}{R_2}\int_0^L$$
$$\cdot\left\{-A_2\theta_1 + (1 - B_2)\theta_2 - C_2\left[\theta_{cav,\infty} - (\theta_{cav,\infty} - \theta_{cav,0})\exp\left(-\frac{z}{b}\right)\right]\right\}dz$$
$$= \frac{(\theta_2 - \theta_{s2\infty})L}{R_2} - \frac{C_2 b_1 (\theta_{sp\infty} - \theta_{sp0})}{R_2}\left[\exp\left(-\frac{L}{b}\right) - 1\right]$$

This gives as average thermal transmittance U over the height L:

$$U = \frac{\Phi}{L(\theta_1 - \theta_2)} = U_o + \frac{C_2 b_1(\theta_{sp\infty} - \theta_{spo})}{LR_2(\theta_1 - \theta_2)}\left[1 - \exp\left(-\frac{L}{b}\right)\right]$$

As the ratio $(\theta_{sp\infty} - \theta_{spo})/(\theta_1 - \theta_2)$ is close to $R_1/R_a = R_1 U_o$, the formula can be simplified to:

$$U = U_o\left\{1 + \frac{C_2 b_1 R_1}{LR_2}\left[1 - \exp\left(-\frac{L}{b}\right)\right]\right\} \tag{2.49}$$

The impact a vented cavity has on the U-value so depends on the thermal resistance of both leaves (R_1, R_2). Is the inside leaf well-insulated thanks to a partial fill mounted perfectly, R_2 thus large, and is the outside leaf heat conducting, R_1 small, then venting hardly lifts the U-value. If instead, the partial fill should contact the outside leaf with the cavity at its back, R_1 large, the increase will become significant. At the air intake, the heat flux anyhow remains:

$$q_{x=0} = \frac{\theta_2 - \theta_{s2,0}}{R_2} \tag{2.50}$$

There, a well-insulated inside leaf is of cutting importance.

2.3 Water Vapour

2.3.1 Water Vapour in the Air

2.3.1.1 In General

That air contains water vapour is seen during cold days when condensate is depositing on single glass. Those wearing glasses have a same experience entering a heated, crowded room from the cold outdoors. Humid air is considered a mixture of two ideal gases: dry air and vapour (the additive water is from now on omitted), albeit 'ideal' is more accurate for dry air than for vapour. The vapour's equation of state so is:

$$pV = m_v RT$$

with p the partial vapour pressure in Pa, V the volume filled with vapour in m³, T the vapour's temperature in K, m_v the amount of vapour filling volume V in kg and R the gas constant of vapour, 461.52 J/(kg·K). More accurate is following non-ideal gas equation:

$$\frac{pV}{n_v R_o T} = 1 + \frac{B_{vv}}{(V/n_v)} + \frac{C_{vvv}}{(V/n_v)^2}$$

with n_v the moles of vapour filling volume V and R_o the general gas constant (8314.41 J/(mol·K)). B_{vv} and C_{vvv} are the vapour's viral coefficients. Most gases present in dry air do not react with vapour, except contaminants such as SO_2 and Cl. With vapour, SO_2 forms H_2SO_3, Cl HCl. Hapilly, their concentrations are mostly too low to have an effect. So, Dalton's law applies, stating that the total air pressure in a volume V is the sum of the partial dry air and the partial vapour pressure: $P_a = p_1 + p$, with P_a the atmospheric pressure, ≈ 101.3 kPa or 1 atm at sea level. Thanks to its limited solubility in water, dry air hardly changes the balance between liquid, vapour and ice (Raoult and Henry's law) or, in air, the diagram of state of water remains (triple point at 0 °C, etc.). Kinetics however change. Evaporation and condensation go slower than in vacuum.

2.3.1.2 Quantities

Following quantities describe the presence of vapour in air:

Partial vapour pressure, symbol p, units Pa	With temperature and total air pressure, a basic variable of state
Vapour concentration, symbol ρ_v, units kg/m³	The mass of vapour per unit volume of dry air. According to the ideal gas law, the quantity can be written as: $$\rho_v = \frac{m_v}{V} = \frac{p}{RT} \qquad (2.51)$$ showing that the vapour concentration is a function of the partial vapour pressure and the temperature
Vapour ratio, symbol x, units: kg/kg	The mass of vapour per unit mass of dry air; a derived variable of state

2.3 Water Vapour

The relation between the vapour ratio x and the partial vapour pressure p is:

$$x = \frac{\rho_v}{\rho_l} = \frac{R_a p}{R_v(P_a - p)} = \frac{0.62\, p}{P_a - p} \qquad (2.52)$$

From now on the additive 'partial' is mostly omitted.

2.3.1.3 Vapour Saturation Pressure

As dry air hardly affects the equilibrium between liquid, gaseous and solid, the vapour pressure present at each temperature could be the highest possible and is therefore provided with the prefix 'saturation' and suffix 'sat': p_{sat}. Idem so for the vapour concentration and vapour ratio: $\rho_{v,sat}$, x_{sat}. The three increase with temperature. At 100 °C, the saturation pressure touches the standard atmospheric pressure at sea level. Figure 2.18 shows and Table 2.4 lists this dependence of temperature, for the table in steps of 0.1 °C between −30 and 41 °C and of 5 °C above 45 °C.

As the figure shows, the dependence follows a kind of exponential curve, between 0 and 50 °C approximated as:

$$p_{sat} = p_{c,sat} \exp\left[2.3026\, \kappa \left(1 - \frac{T_c}{T}\right)\right] \qquad (2.53)$$

with T_c the temperature at the triple point of water and $p_{c,sat}$ related saturation pressure ($T_c = 647.4$ K, $p_{c,sat} = 217.5 \cdot 10^5$ Pa). The parameter κ depends on the temperature in K:

$$\kappa = 4.39553 - 6.2442\left(\frac{T}{1000}\right) + 9.953\left(\frac{T}{1000}\right)^2 - 5.151\left(\frac{T}{1000}\right)^3$$

Less accurate are following formulas:

$-10 \leq \theta \leq 50$ °C	$p_{sat} = \exp(65.8094 - 7066.27/T - 5.976 \ln(T))$
$-30 \leq \theta \leq 0$ °C	$p_{sat} = 611 \exp(82.9 \cdot 10^{-3}\theta - 288.1 \cdot 10^{-6}\theta^2 + 4.403 \cdot 10^{-6}\theta^3)$
$0 \leq \theta \leq 40$ °C	$p_{sat} = 611 \exp(72.5 \cdot 10^{-3}\theta - 288.1 \cdot 10^{-6}\theta^2 + 0.79 \cdot 10^{-6}\theta^3)$
$0 \leq \theta \leq 80$ °C	$p_{sat} = \exp(23.5771 - 4042.9/(T - 37.58))$
$\theta \leq 0$ °C:	$p_{sat} = 611 \exp\left(\frac{22.44\theta}{272.44 + \theta}\right)$ $\quad \theta > 0$ °C: $\quad p_{sat} = 611 \exp\left(\frac{17.08\theta}{234.18 + \theta}\right)$

The last two given allow calculating the temperature linked to a given saturation pressure.

2.3.1.4 Relative Humidity

Air usually contains less vapour (ρ_v) than when saturated. The ratio with the saturated concentration ($\rho_{v,sat}$) or, when vapour pressures are considered, with the saturation pressure at the temperature, the air has, now is called the 'relative humidity', symbol ϕ or RH, a dimensionless number between 0 and 1 or, as a percentage, between 0 and 100%:

$$\phi = 100 \rho_v / \rho_{v,sat} = 100\, p/p_{sat} \quad (\%) \qquad (2.54)$$

Table 2.4 Vapour saturation pressure in Pa.

	(a) Temperatures between 0 and −30 °C									
−1 °C↓−					−0.1 °C→					
	−0.0	−0.1	−0.2	−0.3	−0.4	−05	−0.6	−0.7	−0.8	−0.9
−0	611	606	601	596	591	586	581	576	572	567
−1	562	558	553	548	544	539	535	530	526	522
−2	517	513	509	504	500	496	492	488	484	479
−3	475	471	467	464	460	456	452	448	444	441
−4	437	433	430	426	422	419	415	412	408	405
−5	401	398	394	391	388	384	381	378	375	371
−6	368	365	362	359	356	353	350	347	344	341
−7	338	335	332	329	326	323	321	318	315	312
−8	310	307	304	302	299	296	294	291	289	286
−9	284	281	279	276	274	271	269	267	264	262
−10	260	257	255	253	251	248	246	244	242	240
−11	237	235	233	231	229	227	225	223	221	219
−12	217	215	213	211	209	207	206	204	202	200
−13	198	196	195	193	191	189	188	186	184	183
−14	181	179	178	176	174	173	171	170	168	167
−15	165	164	162	160	159	158	156	155	153	152
−16	150	149	148	146	145	143	142	141	139	138
−17	137	136	134	133	132	131	129	128	127	126
−18	125	123	122	121	120	119	118	116	115	114
−19	113	112	111	110	109	108	107	106	105	104
−20	103	102	101	100	99	98	97	96	95	94
−21	96	92	91	90	90	89	88	87	86	85
−22	84	84	83	82	81	80	80	79	78	77
−23	76	76	75	74	73	73	72	71	70	70
−24	69	68	68	67	66	66	65	64	64	63
−25	62	62	61	60	60	59	59	58	57	57
−26	56	56	55	55	54	53	53	52	52	51
−27	51	50	50	49	49	48	48	47	47	46
−28	46	45	45	44	44	43	43	42	42	41
−29	41	41	40	40	39	39	38	38	38	37
−30	37	36	36	36	35	35	35	34	34	33

Table 2.4 Vapour saturation pressure in Pa. (Continued)

(b) Temperatures between 0 and 40 °C

+1 °C↓	+0.1 °C→									
	0.0	0.1	0.2	0.3	0.4	0.5	0.6	0.7	0.8	0.9
0	611	615	620	624	629	634	638	643	647	652
1	657	662	666	671	676	681	686	691	696	701
2	706	711	716	721	726	731	736	742	747	752
3	758	763	768	774	779	785	790	796	802	807
4	813	819	824	830	836	842	848	854	860	866
5	872	878	884	890	896	903	909	915	922	928
6	935	941	948	954	961	967	974	981	987	994
7	1001	1008	1015	1022	1029	1036	1043	1050	1057	1065
8	1072	1079	1087	1094	1101	1109	1117	1124	1132	1139
9	1147	1155	1163	1171	1178	1186	1194	1203	1211	1219
10	1227	1235	1243	1252	1260	1269	1277	1286	1294	1303
11	1312	1320	1329	1338	1347	1356	1365	1374	1383	1392
12	1401	1411	1420	1429	1439	1448	1458	1467	1477	1487
13	1497	1506	1516	1526	1536	1546	1556	1566	1577	1587
14	1597	1608	1618	1629	1639	1650	1661	1671	1682	1693
15	1704	1715	1726	1737	1748	1760	1771	1782	1794	1805
16	1817	1829	1840	1852	1864	1876	1888	1900	1912	1924
17	1936	1949	1961	1973	1986	1999	2011	2024	2037	2050
18	2063	2076	2089	2102	2115	2128	2142	2155	2169	2182
19	2196	2210	2224	2237	2251	2265	2280	2294	2308	2322
20	2337	2351	2366	2381	2395	2410	2425	2440	2455	2470
21	2486	2501	2516	2532	2547	2563	2579	2595	2611	2627
22	2643	2659	2675	2691	2708	2724	2741	2758	2774	2791
23	2808	2825	2842	2859	2877	2894	2912	2929	2947	2965
24	2983	3001	3019	3037	3055	3073	3092	3110	3129	3148
25	3166	3185	3204	3224	3243	3262	3281	3301	3321	3340
26	3360	3380	3400	3420	3440	3461	3481	3502	3522	3543
27	3564	3585	3606	3627	3649	3670	3692	3713	3735	3757
28	3779	3801	3823	3845	3868	3890	3913	3935	3958	3981
29	4004	4028	4051	4074	4098	4122	4145	4169	4193	4218
30	4242	4266	4291	4315	4340	4365	4390	4415	4440	4466
31	4491	4517	4543	4569	4595	4621	4647	4673	4700	4727
32	4753	4780	4807	4835	4862	4889	4917	4945	4973	5001

(continued)

Table 2.4 Vapour saturation pressure in Pa. (Continued)

(b) Temperatures between 0 and 40 °C										
+1 °C↓					+0.1 °C→					
	0.0	0.1	0.2	0.3	0.4	0.5	0.6	0.7	0.8	0.9
33	5029	5057	5085	5114	5143	5171	5200	5229	5259	5288
34	5318	5347	5377	5407	5437	5467	5498	5528	5559	5590
35	5621	5652	5683	5715	5746	5778	5810	5842	5874	5907
36	5939	5972	6004	6037	6071	6104	6137	6171	6205	6239
37	6273	6307	6341	6376	6410	6445	6480	6516	6551	6587
38	6622	6658	6694	6730	6767	6803	6840	6877	6914	6951
39	6989	7026	7064	7102	7140	7178	7217	7255	7294	7333
40	7372	7412	7451	7491	7531	7571	7611	7652	7692	7733

Temperatures between 45 and 95 °C							
θ (°C)	p' (Pa)	θ (°C)	p' (Pa)	θ (°C)	p' (Pa)	θ (°C)	p' (Pa)
45	9 582	60	19 917	75	38 550	90	70 108
50	12 335	65	25 007	80	47 356	95	84 524
55	15 741	70	31 156	85	57 800		

In point Q [$\theta = 20$ °C, $p = 1169$ Pa] in Figure 2.17, the RH touches 50%. All points with that value lay on a curve through Q having the same shape as the saturation line. The RH cannot exceed 100% since per temperature, the saturation pressure and vapour concentration then are the highest attainable. Any attempt to pass 100% sees the surplus condensing while releasing the latent heat of evaporation, $2.5 \cdot 10^6$ J/kg at 0 °C. Each kg evaporating in turn demands the same amount of latent heat.

A simple relation exists between RH, air pressure and temperature. A given air mass is at RH ϕ_1, air pressure P_{a1} and temperature θ_1. Suddenly, its pressure turns P_{a2} and the air temperature θ_2, while the vapour quantity remains unchanged. The RH (ϕ_2) then becomes:

$$\phi_2 = \phi_1 \frac{P_{sat1} P_{a2}}{P_{sat2} P_{a1}} \tag{2.55}$$

The proof goes as follows. Because the mass of the dry air and vapour does not change, neither the ratio $m_v/(m_l + m_v)$ does. The equilibriums linked to [P_{a1}, θ_1] and [P_{a2}, θ_2] so look:

$$\frac{m_v}{m_l + m_v} = \frac{\dfrac{p_1 V_1}{R T_1}}{\dfrac{P_{a1} V_1}{R_a T_1}} = \frac{p_1 R_a}{P_{a1} R} = \underbrace{\frac{\phi p_{sat,1} R_a}{P_{a1} R}}$$

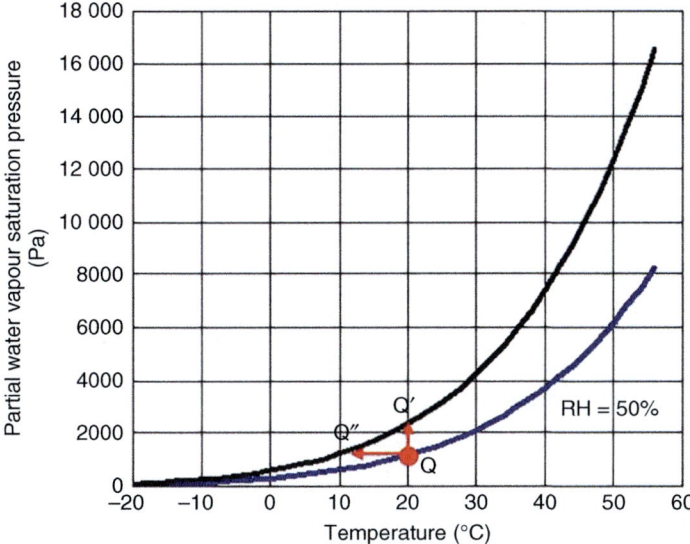

Figure 2.17 The upper curve gives the vapour saturation pressure, the lower the vapour pressure at 50% relative humidity, both as function of temperature.

$$\frac{m_v}{m_l + m_v} = \frac{\dfrac{P_2 V_2}{RT_2}}{\dfrac{P_{a2} V_2}{R_a T_2}} = \frac{P_2 R_a}{P_{a2} R} = \underbrace{\frac{\phi p_{sat,2} R_a}{P_{a2} R}}$$

In both, the gas constant R_a for moist air is $(m_v R + m_l R_l)/m_a$. Equating both therms above the brackets proves ϕ_2 in fact is as just given. Under atmospheric conditions, the named changes of state become isobaric ($P_{a1} = P_{a2}$), converting ϕ_2 into:

$$\phi_2 = \phi_1 p_{sat1}/p_{sat2} \tag{2.56}$$

This simple equation shows what happens with RH when the air temperature changes: an increase when dropping, and a decrease when rising.

To correctly quantify a mixture of two gases, three variables of state have to be known. For moist air, obvious choices are air pressure, vapour ratio and temperature. In most building applications, the air pressure is atmospheric, so two variables remain. Put otherwise, RH does not fully describe the state of moist air, unless lying on a curve of equal RH, see Figure 2.17. To fully define a dry air/vapour mixture also temperature, vapour pressure, vapour saturation pressure or any other variable must be known.

2.3.1.5 Changes of State in Humid Air

Of the different possibilities to reach saturation starting from point Q (humid air), see Figure 2.17, two look characteristics. If without temperature change vapour is added, Q will move parallel to the vapour pressure axis to Q' on the saturation curve. Adding more vapour keeps it there, the vapour pressure saturated, $p_{sat,Q'}$, while the surplus will condense, a change of state called isothermal. Instead, lowering the air temperature while keeping the vapour pressure constant, moves Q parallel to the temperature axis to Q'' on the saturation curve with the temperature there called the dew point (θ_d). Any further decrease in temperature gives condensation again with Q'' descending along the saturation curve and the difference ($p_{sat,Q''} - p_{sat,Q''desc}$) driving the deposit. Related change of state is called isobaric. All points on the line QQ'' through Q, parallel to the temperature axis, have the same dew point or again, the dew point does not fully characterize moist air. A second variable is needed. Fixing the dewpoint is simple. The RH in a point x on the line QQ'' is:

$$\phi_x = 100 \frac{p_Q}{p_{sat,x}}$$

In Q'', the RH touches 100%, meaning:

$$p_Q = \frac{100 p_{sat,Q''}}{100} = p_{sat,Q''}$$

Or, the dew point is the temperature at which the actual vapour pressure becomes the saturation pressure. When the temperature drops below, condensate is deposited.

Real changes of state now combine isothermal and isobaric, so include humidification or dehumidification and temperature change. A graph widely used to analyse the effects of changes of state, is the Mollier diagram, of which an ASHRAE and a VDI version exists.

2.3.1.6 Enthalpy of Humid Air

For air containing x kg vapour per kg dry air, the enthalpy is:

$$h = c_a \theta + x(c_v \theta + l_{bo}) \tag{2.57}$$

with c_a and c_v the isobaric specific heats of dry air (≈ 1008 J/(kg·K)) and vapour (≈ 1860 J/(kg·K)) and l_{bo} the heat of evaporation at 0 °C.

2.3.1.7 Measuring Air Humidity

A direct measurement of the air humidity can be done using a mirror dew point meter or a psychrometer. The first logs the temperature of a mirror cooled until condensate deposits, which happens when its temperature equals the dew point. The second measures the air's dry (θ) and wet bulb temperature (θ_w). To get the last, air with known dry bulb is blown through a thermally well protected, moist tissue, whose adiabatic evaporation saturates the air while the latent heat needed cools the air down to the wet bulb temperature. With both known, all other variables of state can be deduced. So for the vapour pressure:

$$p = p_{sat}(\theta_w) - 66.71(\theta - \theta_w) \tag{2.58}$$

Indirect measurements are done using a hair or a capacitance meter. The first determines the RH from the hygric expansion or contraction of a hair bundle, and the second from the RH-depending change in dielectric constant of a hygroscopic polymer.

Figure 2.18 Vapour balance in a room. Storage in the air is not shown.

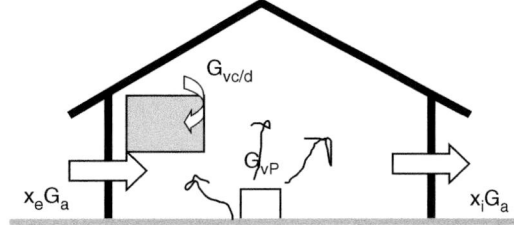

2.3.2 Vapour Balance in Spaces

In non-air-conditioned spaces, the heat, air, and vapour balance is what fixes the vapour pressure and the RH. A room at constant temperature lacking sorption-active surfaces sees the ventilation air from outdoors or from adjacent rooms mixed ideally with the air inside it (Figure 2.18).

Considered representative for the ambient in that room are the air temperature, the vapour pressure and the RH in the centre, 1.8 m above floor level. Conservation of mass now dictates that the vapour entering that room with the inflowing air $(x_{ve}G_a)$, added to the vapour produced or removed in the room (G_{vP}) and occasionally released by surface drying or taken away by surface condensation $(G_{vc/d})$ must equal the vapour stored in the room air, added what the exhaust air $(x_{vi}G_a)$ and diffusion through the enclosure remove. In case the inflow comes from outside and diffusion through the enclosure is negligible, the balance writes as:

$$x_{ve}G_a + G_{vP} + G_{vc/d} = x_{vi}G_a + \frac{d(x_{vi}M_l)}{dt} \tag{2.59}$$

The air mass M_l in the room now is $\rho_a V$ with V its inside volume in m³. The production G_{vP} includes vapour released by people, animals, plants, activities such as cooking, washing, cleaning, exposed water surfaces, etc. The ideal gas law yet allows writing:

$$x_{ve}G_a = \frac{\rho_{ve}}{\rho_{ae}}\rho_{ae}(nV) = \frac{p_e}{RT_e}(nV) \quad x_{vi}G_a = \frac{\rho_{vi}}{\rho_{ai}}\rho_{ai}(nV) = \frac{p_i}{RT_i}(nV) \tag{2.60}$$

The storage term can be rewritten as

$$\frac{d(x_{vi}\rho_{ai}V)}{dt} = \frac{d(\rho_{vi}\rho_{ai}V/\rho_{ai})}{dt} = \frac{V}{RT_i}\frac{dp_i}{dt}$$

In these three equations n, is the ventilation rate in ach, standing for the ratio between the air inflow in m³ per hour and the air volume in the room, p_e and p_i are the vapour pressures out- and indoors, T_i is the temperature indoors in K and R is the gas constant for vapour. The inflowing air so undergoes two changes of state: getting humidified or dehumidified from x_{ve} to x_{vi} and warmed from T_e to T_i, or:

$$x_{ve}G_a = \frac{p_e}{RT_i}(nV).$$

This converts the balance into a first-order differential equation with the vapour pressure indoors as variable. If no surface condensation nor surface drying occurs that equation looks:

$$p_e + \frac{RT_i}{nV} G_{vP} = p_i + \frac{1}{n} \frac{dp_i}{dt}$$

For the average situation over a long period of time, the vapour pressure indoors so becomes:

$$p_i = p_e + \frac{RT_i}{nV} G_{vP} \qquad (2.61)$$

showing that the vapour released keeps it above the value outdoors. The difference between both increases proportionally to the average vapour release and inversely proportional to the average ventilation rate, whose impact consequently quickly loses effect (Figure 2.19).

Related RH becomes:

$$\phi_i = 100 \frac{p_i}{p_{sat,i}} = \frac{p_e}{p_{sat,i}} + \frac{RT_i}{p_{sat,i} nV} G_{vP} \qquad (2.62)$$

Several situations can turn the response non-steady-state. The vapour release indoors may increase suddenly without change in vapour pressure outside and ventilation rate. The initial conditions then become: $t < 0$: $G_{vP} = G_{vP1}$, $t \geq 0$: $G_{vP} = G_{vP2}$, giving as solution:

$$p_i = p_{i\infty} + (p_{i0} - p_{i\infty}) \exp(-nt) \qquad (2.63)$$

In it, p_{i0} is the vapour pressure indoors at time 0 and $p_{i\infty}$ the steady state value for a vapour release G_{vP2}. Figure 2.20 shows how the time constant $1/n$ lets the change lag behind, surely at higher ventilation rates.

Vapour pressure outdoors may also change suddenly while the ventilation rate and the vapour release indoors remain constant ($n = C^t$, $G_{vP} = C^t$). The initial conditions then are $t < 0$ $p_e = p_{e1}$, $t \geq 0$: $p_e = p_{e2}$. The solution, included a lagging behind, resembles the vapour release case with p_{i0} now the vapour pressure indoors at time 0 and $p_{i\infty}$ the steady state value for a vapour pressure p_{e2} outdoors.

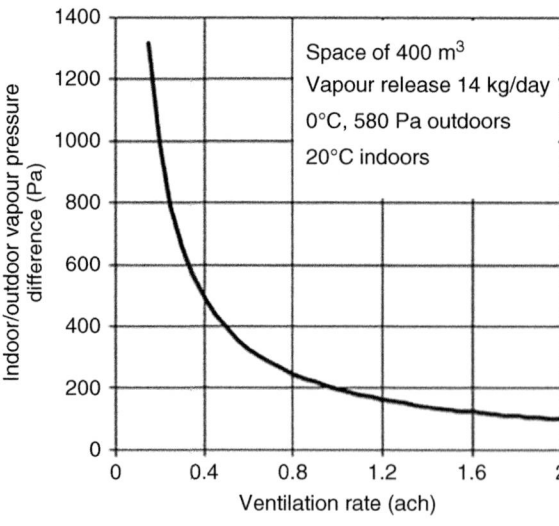

Figure 2.19 The effect of better ventilation on the in-/outdoor vapour pressure difference.

Figure 2.20 Same case as Figure 2.19, sudden increase in vapour release, air buffering does the vapour pressure indoors lag behind.

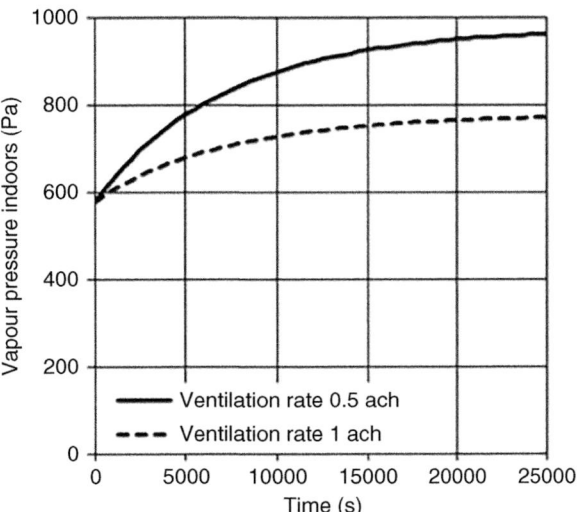

Of course, vapour release in- and vapour pressure outdoors can even change periodically, while the ventilation rate remains constant. The solution then is:

$$p_i = \bar{p}_i + \frac{1}{Z}\left(\mathbf{p_e} + \frac{RT_i \mathbf{G_{vP}}}{nV}\right) \text{Re}\left(\exp\left(\frac{2i\pi t}{T}\right)\right) \qquad (2.64)$$

with $\bar{p}_i$ the vapour pressure indoors for the average values of the vapour release in- and the vapour pressure outdoors. The complex values $\mathbf{p_e}$ and $\mathbf{G_{vP}}$ and the complex damping of the vapour pressure indoors Z in that formula equal:

$$\mathbf{p_e} = \hat{p}_e \exp(i\phi_{p_e}) \quad \mathbf{G_{vP}} = \hat{G}_{vP} \exp(i\phi_{G_{vP}}) \quad Z = \left[1 + \left(\frac{2\pi}{nT}\right)^2\right]^{1/2},$$

$$\arg(Z) = bgtg\left(\frac{2\pi}{nT}\right)$$

with Z, $\hat{p}_e$ and $\hat{G}_{vP}$ the amplitudes, $\phi_{p,e}$ the time lag linked to a cosine linked change of the vapour pressure outdoors and ϕ_{vP} the time lag linked to a cosine linked change of the vapour release indoors. A lower ventilation rate, combined with periodic changes in vapour release indoors and vapour pressure outdoors, so gives a periodically changing vapour pressure indoors, more time shifted and dampened than the one outdoors.

All parameters finally can vary randomly. Then, only a finite difference calculation with small time steps will show the impact on the vapour pressure indoors.

2.3.3 Relative Humidity On Inside Surfaces

A high RH on inner surfaces can have annoying consequences. If $>52 + 1.2(\theta_s - 15)\%$ for a long period of time, with θ_s the surface temperature, the dust mite 'dermatophagoides farinae' will abundantly reproduce on it. Surface condensation occurs each time the RH on touches 100%, i.e. when the surface temperature drops below the dew point indoors (Figure 2.21).

Figure 2.21 Surface condensation in a toilet against the non-insulated low slope roof.

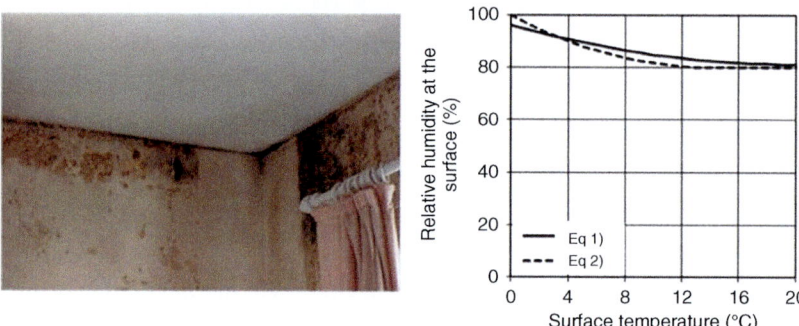

Figure 2.22 Isopleths fixing the lowest mould growth rate of a mould species.

A long-lasting high enough RH on a surface can additionally induce mould growth. It is the mould-linked isopleths that dictate which four weeks means are required for their appearance and growth, see the two in Figure 2.22:

1) $\phi_{crit} = 0.033\theta_s^2 - 1.5\theta_s + 96$

2) $\theta_s < 20°C \quad \phi_{crit} = \max\left(80; -0.00297\theta_s^3 + 0.16\theta_s^2 - 3.13\theta_s + 100\right) \quad (\%)$
$\theta_s \geq 20°C \quad \phi_{crit} = 80\%$

Below these values, mould should not be expected. Globally, surface temperatures between 17 and 27 °C and a four-week mean surface RH of 80% so are sufficient to get mould. This 80% has become the number used to control whether mould will colonize a surface. Besides, a mean RH above 80% shortens the period for germination as indirectly shown by:

$$\phi_{crit} \geq \min\left\{100;\; (0.033\theta_s^2 - 1.5\theta_s + 96)[1.25 - 0.072\ln(t)]\right\} \quad (t \text{ in days})$$

That four weeks mean surface RH in turn is given by the ratio of the four weeks mean vapour pressure p_s indoors to the four weeks mean vapour saturation pressure $p_{sat,s}$ on the surface:

$$p_s \geq 0.8 p_{sat,s} = \text{mould}$$

The four weeks mean vapour saturation pressure in turn depends directly on the four weeks mean inside surface temperature, which for envelope assemblies is given by:

$$\theta_{i,s} = \theta_e + f_{h_i}(\theta_i - \theta_e) \tag{2.65}$$

with: $f_{h_i} = \frac{\theta_{io} - \theta_e}{\theta_i - \theta_e}$ the temperature factor of that surface.

Seen as given are the four weeks mean temperature and four weeks mean vapour pressure outdoors. When cold, the heating habits fix the four weeks mean air temperature indoors. The greater than the ratio between the surface of the envelope and of the partitions, ceiling and floor in a room, the smaller the radiant exchanges and the lower the inside surface film coefficient, so, the envelope's inside surface temperatures will be. Behind cup-boards mounted against an outside wall, radiation and convection are even so limited that the inside surface film coefficient drops to a fraction of the standard 7.7 W/(m²·K), pushing, when cold outside, the wall's inside surface temperature to really low values, what truly heightens the mould risk there. Likewise increasing the mould risk does a high U-value or the presence of thermal bridges (Figure 2.23). A higher inside surface film coefficient (h_i) of course will lift the temperature factor and reduce to some extent the mould risk. The best weapon to minimize any mould risk on inside surfaces is an excellent thermal insulation of the envelope leaving hardly problematic thermal bridges.

At 100% RH on surfaces, surface condensation takes over.

The factors fixing the vapour pressure indoors are a higher vapour pressure outdoors, less ventilation and more vapour produced indoors. In temperate and cold climates the highest vapour pressures outdoors are noted in summer. But, the temperatures indoors are also highest then, so, mostly no problem. Ventilation in homes is often limited to infiltration and accidental window opening, while the amounts of vapour produced mainly depend on the number of residents and their living habits. Among the other factors potentially causing the problems just discussed, first comes a non- or messy insulated envelope resulting in low-temperature factors and low temperatures in unheated rooms. The term messy refers to insulating while leaving many thermal bridges but also to lazy mounted cavity fills, allowing air looping

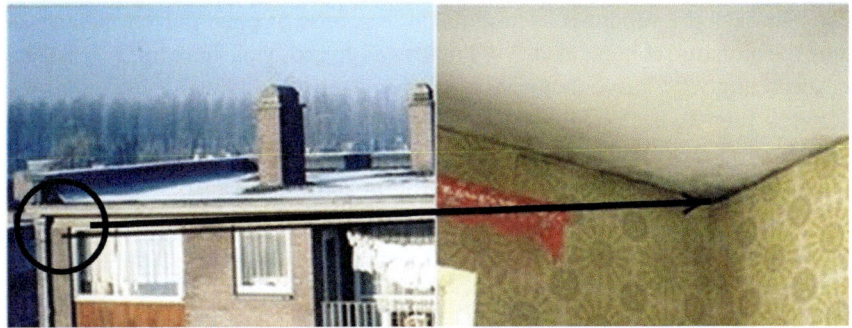

Figure 2.23 Edge between a low-slope roof with concrete slab and two non-insulated outer walls, a thermal bridge that turned mouldy.

around and wind washing behind. Second comes the absence of purpose-designed ventilation systems, since infiltration and accidental airing are often too random to ensure a sufficient air exchange.

2.3.4 Vapour in Open-porous Materials

2.3.4.1 Different from Air?

Under atmospheric conditions, humid air is what fills the open pores in dry porous materials. It so seems logic to evaluate the vapour presence as done for air, now enclosed in a volume $\Psi_0 V$ in m³, Ψ_0 being the material's open porosity. The amount of vapour linked to the RH in the pore air so should vary linearly with that RH and show a temperature dependence that reflects the saturation curve, meaning: the linear relation is steeper when warmer:

$$G_m = \rho_v \Psi_0 V = \frac{p\Psi_0 V}{462\, T} \quad w_H = \frac{G_m}{V} = \frac{p\Psi_0}{462\, T}$$

Concrete for example has 15% open pores. At 20 °C and 65% RH (i) and at 50 °C and 65% RH (ii), 1 m³ so should contain as vapour content:

$$(1)\ w_H = \frac{G_m}{V} = \frac{p\Psi_0}{462\, T} = \frac{0.15 \cdot 2340}{462 \cdot 293.16} = 0.0017 \frac{\text{kg}}{\text{m}^3}$$

$$(2)\ w_H = \frac{G_m}{V} = \frac{p\Psi_0}{462\, T} = \frac{0.15 \cdot 12335}{462 \cdot 293.16} = 0.0081 \frac{\text{kg}}{\text{m}^3}$$

However, in reality, at 20 °C and 65% RH, 1 m³ of concrete shows a moisture content touching 40–50 kg/m³, while for a given temperature the dependence of RH is not linear but S-shaped. The curve also drops a little when warmer. Cause of that huge difference is the sorption/desorption behaviour of open-porous materials.

2.3.4.2 Sorption/Desorption Isotherm
2.3.4.2.1 In General

The S-curve just mentioned is called the 'sorption/ desorption isotherm' or 'hygroscopic curve' (Figure 2.24).

It characterizes the vapour-related moisture response of open-porous materials; whereby the moisture content at 98% RH figures as 'hygroscopic maximum'. At each RH, desorption gives a curve that passes sorption. Due to hysteresis, every point on and between the two is a possible equilibrium. Open-porous materials that only at high RH show a few sorption, as bricks, synthetics and most insulation materials do, are called non-hygroscopic, those that are sorption-active from a RH = 0 on, such as cement- and timber-based materials, hygroscopic.

Several mathematical expressions for the sorption isotherm are in use. None is universal, although the one below applies for RHs <95%, provided the roots of the denominator lie outside the interval [0, 1]:

$$w_H = \frac{\phi}{a_H \phi^2 + b_H \phi + c_H} \tag{2.66}$$

Figure 2.24 Sorption/desorption isotherm.

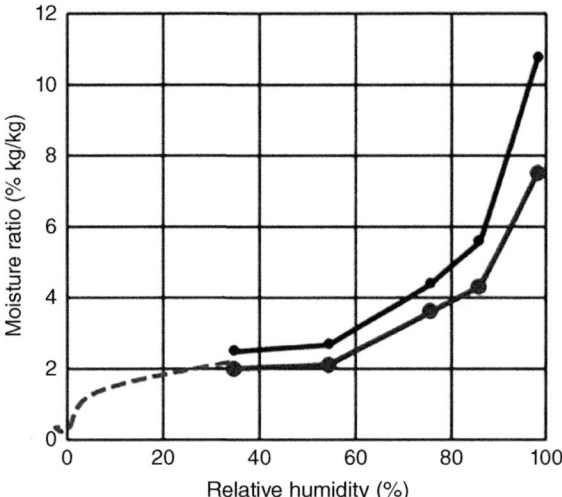

The coefficients a_H, b_H and c_H are material-specific and differ between sorption and desorption. Another formula applied for RHs beyond 20% is:

$$w_H = w_c[1 - \ln(\phi)/b]^{-\frac{1}{c}} \qquad (2.67)$$

In it, w_c is the capillary moisture content and b and c are material-specific parameters.

The sorption/desorption isotherm allows calculating the specific moisture ratio, being the derivative of the moisture ratio in the material to the RH in kg/(kg·RH):

$$\xi_\phi = dX_H/d\phi$$

with X_H the moisture ratio and ϕ the RH. The property quantifies how much moisture a dry material can store per kg at any RH, which is comparable to what the specific heat capacity is for heat. Multiplication with the density gives the specific moisture content, units kg/(m³·RH), comparable to the volumetric specific heat capacity for heat:

$$\rho\xi_\phi = dw_H/d\phi \qquad (2.68)$$

In it, w_H is the hygroscopic moisture content. Both the specific moisture ratio and specific moisture content vary with RH as differentiation of the two sorption formulas underlines. For the specific moisture content:

$$\rho\xi_\phi = \frac{dw_H}{d\phi} = w_H^2\left(\frac{c_H}{\phi^2} - a_H\right) \qquad \rho\xi_\phi = \frac{dw_H}{d\phi} = \frac{w_H}{cb\phi}\left(1 - \frac{\ln\phi}{b}\right)^{-1}$$

That the temporary hygroscopic equilibrium may lay on or in between the sorption and desorption curve also impacts the specific moisture ratio and specific moisture content.

2.3.4.2.2 Physics Involved

A first actor determining the hygroscopic response of open-porous materials is adsorption sticking the water molecules entering the potes to the pore walls. Starting at 0% RH first a monolayer of molecules so forms as described by the Langmuir equation:

$$w_H = \frac{M_w A_P}{A_w N} \frac{C\phi}{1+C\phi} = 2.62 \cdot 10^{-7} A_P \frac{C\phi}{1+C\phi} \quad (\phi \text{ on a scale from 0 to 1}) \quad (2.69)$$

In it, ϕ is RH, M_w the mass of 1 mol of water (0.018016 kg), A_w the area a water molecule occupies ($11.4 \cdot 10^{-20}$ m^2), N the Avogadro number ($6.023 \cdot 10^{23}$ molecules/mol), A_p the material's specific pore surface (m^2/m^3) and C the heat the adsorption releases (J/kg):

$$C = k \exp[(l_a - l_b)/RT]$$

with k the adsorption constant, l_a the heat of adsorption (J/kg), l_b the heat of evaporation (J/kg) and R the gas constant of vapour. Beyond 20% RH, multi-layer adsorption starts, explaining why until 40% RH the hygroscopic curve looks convex and reflects the Brunauer–Emmett–Teller or BET-equation:

$$w_H = 2.62 \cdot 10^{-7} A_P \frac{C\phi}{1-\phi} \left[\frac{1 - (n+1)\phi^n + n\phi^{n+1}}{1 + (C-1)\phi - C\phi^{n+1}} \right] \quad (2.70)$$

where n is the number of water molecule layers formed.

Both the Langmuir and BET-equations underline that at low RH the hygroscopic moisture content increases with the specific pore surface available as also the heat exchanged does, albeit higher temperatures temper that more. A high open porosity and all open pores very small gives a large specific pore surface. Take sand-lime stones and bricks. Their open porosity hardly differs ($\approx$33%) but the average pore diameter of sand-lime stone is 0.1 μm versus 8 μm for bricks. Sand-lime stone so has a 6000 times larger specific pore surface than bricks, so will be far more hygroscopic at low RH.

Moreover, somewhere between 20% and 40% capillary condensation on the adsorbed water layers already starts. In fact, once in the smallest pores, these touch each other, surface tension turns them into stable water meniscuses, allowing capillary condensation on both their sides, a phenomenon lasting until RH $\approx$ 100% and turning the curve concave. Why already at 40%, is related to the vapour saturation pressure depending on the form of a meniscus. For a molecule, escaping from a concave meniscus is harder than from a flat one. From a convex meniscus, it's easier. Because of that, the vapour phase above a concave meniscus contains less, above a convex one more water molecules than above a flat one. Conversely, condensation on a concave meniscus requires less, on a convex one more water molecules in the air above than on a flat one. Water so is already depositing on a concave at RHs below 100%, while on a convex, the RH must pass 100%, a fact Thompson's law underlines:

$$p'_{sat} = p_{sat} \exp\left[-\frac{\sigma_w \cos\vartheta}{\rho_w RT} \left(\frac{1}{r_1} + \frac{1}{r_2} \right) \right] \quad (2.71)$$

In it, p'_{sat} is the saturation pressures above a curved and p_{sat} the saturation pressure above a flat meniscus, ρ_w the density of water, σ_w the surface tension of water and ϑ the contact angle between meniscus and the pore wall having as curvature radii r_1 and r_2. If all open pores are assumed circular, equivalent diameter d_{eq}, the law simplifies to:

$$p'_{sat} = p_{sat} \exp\left(-\frac{4\sigma_w \cos\vartheta}{\rho_w RT d_{eq}}\right) \tag{2.72}$$

or, with the ratio p'_{sat}/p_{sat} the RH (ϕ):

$$\ln(\varphi) = -\frac{4\sigma_w \cos\vartheta}{\rho_w RT d_{eq}} \quad (0 \le \phi \le 1). \tag{2.73}$$

Also see Figure 2.25.

The smaller so the equivalent pore diameter (d_{eq}) and the higher the temperature, the lower the RH above which capillary condensation starts. However, below 20% RH, d_{eq} turns smaller than 10^{-9} m, the radius of a water molecule's sphere of influence, making capillary condensation meaningless. At 100% RH, d_{eq} in turn touches ∞, meaning all open pores should be filled with water then, although air bubbles left in the pores will limit the water storage to a value prohibiting them to escape, so fixing the 'capillary moisture content'. Of course, permanently 100% RH will lead to a slow dissolving of the air in the water present, pushing the moisture content up to saturated.

On a concave meniscus in a pore with equivalent diameter d_{eq}, capillary condensation starts when the RH in the surrounding air passes:

$$\phi = 100 \, p'_{sat}(d_{eq})/p_{sat}$$

In pores with equivalent diameter below d_{eq}, condensate is then already depositing. In these wider than $2d_{eq}$, multilayer adsorption continues while in these with diameter between d_{eq} and $2d_{eq}$, condensation will thicken the already adsorbed layers. Many wide pores in a pore volume so explains why above 90% RH the hygroscopic moisture content increases really fast.

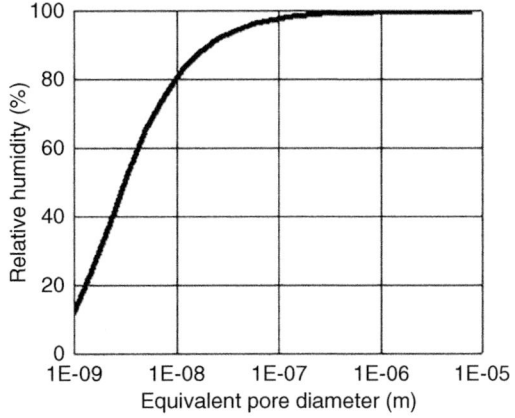

Figure 2.25 Thompson's law at 20 °C.

The hysteresis mentioned between sorption and desorption has many causes. Under atmospheric conditions, the hygroscopic water uptake or release progresses so slowly that, when a test stops too early, a too low sorption or desorption moisture content is noted. Vacuum reduces the hysteresis, learning that the atmospheric pressure slows down the interactions between liquid and air in the pores and changes it between sorption and desorption, because, due to a changing contact angle, meniscuses growth differs from their shrinkage.

2.3.4.2.3 Impact of Salts

The presence of salts in the pores lifts sorption, a consequence of what saturated salt solutions do: lowering the equilibrium RH:

Saturated solution	Equilibrium RH (%)
$MgCl_2$	33
NaCl	75
KCl	86

With NaCl in the pores, a sudden jump in hygroscopic moisture content will show up at 75%. RH. One of the consequences is that masonry wetted by seawater hardly dries in climates with RH passing on average 75%. A mixture of salts even lifts the whole hygroscopic curve, a fact also proving their presence.

2.3.4.2.4 Consequences

Sorption and desorption play a dominant role in the hygric response of assemblies made of open-porous materials, while, linked to their sorption/desorption isotherms, the RH becomes the factual moisture potential. Both add inertia to the hygric response with the specific moisture content of the materials used acting the way the volumetric specific heat capacity does thermally. Their influence often limits interstitial condensation to an increase in hygroscopic moisture content. Of course, also undesired influences surface. Due to sorption and desorption, materials see their dimensions change with RH, a drop giving shrinkage, an increase expansion. Very sensitive are thin wood-based layers, but also cement-based materials, even bricks, be it hardly, are. Above 75–80% RH, hygroscopicity may also induce biological activity.

2.3.5 Vapour Transfer in the Air

Back to moist air now. With ρ_{da} $(=m_1/V)$ the dry air and ρ_v $(=m_v/V)$ the vapour concentration, the moist air concentration becomes:

$$\rho_a = (m_{da} + m_v)/V = \rho_{da} + \rho_v$$

A main cause of air moving is convection. Let $\mathbf{v_a}$ be the moist air speed. In the mixture, dry air and vapour can have speeds $\mathbf{v_{da}}$ and $\mathbf{v_v}$ differing from $\mathbf{v_a}$, the link being:

$$\mathbf{v_a} = \frac{\rho_{da}\mathbf{v_{da}} + \rho_v\mathbf{v_v}}{\rho_{da} + \rho_v} = \frac{\rho_{da}\mathbf{v_{da}} + \rho_v\mathbf{v_v}}{\rho_a}$$

Related dry air, vapour and moist air fluxes so are:

$$\mathbf{g_{da}} = \rho_{da}\mathbf{v_{da}} \quad \mathbf{g_v} = \rho_v\mathbf{v_v} \quad \mathbf{g_a} = \mathbf{g_{da}} + \mathbf{g_v} = \rho_a\mathbf{v_a}$$

What makes these speeds different is, in addition to convection, diffusion. Brownian motion in fact tends to equalize concentration differences in liquid and gaseous mixtures. Diffusion for example drives the carbonisation of lime (CO_2/moist air), the lime to gypsum transformation (SO_2/moist air) and chlorine corrosion in concrete (Cl_2/moist air). In moist air, it acts when gradients in the vapour to dry air ratio exist. In a right-angled coordinate system that moves with the moist air, the total vapour flux ($\mathbf{g_v}$) by convection ($\mathbf{g_{v1}}$) and diffusion ($\mathbf{g_{v2}}$), which for dry air and vapour obeys Fick's empirical diffusion law, so becomes:

$$\text{Vapour: } \mathbf{g_{v2}} = -\rho_a D_{v,da} \mathbf{grad}\left(\frac{\rho_v}{\rho_a}\right) \quad \text{Dry air: } \mathbf{g_{da2}} = -\rho_a D_{v,da} \mathbf{grad}\left(\frac{\rho_{da}}{\rho_a}\right)$$

$$\mathbf{g_v} = \mathbf{g_{v1}} + \mathbf{g_{v2}} = \rho_v \mathbf{v_a} - \rho_a D_{v,da} \mathbf{grad}\frac{\rho_v}{\rho_a} \tag{2.74}$$

In these three equations, $D_{v,da}$, units m²/s, is the binary diffusion coefficient between vapour and dry air, which obeys Schirmer's equation:

$$D_{v,da} = \frac{2.26173}{P_a}\left(\frac{T}{273.15}\right)^{1,81} \tag{2.75}$$

For moist air at 1 atmosphere (1 Bar, 101 300 Pa) this equation converts to:

$$D_{v,da} = 8.69 \cdot 10^{-10} T^{1.81}$$

Usually, convective flows dominate, but a diffusive vapour flow in one direction will always generate an equally large diffusive dry airflow in the opposite direction. When only diffusion plays, no moist air movement nor changes in its overall concentration will be noted.

Under terrestrial conditions and at temperatures below 50 °C, the atmospheric pressure P_a largely exceeds the vapour pressure p, a fact turning $p/R + p_{da}/R_{da}$ into nearly a constant. The general expression for the total vapour flux in moist air then turns into:

$$\mathbf{g_v} = \frac{p}{\rho_a RT}\mathbf{g_a} - \frac{D_{v,da}}{T}\left(\frac{p}{R} + \frac{p_{da}}{R_{da}}\right)\mathbf{grad}\frac{\frac{p}{R}}{\frac{p}{R} + \frac{p_{da}}{R_{da}}} \approx \frac{\mathbf{g_a}}{\rho_a RT}p - \frac{D_{v,da}}{RT}\mathbf{grad}\ p$$

$$\tag{2.76}$$

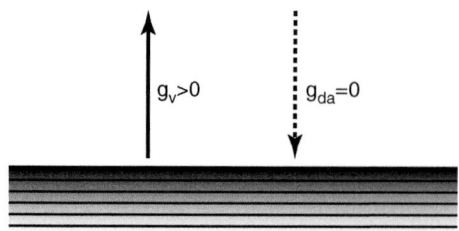

Figure 2.26 The diffusion law applied to evaporation from a water surface.

This equation turns the vapour pressure and its gradient into the driving forces, so reflects the role of temperature in the heat flux by conduction and convection.

A first case now is a vapour flux in one direction without opposing dry air flux, the case for evaporation from a water surface (Figure 2.26). For $R_a \approx R_{da}$, the vapour flux then converts to:

$$\mathbf{g_v} = -\frac{1}{1 - \frac{\rho_v}{\rho_a}} \frac{D_{v,da}}{RT} \mathbf{grad}\, p \approx -\frac{D_{v,da}}{RT} \frac{P_a}{P_a - p} \mathbf{grad}\, p$$

A second is diffusion in stagnant moist air. The total vapour flux then is diffusive, or:

$$\mathbf{g_v} \approx -\frac{D_{v,da}}{RT} \mathbf{grad}\, p \tag{2.77}$$

The ratio $D_{v,da}/(RT)$ is called the vapour permeability of air, symbol δ_a, units s. This allows, if no opposing dry air flux, rewriting the vapour flux by diffusion and convection and by diffusion only as:

$$\mathbf{g_v} = \frac{\mathbf{g_a}}{\rho_a RT} p - \delta_a \mathbf{grad}\, p$$

$$\mathbf{g_v} = -\delta_a \frac{P_a}{P_a - p} \mathbf{grad}\, p \approx -\delta_a \mathbf{grad}\, p \left(\frac{P_a}{P_a - p} \approx 1 \text{ for } \theta < 50\ °C\right)$$

The last equation resembles the one for heat conduction, with the vapour permeability replacing the thermal conductivity and vapour pressure temperature. However, could the thermal conductivity mostly be handled as a constant, the vapour permeability is by definition temperature dependent.

2.3.6 Vapour Flow by Diffusion in Open-porous Materials and Building Assemblies

2.3.6.1 Flow Equation

At first sight, vapour flow in open-porous materials must develop as in air. Anyhow, its flowing requires pores accessible for water molecules, while for moist air to move through large enough open pores are needed. The moist air-related convective vapour flux anyhow looks:

$$\mathbf{g_{v1}} = \frac{\mathbf{g_a}}{\rho_a RT} p \tag{2.78}$$

Below 50 °C, vapour diffusion obeys:

$$\mathbf{g_{v2}} = -\delta\ \mathbf{grad}\, p \tag{2.79}$$

with δ the vapour permeability of the material in s. Because the area taken by the open pores per m² of material is much smaller than that m², the vapour flux through a material must be much lower than through 1 m² of air, so, its vapour permeability correspondingly much smaller. Krischer therefore proposed using a dimensionless material characteristic, called the vapour resistance factor μ, which tells how larger the vapour permeability of stagnant air is compared to the one of a material at a same temperature and total pressure:

$$\mu = \delta_a/\delta \quad (0 \leq \mu \leq \infty) \quad (-)$$

2.3.6.2 Vapour Resistance Factor μ

First parameter of influence fixing its value is the open pore area A_o per m² of material. Larger means a lower, smaller a higher μ-value (Figure 2.27a):

$$\mu \propto 1/A_o$$

In materials without open pores, $\mu = \infty$, which excludes diffusion. Of course, also convection is impossible then.

Second parameter of influence is the ratio between the path length l_o, equal to the average distance (l_o) the molecules are travelling through the pores, and the material's thickness (d_o). The higher that ratio, the larger the μ-factor (Figure 2.27b):

$$\mu \nearrow \text{ if } (l_o/d_o) \nearrow$$

Third parameter of influence is the deviousness Ψ_1 of the pore system. Generally, the vapour resistance factor increases when the ratio between the deviousness and the total open porosity (Ψ_o) increases (Figure 2.27c):

$$\mu \propto \Psi_1/\Psi_o \mu \quad \nearrow \text{ if } \Psi_1/\Psi_o \nearrow$$

Fourth parameter of influence are the other transport mechanisms in hygroscopic materials. In pores with diameter close to the free path length of a water molecule,

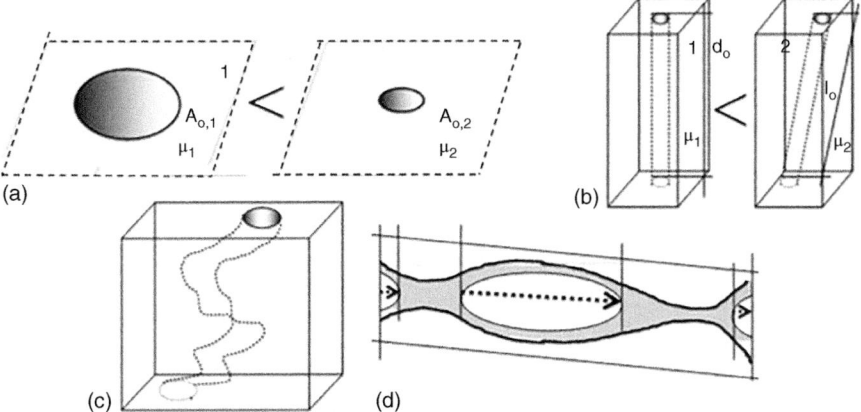

Figure 2.27 Vapour resistance factor (μ): higher with lower open pore area (a), with larger path length l_o (b) and with more deviousness (c), lower due to capillary condensation (d).

friction diffusion will replace Fickian diffusion. At high RH, adsorbed water layers and pores filled with capillary condensate start adding some liquid transfer, while diffusion in the pore air left turns into vapour movement between the water islets present. This shortens the path length (Figure 2.27d). The result is a higher temperature impact but above all a lower vapour resistance factor at higher RHs, better, at higher moisture content. Due to this mix of water presence and vapour diffusing, the name 'equivalent vapour resistance factor μ' is often used. Fick's law still dominates, but with a μ-value that turns lower at higher RH.

Because the vapour permeability of air is already so small ($\approx 1.87 \cdot 10^{-10}$ s at 20 °C) that diffusion looks unimportant, the inverse, an impressive number, named the diffusion constant (N) in s^{-1}, see Figure 2.28, is used.

This and the equivalent vapour resistance factor μ redresses Fick's law into:

$$\mathbf{g}_{v2} = -\frac{1}{\mu N}\text{grad } p \tag{2.80}$$

2.3.6.3 Mass Conservation

Mass conservation, with the vapour content represented by the vapour pressure, dictates that:

$$\text{div } \mathbf{g}_v \pm G'_c = -\frac{\partial}{\partial t}\left(\frac{\Psi_o p}{RT}\right) \tag{2.81}$$

Each time vapour condenses, evaporates, or sublimates, the source term G'_c differs from 0. In the buffer term, the open porosity Ψ_o decreases when vapour condenses and increases when liquid evaporates. In non-hygroscopic materials, a source term G'_c is non-existent till 100% RH. In hygroscopic materials, changes in sorption moisture replace the source and buffering terms. Under isothermal conditions and for a RH below ϕ_M, ϕ_M being the hygroscopic pivot, 95–98%, that marks the start of combined liquid and vapour flow, mass equilibrium rewrites as:

$$\text{div } \mathbf{g}_v = -\frac{\partial w_H}{\partial t} = -\rho \xi_\varphi \frac{\partial \phi}{\partial t} \tag{2.82}$$

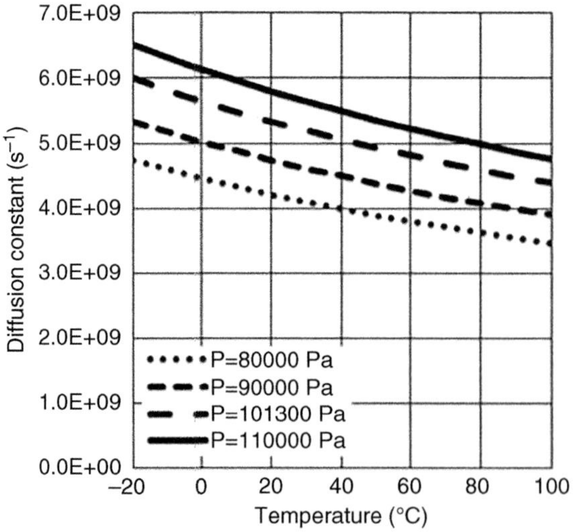

Figure 2.28 Diffusion constant N as function of temperature and air pressure.

2.3.6.4 Applicability of the <Equivalent> Diffusion Concept

The applicability lasts as long as no air moves in and through an assembly. For that, neither air-permeable layers nor cracks, leaks, etc, are allowed, which restricts the applicability of the <equivalent> diffusion concept to compact assemblies with at least one layer airtight, i.e. consisting of a material having pores too small for air to pass through. Such materials can also be vapour tight. Another possibility is no air pressure differences, which is fiction as even different temperatures at both sides of a vertical or inclined assembly are activating stack with as unavoidable consequence airflow through that assembly if lacking an airtight layer. Equivalent diffusion only so is far from guaranteed. Next paragraphs therefore mainly have a theoretical value. By the way, due to the presence of construction moisture, even an assumption as assemblies start dry is fiction.

2.3.6.5 Steady State: Flat Assemblies

2.3.6.5.1 The Equations Adapted

In steady state the derivative $\partial w_H/\partial t$ turns 0, which converts mass conservation into:

$$\mathrm{div}(\mathbf{g_v}) = 0$$

Inclusion of the flux equation gives:

$$\mathrm{div}\left(\frac{1}{\mu N}\mathbf{grad}\,p\right) = 0$$

In flat one-dimensional assemblies, this relation simplifies to:

$$\frac{d}{dx}\left(\frac{1}{\mu N}\frac{dp}{dx}\right) = 0.$$

Even with the vapour resistance factor μ assumed constant, the diffusion constant N remains temperature and air pressure dependent. Under atmospheric conditions, a distinction should so be made between isothermal, N constant, and non-isothermal, N variable. N constant reduces the mass balance to

$$\frac{d^2 p}{dx^2} = 0$$

with as solution $p = C_1 x + C_2$

2.3.6.5.2 Isothermal, Single-layered

With the vapour pressures on both end faces (p_{s1}, p_{s2}), the thickness (d), the vapour resistance factor (μ) of the material used and the boundary conditions known ($x = 0, p = p_{s1}$; $x = d, p = p_{s2}, p_{s1} < p_{s2}$), the integration constants become $C_2 = p_{s1}$, $C_1 = (p_{s2} - p_{s1})/d$ and the vapour pressure course and vapour flux through look:

$$p = \frac{p_{s2} - p_{s1}}{d}x + p_{s1} \tag{2.83}$$

$$g_v = -\frac{1}{\mu N}\frac{dp}{dx} = -\frac{p_{s1} - p_{s2}}{\mu N d} \tag{2.84}$$

Or, isothermally, the vapour pressure in an airtight single-layer assembly changes linearly, see Figure 2.29, while the vapour flux is proportional to the difference in

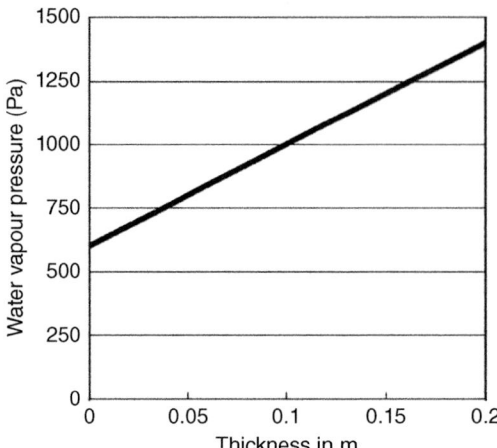

Figure 2.29 Single-layer assembly, isothermal regime, vapour pressure across.

vapour pressure between both end faces and inversely proportional to the product of the thickness, the diffusion constant and the vapour resistance factor.

The quantity μNd with $N \approx 5.4 \cdot 10^9$ s^{-1} is called the diffusion resistance, symbol Z, units m/s. The larger the diffusion resistance is, the lower the flux. A large value requires either a large thickness or a, if to remain small, a material with high vapour resistance factor. Anyhow, the diffusion constant is such a large number that even fluxes through vapour-permeable assemblies remain really small. To give an example, a museum in an old historic building is built with massive outer brick walls, 60 cm thick. The masonry has a diffusion resistance factor 7. Inside, the temperature is on average 20 °C and the RH 60%, giving a vapour pressure of 1402 Pa. The January mean vapour pressure outside touches 660 Pa. The amount of vapour diffusing through 1 m^2 of outer wall that month then equals: $(1402 - 660)/(7 \cdot 0.6 \cdot 5.4 \cdot 10^9) \cdot 3600 \cdot 24 \cdot 31 \cdot 1000 = 88$ g/m^2. To compare with, perspiration by a moderately active adult equals 50–120 g per hour!

2.3.6.5.3 Isothermal, Multi-layered

Are known, the vapour pressures on both end faces ($p_{s1} < p_{s2}$), the layer thicknesses (d_i) and the vapour resistance factors (μ_i) of all materials used. Typical for diffusion is the existence of contact resistances Z_{ci} between layers. If for example glued together, high values for these can be expected. Because of that, each interface generates two vapour pressures:

$$\text{Layer 1}: g_v = \frac{p_{11} - p_{s1}}{\mu_1 N d_1} \quad \text{Contact resistence 1}: g_v = \frac{p_{12} - p_{11}}{Z_{c1}}$$

$$\text{Layer 2}: g_v = \frac{p_{21} - p_{12}}{\mu_2 N d_2} \quad \text{Contact resistence 2}: g_v = \frac{p_{22} - p_{21}}{Z_{c2}}$$

....

$$\text{Layer } n: g_v = \frac{p_{s2} - p_{n-1,2}}{\mu_n N d_n} \quad \text{Contact resistence } n-1: g_v = \frac{p_{n-1,2} - p_{n-1,1}}{Z_{c,n-1}}$$

In these, $p_{11}, p_{12}, \ldots, p_{n-1,1}, p_{n-1,2}$ are as interface values unknown and g_v is the unknown vapour flux. Rearranging and adding gives:

$$g_v = \frac{p_{s2} - p_{s1}}{N \sum(\mu_i d_i) + \sum Z_{ci}} \qquad (2.85)$$

The denominator stands for the total diffusion resistance of the multi-layer assembly, symbol Z_T, units m/s, while $\mu_i N d_i$ and $\mu_i d_i$ are the diffusion resistance in m/s and the diffusion thickness in m of layer i.

The higher the total, for example by inserting a layer with high diffusion thickness or by caring for a high contact resistance, the lower the vapour flux. If that layer with high diffusion thickness is a thin foil, the name vapour retarder or vapour barrier is given. As the same vapour flux must traverse all layers and all contact resistances, following algorithms give the vapour pressure course through the assembly:

$$p_x = p_{s1} + g_v Z_{s1x} \quad p_x = p_{s2} - g_v Z_{s2x} \qquad (2.86)$$

In the $[Z, p]$ axis system, the result is a straight line with the vapour flux as slope, linking the vapour pressures on both end faces $(0, p_{s1}; Z_T, p_{s2}, p_{s1} < p_{s2})$. How that course looks in the $[x,p]$ axis system follows from transposing each intersection with an interface and contact resistance in the correct order to the assembly and connecting the successive values with line segments (Figure 2.30).

The most vapour-tight layer or highest contact resistance, which then acts as unplanned vapour retarder, gives the largest drop in vapour pressure. A problem is that most contact resistances remain unknown, reason why a free contact between layers is commonly assumed.

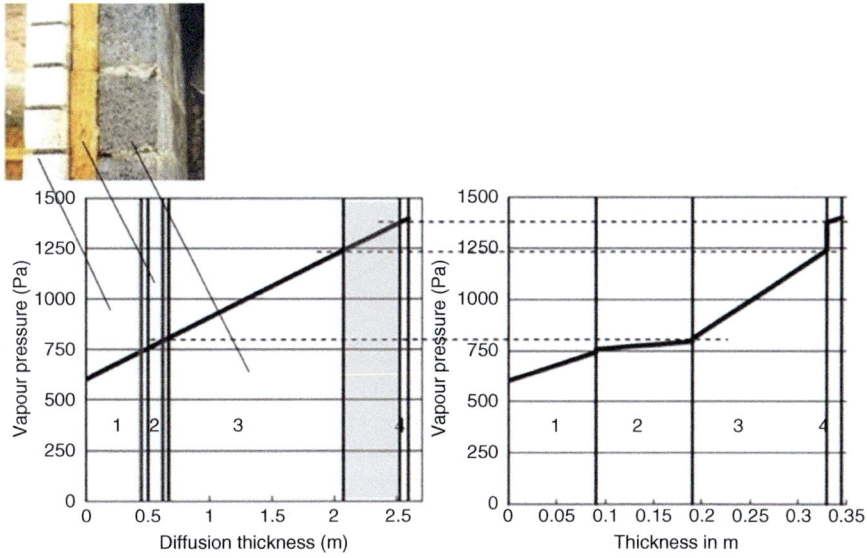

Figure 2.30 Multi-layered assembly, isothermal regime, vapour pressure course. Notice the large contact resistance in grey between layers 3 and 4.

2.3.6.5.4 Non-isothermal, Single-layered

In steady state, the temperature in the single-layer assembly varies linearly along the thickness (x), which makes the diffusion constant depending on x. With the assembly split in infinitesimal small layers dx, the temperature $\theta(x) = \theta_1 + (\theta_2 - \theta_1)x/d$ allows writing the vapour flux as:

$$g_v = -\frac{1}{\mu N(\theta(x))}\frac{dp}{dx}$$

Integration over the thickness d so gives:

$$g_v\mu \int_0^d N\left[\theta_1 + (\theta_2 - \theta_1)\frac{x}{d}\right] dx = -\int_{p_{s1}}^{p_{s2}} dp \tag{2.87}$$

The diffusion constant N now equals:

$$N = \frac{RT}{D_{vl}} = (5.25 \cdot 10^6 P_a)T^{-0.81}$$

Implementing it in the integral equation and solving gives:

$$g_v = -\frac{p_{s2} - p_{s1}}{\mu d\, 5.25 \cdot 10^6 P_a\, F(T)} \quad \text{with } F(T) = \left(T_{s2}^{0.19} - T_{s1}^{0.19}\right)/[0.19(T_{s2} - T_{s1})] \tag{2.88}$$

Numerically, $F(T)$ hardly differs from the integral of a linear change in temperature between θ_{s1} on the one and θ_{s2} on the other end face, divided by their difference (Figure 2.31).

With the temperature a linear function of x, the term $5.25 \cdot 10^6 P_a\, F(T)$ equals the value the diffusion constant N has for the mean temperature $\theta_m = (\theta_{s1} + \theta_{s2})/2$. The vapour flux so becomes:

$$g_v = \frac{p_{s2} - p_{s1}}{\mu d N(\theta_m)} \tag{2.89}$$

This equation resembles isothermal. The vapour pressures follow from:

$$p = p_{s1} + \mu g_v \int_0^x N(x)dx$$

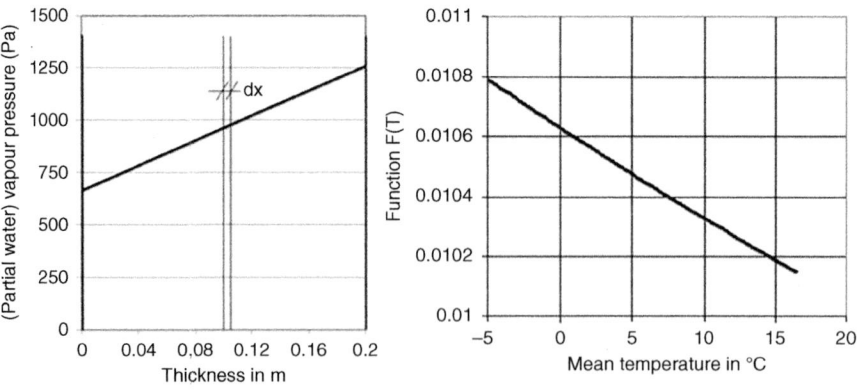

Figure 2.31 Single-layered assembly, diffusion under non-isothermal steady state conditions, $F(T)$ as function of the mean temperature: a straight line.

where:

$$\int_0^x N(x)dx = \frac{\left\{(5.25 \cdot 10^6 P_a)d\left\{\left[T_{s1} + (T_{s2}-T_{s1})\frac{x}{d}\right]^{0.19} - T_{s1}^{0.19}\right\}\right\}}{[0.9(T_{s2}-T_{s1})]}$$

This expression closely matches the product $N(\theta_m)x$, or:

$$p = \frac{p_{s2}-p_{s1}}{d}x + p_{s1} \qquad (2.90)$$

As in isothermal regime, the vapour pressures in a non-isothermal single-layer assembly change almost linearly from the end face with highest down to the end face with lowest vapour pressure.

2.3.6.5.5 Non-isothermal: Multi-layer

With the vapour pressures and temperatures on both end faces, the layer thicknesses (d_i), the vapour resistance factors (μ_i) and the thermal conductivities of all materials (λ_i) known, then the diffusion constant per layer has a value associated with its average temperature, per contact resistance a value associated with the contact temperature. The vapour flux so becomes:

$$g_v = \frac{p_{s2}-p_{s1}}{\sum[\mu_i d_i N(\theta_{mi})] + \sum Z_{ci}(\theta_{i,i+1})} \qquad (2.91)$$

Again, the denominator represents the total diffusion resistance Z_T of the multi-layer assembly with as first term the sum of the diffusion resistances Z_i of all layers and as second term the sum of the contact diffusion resistances. The vapour pressure course hardly differs from isothermal, provided the diffusion constant is calculated as advanced:

$$p_x = p_{s1} + g_v Z_{s1x} \qquad p_x = p_{s2} - g_v Z_{s2x} \qquad (2.92)$$

In the $[Z, p]$ axis system, the vapour pressure course so remains a straight line linking the vapour pressure values on both end faces $(0, p_{s1}; Z_T, p_{s2}, p_{s1} < p_{s2})$ with the flux as slope. Transposing it to the $[x,p]$ axis system follows again from moving the intersections with all interfaces in the correct order to that axis system and connecting their successive values by line segments. As the contact resistances Z_{ci} are mostly unknown, they are mostly set zero.

But, that vapour pressure curve is only possible if it does not intersect the saturation pressure curve in the assembly. If it does, interstitial condensation is a fact. To control, the saturation pressure curve deduced from the temperature course is traced in the $[Z, p]$ axis system (Figure 2.32). When crossing the vapour pressure course, condensation is a fact in the zones in between two successive intersections. How to correct the vapour pressure course then, requires a return to mass conservation. Where in a dry assembly the vapour pressure passes saturation, the vapour balance turns to:

$$\frac{d^2 p_{sat}}{dZ^2} = \pm \frac{G'_c}{\mu N} \qquad (2.93)$$

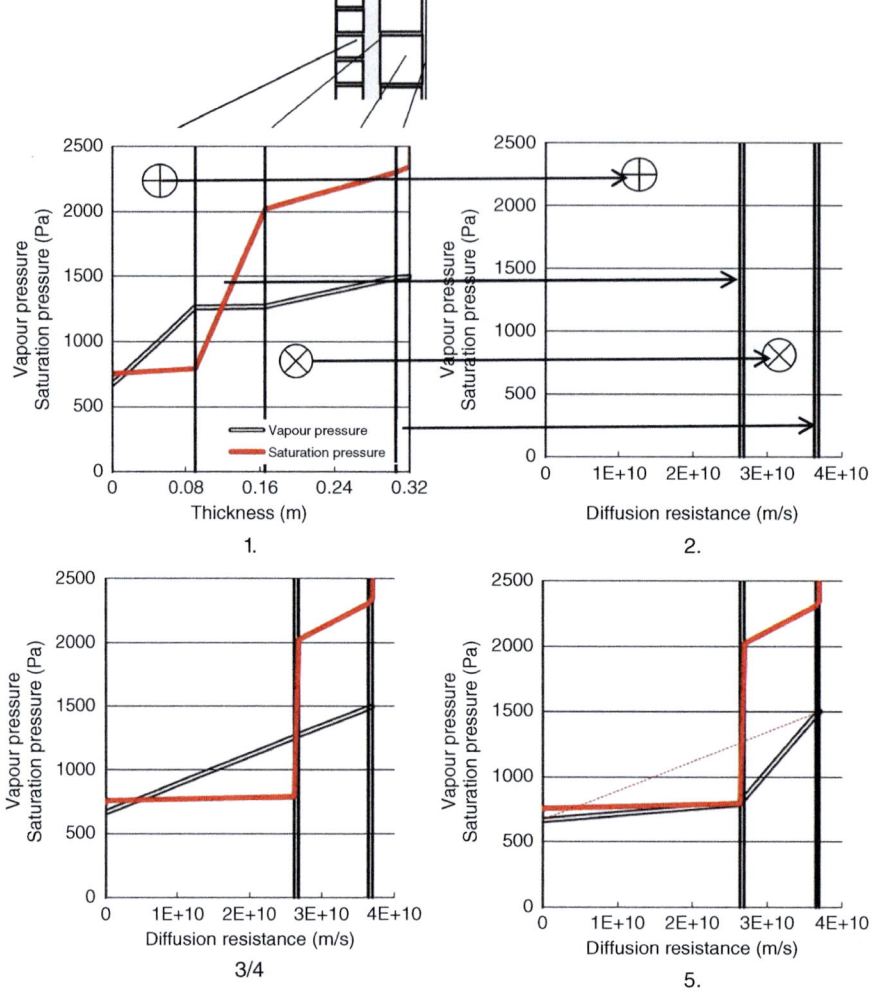

Figure 2.32 Non-isothermal steady state diffusion through a multi-layered assembly, the tangent method: (1) vapour pressure passes saturation, so, interstitial condensation, (2) transposing the whole to the [Z,p]-plane, (3/4) vapour (p) and saturation pressure (p_{sat}) intersect, physically impossible, (5) correction: tangents replace the vapour pressure line.

with G'_c the condensate in kg/(m·s) per m² and p_{sat} the saturation pressure in that zone between two successive intersections. Outside such zone, the balance remains:

$$\frac{d^2p}{dZ^2} = 0 \qquad (2.94)$$

Or, there, the vapour pressure course remains linear. In the condensing zones instead, the deposit in kg/(m³·s) is proportional to the second derivative of the saturation curve. Anyhow, in the intersections in and out a condensation zone, whether approached from the vapour pressure or the saturation pressure side, the vapour fluxes must have the same value. Otherwise, both intersections would contain a

2.3 Water Vapour

vapour source, what conflicts with 'initially dry'. Therefore, the lines to, the lines from and those between saturated zones must see their slope adapted. As the derivatives of the vapour pressure line and the saturation curve at an intersection so must stand for the same vapour flux, there must hold:

$$(dp_{sat}/dZ)_{c\to} = (dp/dZ)_{c\leftarrow} \tag{2.95}$$

Or, in such intersection, the slope of the saturation and the vapour pressure course must be the same, meaning the vapour pressure lines in the zones without deposit have to be tangents to the saturation curve in the condensation zones.

At the point of tangency of the vapour pressure line coming from the end face 2, the one at the highest vapour pressure side, vapour and saturation pressure must be equal, or:

$$p_{sat,c2} = p_{s2} - g_{v2}(Z_T - Z_{c2})$$

with g_{v2} the vapour flux, still unknown, Z_T the total diffusion resistance and Z_{c2} the diffusion resistance from the point of tangency to end face s1. Since the vapour pressure line in the no condensation and the saturation curve in the condensation zone have the same slope in their contact point, the following holds:

$$g_{v2} = \left[\frac{dp_{sat}}{dZ}\right]_{c2} = \left[\frac{dp_{sat}}{d\theta}\frac{q}{\lambda\mu N}\right]_{c2}$$

In it, q is the heat flux through the assembly, λ the thermal conductivity and μ the vapour resistance factor of the layer containing the contact point. For the vapour pressure tangent going out, a same logic applies. In many cases, the points of tangency of the incoming and outgoing vapour pressure lines coincide with an interface between two layers. There, it condenses.

This tangent method answers several questions:

(1) *How does the vapour pressure course looks in case of interstitial condensation?* Outside the condensation zones, it consists of the vapour pressure tangents, inside it follows the saturation curve.
(2) *Where does condensate deposit?* In all zones where the vapour saturation curve reigns. Is such zone limited to an interface, it only condenses there.
(3) *How much vapour condenses?* In general, the amounts follow from the difference in slope between each couple of in- and outgoing tangents. Is Z_{c1} the diffusion resistance between the end face with lowest vapour pressure and temperature ($Z = 0$) and the point of tangency of the outgoing tangent, is Z_{c2} the diffusion resistance between the end face with highest vapour pressure and temperature ($Z = Z_T$) and the point of tangency of the incoming tangent and are $p_{sat,c1}$ and $p_{sat,c2}$ the saturation pressures in both points of tangency, then the flux condensing equals:

$$g_c = \frac{p_2 - p_{sat,c2}}{Z_T - Z_{c2}} - \frac{p_{sat,c1} - p_1}{Z_{c1}} \tag{2.96}$$

In case condensate is deposited in several interfaces, calculating the distribution goes the same way, starting at the end face where the vapour pressure and the temperature are highest, then, jumping from interface to interface where condensation happens, and calculating per interface the difference between the vapour flux in and out:

$$g_c = \frac{p_{sat,i-1} - p_{sat,i}}{Z_{i-1} - Z_i} - \frac{p_{sat,i} - p_{i+1}}{Z_i - Z_{i+1}}$$

Is the deposit spread over a zone served by an in- and outgoing tangent, the distribution on the $[x, p]$ graph, showing the assembly, is:

$$\frac{dg_c}{dx} = G'_c = \mu N \frac{d^2 p_{sat}}{dZ^2} = \frac{1}{\mu N} \frac{d^2 p_{sat}}{dx^2} = \frac{1}{\mu N} \frac{d^2 p'}{d\theta^2} \left(\frac{d\theta}{dx}\right)^2 \quad (2.97)$$

A same approach applies to single-layer assemblies with a vapour resistance factor varying along the thickness. They behave as if composed of n layers with thickness Δx.

(4) *How to avoid interstitial condensation?* A first possibility, suggested by the $[Z, p]$-graph, consists of lowering the vapour pressure at the warm side until no intersections remain. For that, a better ventilation or a lower vapour release indoors are the actions needed (Figure 2.33a).

A second possibility consists of increasing the vapour resistance at the warm side to a level no intersections are left. This demands adding a vapour resistance ΔZ ($= Z_{vapour\ retarder}$) to the existing $Z_T - Z_{c2}$ value between the interface with condensation closest to and the warm end face. ΔZ as read on the Z-axis is equal to the diffusion resistance stretching from the outgoing end face to the intersection of the outgoing tangent with a horizontal through the vapour pressure at the warm end face minus the diffusion resistance of the assembly (p_2, Figure 2.33b):

$$\Delta Z = Z_{vapor\ retarder} \geq \frac{p_2 - p_1}{p_{sat,c1} - p_1} Z_{c1} - Z_T \quad (2.98)$$

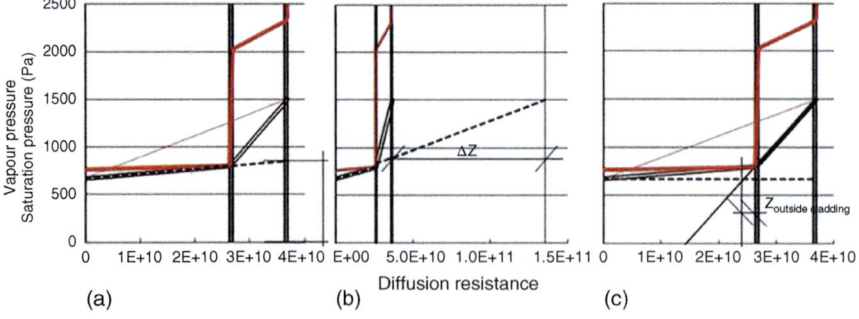

Figure 2.33 Tangent method (a) highest vapour pressure and warmest indoors. Two measures to avoid interstitial condensation valued graphically: (b) diffusion resistance to be added at the warm side, (c) diffusion resistance the outer finish should have.

Thus, the warm side of the thermal insulation is what needs a vapour retarder, which in cold and temperate climates is the inner side, in warm and humid climates the outer side.

A third possible intervention consists of lowering the diffusion resistance at the cold side (Z_{c1}) till all intersections are gone. How low, gives the difference on the Z-axis between the diffusion resistance at the point of tangency of the outgoing tangent and the diffusion resistance at the intersection of the ingoing tangent with a horizontal through the vapour pressure at the cold end face (p_1) (Figure 2.33c):

$$Z'_{c1} = Z_{\text{outside cladding}} \leq \frac{p_{\text{sat,c1}} - p_1}{p_2 - p_{\text{sat,c1}}}(Z_T - Z_{c1}) \qquad (2.99)$$

In cold and moderate climates, this implies the need for a vapour permeable outer, in warm and humid climates a vapour permeable inner finish.

In practice, evaluating diffusion as cause of interstitial condensation in multi-layer assemblies does not necessarily require the calculations commented. Simple rules exist. Of course, which end face is warmest and sees the highest vapour pressure differs between climates. In cold and moderate climates it's the inner, in warm and humid climates, where active cooling is needed, it's the outer end face.

Rule 1: no interstitial condensation? Presupposes no p/p_{sat} intersections. Demands a convex vapour saturation and concave vapour pressure course from warm to cold. Since the saturation pressure and temperature course are similar, the temperature concave suffices for that, what requires thermal conductivities and vapour resistance factors of the materials used decreasing from warm to cold. Or, the best insulating, most vapour permeable layer must sit at the cold, the less insulating, most vapour retarding layer at the warm side (Figure 2.34a).

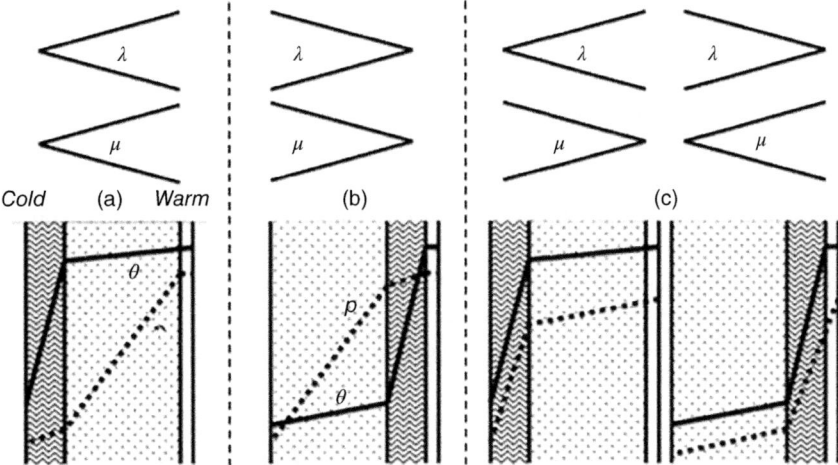

Figure 2.34 Interstitial condensation, risk evaluation: (a) temperature convex and vapour pressure concave, risk zero, (b) temperature concave and vapour pressure convex, high risk, (c) temperature and vapour pressure convex or concave, risk unclear.

Rule 2: interstitial condensation hardly avoidable? Presupposes that a p/p_{sat} intersection is most probable. Needed for that are a concave temperature and convex vapour pressure course from warm to cold, requiring that both the thermal conductivity and the vapour resistance factor of the materials used increase from warm to cold. Or, the best insulating, more vapour permeable layer should sit at the warm, the least insulating, more vapour retarding layer at the cold side (Figure 2.34b). Whether interstitial condensation will occur, then solely depends on the boundary conditions.

Rule 3: perhaps interstitial condensation? Presupposes that a p/p_{sat} intersection is unlikely but possible, which requires either a concave or a convex vapour pressure and temperature course. Whether condensate will deposit, then depends solely on the boundary conditions and how the sequence of materials with their thermal conductivity and vapour resistance factor looks (Figure 2.34c).

Of course, such simple rules do not guarantee an overall moisture tolerance. Many vapour permeable layers are not fit as outer finish because of being rain permeable, too capillary or too weak and flabby. The warm and cold sides may switch between seasons. In moderate climates, in winter the outer is the cold but in summer, the inner can be. This may give problems with assemblies, where the outside finish acts as rain buffer. Many assemblies are also air permeable, inducing airflows in and through that overrule an 'only diffusion' approach.

2.3.6.6 Steady State: Two and Three Dimensions

In isothermal regime, conservation of mass gives as vapour flows ($G_{v,i,j}$) from the adjacent to each central control volume:

$$\sum_{\substack{i=l,m,n \\ j=l\pm1, m\pm1, n\pm1}} G_{v,i,j} = 0$$

Each flow thereby equals the vapour pressure difference between the adjacent and central control volume, multiplied with the vapour permeance (P'_d) connecting both.

In non-isothermal regime, temperatures and their conversion to saturation pressures must be calculated first. Then, using the same mesh, the vapour pressures are quantified. If none passes saturation, interstitial condensation is no issue. If some do, to find the amounts deposited and the correct vapour pressure distribution, in all control volumes, where passing is a fact, the saturation pressure becomes the known and the condensation deposit ($G'_c \Delta V$) the unknown variables. This changes the vapour flows between the adjacent and central control volume to:

$$\sum_{\substack{i=l,m,n \\ j=l\pm1, m\pm1, n\pm1}} G_{v,i,j} = G'_c \Delta V \tag{2.100}$$

That adapted system of n equations has $n - y$ unknown vapour pressures now and y unknown condensation deposits (p and G'_c). When solved, the control volumes with negative deposit get the vapour pressure reintroduced as unknown and the system is solved again. Iterating this way goes on until all control volumes show either a deposit or a vapour pressure below saturation.

2.3.6.7 Non-steady State
2.3.6.7.1 The Equations Adapted
In transient state, conservation of mass gives:

Hygroscopic materials	Non-hygroscopic materials
$\text{div } \mathbf{g}_v = -\dfrac{\partial w_H}{\partial t} = -\rho \xi_\phi \dfrac{\partial \phi}{\partial t}$	$\text{div } \mathbf{g}_v \pm G'_c = \dfrac{\partial}{\partial t}\left(\dfrac{\Psi_0 p}{RT}\right)$

The vapour flux remains:

$$\mathbf{g}_v = -\frac{1}{\mu N}\text{grad } p$$

2.3.6.7.2 Hygroscopic Materials
If the RH does not pass the pivot ϕ_M, see above, then, with temperature and RH replacing vapour pressure as driving forces, the vapour flux becomes:

$$\mathbf{g}_v = -\frac{1}{\mu N}\text{grad }(p_{sat}\phi) = -\frac{1}{\mu N}\left(p_{sat}\text{grad }\phi + \phi\frac{dp_{sat}}{d\theta}\text{grad }\theta\right) \quad (2.101)$$

Without contact resistances, the RH has one value in all interfaces, making it a real potential. Mass conservation turns into:

$$\text{div}(D_\phi \text{grad }\phi + D_\theta \text{grad }\theta) = \frac{\partial(\rho\xi_\phi \phi)}{\partial t} \quad \left(D_\phi = p_{sat}/(\mu N), D_\theta = \frac{\phi \cdot (dp_{sat}/d\theta)}{\mu N}\right)$$
$$(2.102)$$

The functions D_ϕ and D_θ but also the specific moisture content $\rho\xi_\phi$, so depend on the RH and the temperature. Or, solving with FEM or CVM the mass balance is only doable in conjunction with the heat balance. Although isothermal regime is only possible in materials with infinite thermal conductivity as otherwise the conversion of latent to sensible heat must result in temperature differences, assuming such regime changes mass conservation into:

$$\text{div}(D_\phi \text{grad }\phi) = \frac{\partial(\rho\xi_\phi \phi)}{\partial t}$$

If the transport coefficient D_ϕ and the specific moisture content $\rho\xi_\phi$ are constants, this equation further simplifies to:

$$\nabla^2 \phi = \frac{\rho\xi_\phi}{D_\phi}\frac{\partial \phi}{\partial t} \quad (2.103)$$

A constant specific moisture content presumes a sorption isotherm forming a straight line between RH = 0 and RH = ϕ_M. A usable value ensues from calculating the least square line between the 30 and 86% RH moisture contents, see Figure 2.35.

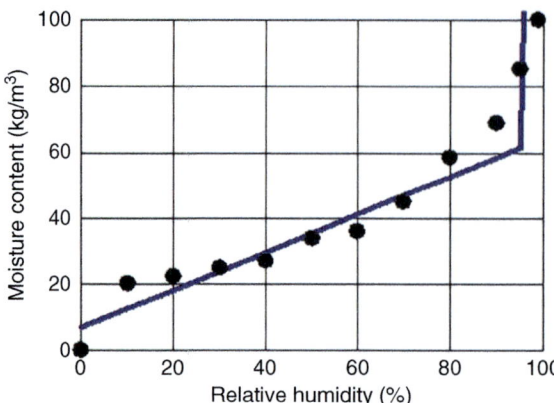

Figure 2.35 Simplified sorption curve. At 0% RH, moisture content jumps from 0 to b, then increases linearly to f_M where it jumps to capillary. The dots show what's measured.

In reality, however, only non-hygroscopic materials have constant transport coefficients! Indeed, the vapour resistance factor of hygroscopic materials turns lower, the higher the RH, or, the transfer coefficient D_ϕ will increase with RH, standing for moisture content. This anyhow limits the applicability of the simple equation given above to small differences in RH. But even then, just like the saturation pressure and the diffusion constant N, the transfer coefficient D_ϕ still depends on temperature. Anyhow, if usable, the simple balance equation resembles Fourier's second law. So, solving the same way can.

For a periodic change in RH, the hygric damping D_ϕ^n with phase shift ϕ_ϕ^n, the dynamic moisture resistance D_g^n with phase shift ϕ_g^n and the hygric admittance Ad_v^n with phase shift ϕ_{Adv}^n. form together the assembly's matrix for diffusion W_{nd}. For a single-layer, the three look:

$$D_\phi^n = \cosh\left(\frac{\omega_{vn} d}{D_\phi}\right) \qquad D_g^n = \frac{\sinh\left(\frac{\omega_{vn} d}{D_\phi}\right)}{\omega_n} \qquad Ad_v^n = \frac{D_\phi^n}{D_g^n}$$

$$\phi_\phi^n = \arg\left(\cosh\frac{\omega_{vn} d}{D_\phi}\right) \qquad \phi_g^n = \arg\left(\frac{\sinh\frac{\omega_{vn} d}{D_\phi}}{\omega_n}\right) \qquad \phi_{Adv}^n = \phi_\phi^n - \phi_g^n$$

with ω_{vn} the complex hygroscopic pulsation:

$$\omega_{vn} = \sqrt{\frac{2 D_\phi \rho \xi_\phi i n \pi}{T}}$$

For composite assemblies, the same multiplication rules as for heat conduction apply.

A sudden change in RH against the surface of a semi-infinite solid ($\Delta\phi_s$) in turn gives as RH response:

$$\phi = \phi_{s0} + \Delta\phi_s \left[1 - \mathrm{erf}\frac{x}{\sqrt{4 a_H t}}\right]$$

with a_H the moisture diffusivity in m²/s, equal to $D_\phi/\rho\xi_\phi$. The ingoing vapour flux becomes:

$$g_v = -D_\phi(\text{grad }\phi)_{x=0} = \Delta\phi\sqrt{\frac{\rho\xi_\phi D_\phi}{\pi t}}$$

Here, $\sqrt{\rho\xi_\phi D_\phi}$ is the water vapour sorption coefficient or vapour effusivity, units kg/(m²·s^(1/2)).

Solving two and three dimensional problems requires FEM or CVM.

2.3.6.7.3 Non-hygroscopic Materials

Inserting the flux equation in the vapour balance gives:

$$\text{div}\left(\frac{1}{\mu N}\text{grad }p\right) \pm G'_c = \frac{\partial}{\partial t}\left(\frac{\Psi_0 p_{sat}}{RT}\right)$$

This differential equation must be solved together with the heat balance, using the equation of state between saturation pressure and temperature. In cases where the vapour pressure equals saturation, the condensation or evaporation term (G'_c) becomes:

$$\pm G'_c = -\text{div}\left(\frac{1}{\mu N}\text{grad }p_{sat}\right) + \frac{\partial}{\partial t}\left(\frac{\Psi_0 p_{sat}}{RT}\right)$$

If the assembly is isothermal and dry, the mass balance simplifies to:

$$\nabla^2 p = -\left(\frac{\Psi_0 \mu N}{RT}\right)\frac{\partial p}{\partial t} \tag{2.104}$$

This relation is identical to the simple isothermal vapour balance for hygroscopic materials. Analogous characteristics emanate, now called the vapour diffusivity (a_v) and effusivity (b_v):

$$a_v = RT/(\Psi_0 \mu N) \quad b_v = \sqrt{\Psi_0/(RT\mu N)}$$

In periodic regime, the complex properties defining the response of any flat assembly are called the vapour pressure damping, the dynamic diffusion resistance and the vapour admittance.

2.3.7 Vapour Flow by Diffusion and Moist Air Moving Through Open-porous Assemblies

2.3.7.1 In General

As stated, equivalent diffusion only is the exception. In many cases, moist airflow in and through adds. In fact, assemblies showing some air permeance see wind, stack and fans force moist air to flow through, changing the vapour balance into:

$$\text{div}(\mathbf{g}_{v1} + \mathbf{g}_{v2}) \pm G'_c = -\frac{\partial}{\partial t}\left(\frac{\Psi_0 p}{RT}\right) \tag{2.105}$$

with $\mathbf{g}_{v1}$ the convective and $\mathbf{g}_{v2}$ the diffusive vapour flux. Including the flux equations gives:

$$\text{div}\left(\frac{1}{\mu N}\text{grad }p - \frac{\mathbf{g}_a}{\rho_a RT}p\right) \pm G'_c = \frac{\partial}{\partial t}\left(\frac{\Psi_0 p}{RT}\right) \tag{2.106}$$

Solving is not straightforward as CVM requires a combination with the air and heat balances.

Further discussion therefore remains restricted to a steady state air in- or outflow, perpendicular to the end faces of flat assemblies, that lack contact resistances and whose materials have constant vapour resistance factors. Although depending on temperature, for reasons of simplicity, also assumed is that the diffusion constant N can be taken constant, $5.4 \cdot 10^9$/s. The vapour balance so looks:

$$\frac{d}{dx}\left(\frac{1}{\mu N}\frac{dp}{dx} - \frac{g_a}{\rho_a RT}p\right) = 0 \tag{2.107}$$

2.3.7.2 Isothermal, Single- and Multi-layered Assemblies

In both cases, transposing the equation above to the $[Z, p]$ axis system gives:

$$\frac{d^2 p}{dZ^2} - \frac{g_a}{\rho_a RT}\frac{dp}{dZ} = \frac{d^2 p}{dZ^2} - a_P \frac{dp}{dZ} = 0 \quad \text{with} \quad a_P = \frac{g_a}{\rho_a RT} \approx \frac{0.62 g_a}{P_a}$$

This differential equation of second order has as solution:

$$p = C_1 + C_2 \exp(a_P Z) \tag{2.108}$$

The diffusive, convective and total vapour fluxes so equal:

Diffusive $\quad g_v = -\dfrac{dp}{dZ} = -a_P C_2 \exp(a_P Z)$

Convective $\quad g_{vc} = a_P p = a_P[C_1 + C_2 \exp(a_P Z)]$

Total $\quad g_{vT} = g_v + g_{vc} = a_P C_1$

While the diffusive and convective vapour fluxes vary while moving through the assembly, the total vapour flux, their sum, remains constant. The integration constants C_1 and C_2 ensue from the boundary conditions. The vapour pressures on both end faces known ($Z = 0, p = p_{s1}; Z = Z_T, p = p_{s2}$), $p_{s1} = C_1 + C_2$ and $p_{s2} = C_1 + C_2 \exp(a_P Z_T)$ gives:

$$C_1 = \frac{p_{s2} - p_{s1}\exp(a_P Z_T)}{1 - \exp(a_P Z_T)} \quad C_2 = \frac{p_{s2} - p_{s1}}{1 - \exp(a_P Z_T)}$$

Vapour pressures and vapour fluxes through the assembly then look:

$$p = p_{s1} + (p_{s2} - p_{s1})F_{v1}(Z) \quad g_v = F_{v2}(Z)(p_{s2} - p_{s1})$$
$$g_{vc} = a_P p \quad g_{vT} = p_{s2}F_{v2}(0) - p_{s1}F_{v2}(Z_T)$$

with:

$$F_{v1}(Z) = \frac{1 - \exp(a_P Z)}{1 - \exp(a_P Z_T)} \quad F_{v2}(Z) = \frac{a_P \exp(a_P Z)}{1 - \exp(a_P Z_T)} \tag{2.109}$$

In the $[Z, p]$ axis system, the vapour pressure course between both end faces $[0, p_{s1}]$ to $[Z_T, p_{s2}]$ so changes from straight to convex exponential when the air is moving from the end face at higher to the end face at lower vapour pressure and to concave exponential in the opposite case, see Figure 2.36). The higher the factor a_P in the equation, the more air moves through and the more concave or convex the curve.

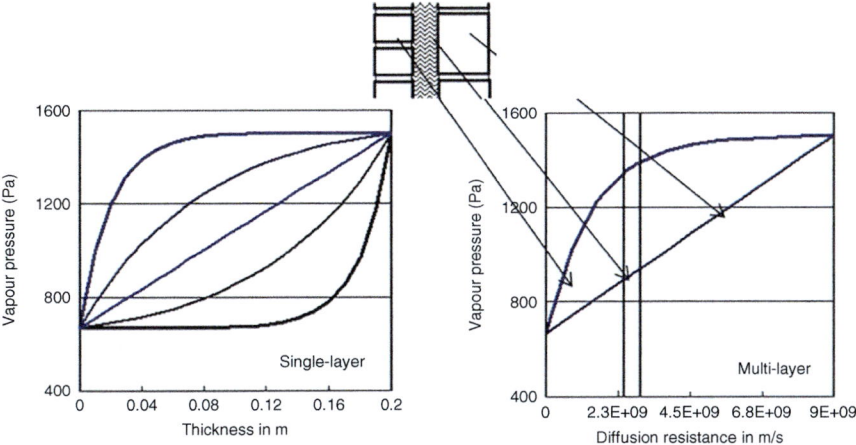

Figure 2.36 Steady state convection and diffusion in isothermal regime. Vapour pressure course in a single-layer in the [x, p] and in a multi-layer in the [Z, p]-axis system.

The vapour pressure course through a multi-layer assembly drawn in the [x,p] axis system follows from transposing the intersections with all successive interfaces in the right order to the interfaces and linking them with exponential segments.

At first sight, the results resemble combined heat conduction and enthalpy flow. But there's a difference. A 10 cm thick mineral fibre layer has a λ-value 0.036 W/(m·K), an air permeability $8 \cdot 10^{-5}$ s and a μ-value 1.2. For 1 Pa air pressure difference between both faces, the air flux so touches 0.0008 kg/(m²·s) or 2.4 m³/(m²·h), giving an exponent $c_a g_a d/\lambda = 2.2$ for the temperature and $g_a \mu N d/(\rho_a R T) = 3.2$ for the vapour pressure curve. So, even for an insulating, extremely vapour-permeable layer, the air flowing through makes the vapour pressure course more convex or concave than the temperature course. For less insulating, air-permeable assemblies, that difference becomes even more pronounced.

The increase in total vapour flux compared to diffusion only (g_{vo}) is:

$$100(g_{vT} - g_{vo})/g_{vo} = 100[Z_T \text{abs}(F_{v2}(Z_T)) - 1] \qquad (2.110)$$

When the air migrates from high to low vapour pressures, $\text{abs}(F_{v2}(Z_T))$ passes 1 and the total vapour flux increases compared to diffusion only. In the opposite case, even a limited air inflow will already force the vapour flow to change direction, from low to high vapour pressure.

2.3.7.3 Non-isothermal, Single- and Multi-layered Assemblies

Here, the vapour saturation (p_{sat}) and vapour pressure courses are fixing what happens. If not intersecting, the isothermal exponential remains correct as vapour pressure response. If intersecting, interstitial condensation will be a fact. As the amounts will usually be too small to see the latent heat of evaporation having impact, only the vapour pressure curve will change. Indeed, where condensation begins and ends, the vapour and saturation pressure sources must coincide with the slope representing the vapour flux, or:

$$p = p_{sat} \frac{dp_{sat}}{dZ} - a_p p_{sat} = \frac{dp}{dZ} - a_p p$$

Unknown in both relations is at which saturation pressure and diffusion resistance condensation begins and stops. Combining both equations gives:

$$\left[\frac{dp}{dZ}\right] = \left[\frac{dp_{sat}}{dZ}\right] \tag{2.111}$$

This shows that a same slope in the [Z,p] axis system means that the in- and outgoing vapour pressure curves demarcating a condensation zone become tangents to the vapour saturation course. In practice, first, the saturation and vapour pressure courses as if no condensate deposits are traced. Then, all interfaces where the vapour pressure exceeds saturation are tested on condensation, starting at the interface where the difference between the vapour saturation and the vapour pressure is highest. If this is interface j, then the exponential vapour pressure course coming from the end face at highest temperature and vapour pressure ($s1$) to the saturation pressure $p_{sat,j}$ in j, turns into:

$$p = p_{s1} + (p_{sat,j} - p_{s1})F_{v1}(Z_{s1}) \quad \text{with} \quad F_{v1}(Z_{s1}) = \frac{1 - \exp\left[a_p Z_{s1}^z\right]}{1 - \exp\left[a_p \left(Z_{s1}^j\right)\right]}$$

In it, Z_{s1}^j is the diffusion resistance between that end face $s1$ and the point of tangency j and Z_{s1}^z the diffusion resistance between $s1$ and every plane z between $s1$ and j. If not intersecting in any other plane z, then plane j is the one where condensate starts depositing. A same reasoning applies for the outgoing vapour pressure exponential, now with the saturation pressure $p_{sat,j}$ in interface j and the vapour pressure at end face $s2$ as bounding values:

$$p = p_{sat,j} + (p_{s2} - p_{sat,j})F_{v1}(Z) \quad \text{with} \quad F_{v1}(Z) = \frac{1 - \exp\left(a_p Z_j^z\right)}{1 - \exp\left(a_p Z_j^{s2}\right)}$$

In it, Z_j^{s2} is the diffusion resistance between j and end face $s2$, Z_j^z the diffusion resistance between j and every plane z between j and $s2$. If no other intersection jumps up, j then remains the only condensation interface.

When there are additional intersections between end face $s1$ and j, testing all interfaces in between will give the one or those where condensate is also deposited. If the zone with higher vapour than vapour saturation pressure stretches till between interfaces, then the point of tangency (l) on the saturation curve where condensation starts will follow from:

$$(p_{sat,1} - p_{s1}) \frac{a_p \exp\left[a_p Z_{s1}^l\right]}{1 - \exp\left[a_p Z_{s1}^l\right]} = \left[\frac{dp_{sat}}{dZ}\right]_l$$

Additional intersections between j and end face $s2$ could also be a fact for the outgoing vapour pressure exponential. Then, either extra condensation interfaces exist or, with *l* the starting point of the condensation zone, the end point *k* where condensation stops will obey:

$$(p_{s2} - p_{sat,k}) \frac{a_p \exp(a_p Z_k^{s2})}{1 - \exp(a_p Z_k^{s2})} = \left[\frac{dp_{sat}}{dZ}\right]_k$$

In case *k* and *l* delimit one zone, only this will see condensate depositing. If nonetheless more vapour pressure tangents appear, more interfaces or zones will suffer.

When the air moves through a multi-layer assembly from low to high temperature and low to high vapour pressure, interstitial condensation is definitely excluded. In fact, a more concave exponential vapour pressure course cannot intersect a less concave vapour saturation pressure course. On the contrary, with air mitigating from high to low temperature and high to low vapour pressure, then, as Figure 2.37 illustrates, with the vapour pressure at the warm side high enough, interstitial condensation becomes unavoidable.

Airflow additionally changes the speed by which deposit accumulates, from very slow for diffusion only to much faster when combined with convection.

It must be clear that diffusion and convection combined have consequences.

First Air ingress may increase the number of condensation zones compared to diffusion only. If nonetheless, only one interface sees condensate depositing, then with *s2* the outer and *s1* the inner end face, that deposit becomes:

$$g_c = \left[p_{sat,x} F_{v2}(0) - p_{s2} F_{v2}\left(Z_{s2}^{sat,x}\right)\right] \\ - \left[p_{s1} F_{v2}\left(Z_{s2}^{sat,x} - Z_{s2}^{sat,x}\right) - p_{sat,x} F_{v2}\left(Z_{s1}^{s2} - Z_{s2}^{sat,x}\right)\right]$$

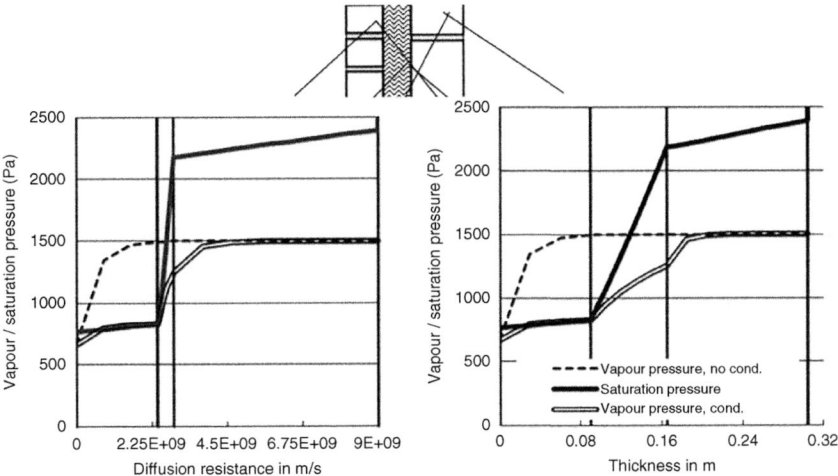

Figure 2.37 Composite assembly, combined diffusion and convection under non-isothermal steady state conditions. Vapour pressure (thin line) intersects saturation, so, interstitial condensation. The correct line gives deposit at the the veneer's backside.

with $p_{sat,x}$ the saturation pressure in this interface and the functions $F_{v2}(...)$ equal to:

$$F_{v2}(0) = \frac{a_P}{1 - \exp(a_P Z_{s2}^{sat,x})} \qquad F_{v2}\left(Z_{s2}^{sat,x}\right) = \frac{a_P \exp(a_P Z_{s2}^{sat,x})}{1 - \exp(a_P Z_{s2}^{sat,x})}$$

$$F_{v2}\left(Z_{s2}^{sat,x} - Z_{s2}^{sat,x}\right) = \frac{a_P}{1 - \exp\left[a_P\left(Z_{sl}^{s2} - Z_{s2}^{sat,x}\right)\right]}$$

$$F_{v2}\left(Z_{sl}^{s2} - Z_{s2}^{sat,x}\right) = \frac{a_P \exp\left[a_P\left(Z_{sl}^{s2} - Z_{s2}^{sat,x}\right)\right]}{1 - \exp\left[a_P\left(Z_{sl}^{s2} - Z_{s2}^{sat,x}\right)\right]}$$

Second Compared to diffusion only, the deposit could increase dramatically. Of course, more air outflow from the warmer end face also heats the condensation interface, which will somewhat slow down the increase in deposit. In fact, after a maximum reached, further warming due to more air flowing out, lowers the deposit again until it turns zero once the temperature in the condensation interface touches the dew point of the air at the warm side (Figure 2.38).

Third The most reliable measure to avoid interstitial condensation in an assembly, also optimal from an energy efficiency point of view, is caring for airtightness and ordering the layers correctly in terms of λ- and μ-value. As it negates energy efficiency and thermal comfort, stupid should be to construct assemblies leaky enough for air inflowing from the warm end face to lift the temperature in the interfaces where condensate could deposit above the dew point at the warm side. Of course, in cold and temperate climates, combining a good ventilation with a moderate vapour release may keep the vapour pressure indoors below saturation at all likely condensation interfaces. Depressurising the indoors to get a few air inflow through the opaque envelope parts is an alternative in winter, while in warm, humid climates, a moderate pressurization will help avoiding interstitial condensation.

In diffusion-only mode, the tangent method allowed checking below which vapour pressure at the warm side interstitial condensation was excluded. Also, the vapour retarder needed at the warm side or how vapour permeable the outside finish should be could be evaluated. With convection added, the measures left are either limiting the vapour pressure at the warm side to a value excluding interstitial condensation or caring for a really airtight envelope. Vapour retarders and

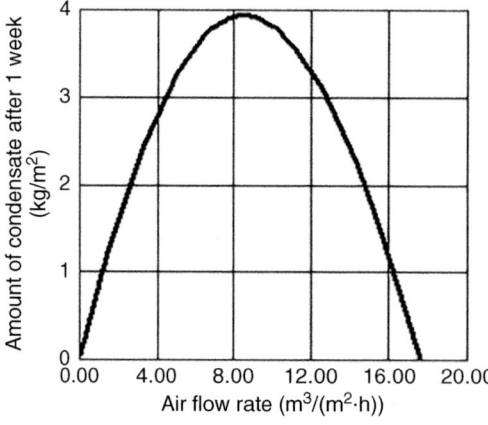

Figure 2.38 Interstitial condensation by combined diffusion and convection in a multi-layer assembly, deposit in kg/week as function of air moving through from the warm end face.

vapour permeable outside finishes, in fact, lose their effect when, them included, airtightness fails.

2.3.8 Surface Film Coefficients for Diffusion

2.3.8.1 Derivation

Till now, the vapour pressure on the end faces was assumed known, although measuring is hardly doable. Instead, logging the air temperature and RH in the ambient is easy. Preferred so are these as boundary condition. The question left then is: how is vapour moving from the air to a surface and vice versa. Must it overcome a resistance? Yes, as diffusion is the mover left in the laminar air layer against a surface, while in the free air convection regns (Figure 2.39). Vapour diffusion through that laminar air layer now is written as:

$$g_v = \beta(p - p_s) \tag{2.112}$$

with p the vapour pressure in the air outside that layer, the reference indoors being its value in the centre of the space, 1.7 m above floor level and outdoors the value measured in the nearest weather station, p_s the vapour pressure on the surface and β the surface film coefficient for diffusion, units s/m. $1/\beta$ is called the surface resistance for diffusion, symbol Z, units m/s.

The suffix i indicates indoors (β_i, Z_i), e outdoors (β_e, Z_e). Both are calculated as the thermal were. Basis is the mass balances for air and vapour, the Navier–Stokes momentum equilibrium, and the turbulence equations. Solving that system of seven partial differential equations numerically or by similarity gives a series of dimensionless numbers, allowing quantification:

Forced convection	Mixed convection	Free convection
Reynolds number: $\text{Re} = \dfrac{vL}{v}$	$\text{Re}/\text{Gr}^{1/2}$	Grashoff number: $\text{Gr} = \dfrac{\Delta \rho_a g L^3}{v^2}$
←———— The Sherwood number, $\text{Sh} = \dfrac{\beta L}{\delta_a}$, replacing the Nusselt number ————→		
←———— The Schmidt number, $\text{Sc} = \dfrac{v}{RT\delta_a}$, replacing the Prandl number ————→		

Figure 2.39 Surface film coefficient for diffusion, boundary layer.

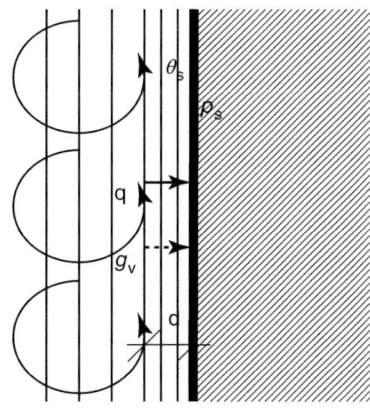

The Sherwood number links the vapour transfer to diffusion, while the Schmidt number underlines the similarity between the flow velocities and the field of vapour pressures. Between the numbers, identical relations exist as for convective heat transfer. Forced laminar flow along a horizontal flat assembly so gives as surface film coefficient for diffusion:

$$Sh_L = 0.664 \, Re_L^{1/2} \, Sc^{1/3}$$

That similarity between the vapour and heat transfer against surfaces led to the Lewis equation:

$$\beta = \frac{h_c}{\rho_a R_a T \, c_p} \left(\frac{R_a T \, c_p \delta}{\lambda_a} \right)^{0.67} \tag{2.113}$$

In it, h_c is the convective surface film coefficient for heat, while ρ_a, R_a, c_p, and λ_a are the density, the gas constant, the specific heat at constant pressure and the thermal conductivity of air. Under atmospheric conditions and for the Schmidt number equal to the Prandl number, the laminar boundary layers for heat and vapour have the same thickness d, giving:

$$q = \lambda_a \frac{\theta - \theta_s}{d} = h_c(\theta - \theta_s) \text{ or } d = \lambda_a / h_c \quad g_v = \frac{p - p_s}{Nd} = \beta(p - p_s) \text{ or } d = \frac{1}{N\beta}$$

So:

$$\beta = h_c \frac{1}{\lambda_a N} \tag{2.114}$$

In air, with $N \approx 5.4 \cdot 10^9$/s and $\lambda = 0.025$ W/(m·K), β becomes (see Table 2.5):

$$\beta \approx 7.7 \cdot 10^{-9} h_c \tag{2.115}$$

In practice, following values are commonly used:

Table 2.5 Indoor and outdoor surface film coefficient for diffusion.

Indoors, β_i (P_a = 1 Atm, $0 \leq \theta_i \leq 20$ °C)		Outdoors, β_e (P_a = 1 Atm, $-20 \leq \theta_i \leq 30$ °C)	
$\theta_i - \theta_{is}$ (K)	β (·10^{-9} s/m)	v_a (m/s)	β_e (·10^{-9} s/m)
2	28.6	<1	≤110
4	30.0	5	212
6	31.4	5–10	280
8	32.8	25	849
10	34.2	—	—
12	36.0	—	—
$\beta_i = [27 + 0.73(\theta_i - \theta_{si})] \cdot 10^{-9}$, $r^2 = 1$		$\beta_e = 49.9 \cdot 10^{-9} v_a^{0.875}$, $r^2 = 0.998$	

	Indoors	Outdoors
β (s/m)	$18.5 \cdot 10^{-9}$	$140 \cdot 10^{-9}$
Z (m/s)	$54 \cdot 10^{6}$	$7.2 \cdot 10^{6}$

The diffusion resistance ambient to ambient through a flat assembly so looks:
In- to outdoors:

$$Z_a = Z_i + Z_T + Z_e$$

In- to indoors:

$$Z_a = 2Z_i + Z_T$$

giving as steady-state vapour fluxes:
In- to outdoors:

$$g_v = \frac{p_i - p_e}{Z_i + Z_T + Z_e}$$

In- to indoors:

$$g_v = \frac{p_{i1} - p_{i2}}{2Z_i + Z_T}$$

The vapour pressures on the end faces of an envelope assembly become:
Indoors:

$$p_{si} = p_i - g_v Z_i$$

Outdoors:

$$p_{se} = p_e + g_v Z_e$$

Compared to the thermal surface film resistances (R_i and R_e), true differences anyhow exist. An 0.3 m thick air-dry aerated concrete wall (ρ = 480 kg/m³, λ = 0.15 W/(m·K), μ = 5) has as thermal resistance and as diffusion resistance: R = 0.3/0.15 = 2 m²·K/W, Z = 0.3 · 5 · 5.4 · 10⁹ = 8.1 · 10⁹ m/s. From ambient to ambient, these values become: R_a = 2 + 0.17 = 2.17 m²·K/W, Z_a = 8.1 · 10⁹ + 61 · 10⁶ = 8.17 · 10⁹ m/s. Or, the thermal surface film resistances represent 7.7%, the surface film resistances for diffusion only 0.7% of the total. Or, except when fluxes at surfaces are of interest, the error by negating the surface film resistances for diffusion when calculating vapour fluxes air to air through assemblies has hardly impact.

2.3.8.2 Applications

2.3.8.2.1 Diffusion Resistance of Unvented Cavities

In cases the air in a cavity has the same temperature and air pressure everywhere, only diffusion can move vapour through, giving as diffusion resistance ($\mu_a = 1$):

$$Z_c = Nd \tag{2.116}$$

If, instead, the temperature and air pressure vary, convection will reduce the diffusion resistance to the sum of the surface film resistances for diffusion at both cavity faces:

$$Z_c = Z_1 + Z_2 = \frac{\beta_1 + \beta_2}{\beta_1 \beta_2} \tag{2.117}$$

Related value is so small compared to the diffusion resistances of the other layers in an assembly with cavity that ignoring the cavity when looking to equivalent diffusion through hardly matters. Not so of course when the vapour fluxes at both cavity faces are the subject of interest.

2.3.8.2.2 Does Ventilating Cavities Enhance Veneer Drying?

A cavity wall has a veneer with open head joints above and below, allowing wind and stack to vent the cavity. Will this accelerate drying? To answer, the vapour flow by diffusion through both leaves is set equal to the vapour removal by venting (Figure 2.40).

The cavity is d_{cav} m wide and L m high, z being the ordinate along. The air moves with velocity v (m/s), β (s/m) is the surface film coefficient for diffusion at both cavity faces, Z_1 (m/s) the diffusion resistance from ambient 1 (p_1) to cavity face 1 and Z_2 (m/s) the diffusion resistance from ambient 2 (p_2) to cavity face 2. Vapour now diffuses from both ambients to the linked cavity faces, or:

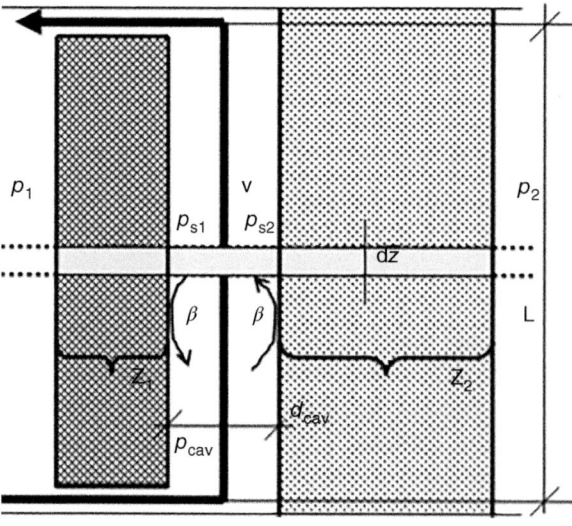

Figure 2.40 Cavity wall, diffusion through veneer and inner leaf, venting the cavity.

Face 1 : $\dfrac{p_1 - p_{s1}}{Z_1} + \beta(p_{cav} - p_{s1}) = 0$

Face 2 : $\dfrac{p_2 - p_{s2}}{Z_2} + \beta(p_{cav} - p_{s2}) = 0$

with p_c the vapour pressure in the cavity. Solving gives:

$$p_{s1} = \left(\dfrac{p_1}{Z_1} + \beta p_c\right) \bigg/ \left(\dfrac{1}{Z_1} + \beta\right) \quad p_{s2} = \left(\dfrac{p_2}{Z_2} + \beta p_c\right) \bigg/ \left(\dfrac{1}{Z_2} + \beta\right) \quad (2.118)$$

The first term stands for the vapour transfer through the veneer, respectively the inside leaf, the second term for the flow between related cavity face and the air moving along the cavity, giving as vapour balance there:

$$[\beta(p_{s1} - p_{cav}) + \beta(p_{s2} - p_{cav})]dz = \dfrac{d_{cav}v}{RT_{cav}} dp_{cav}$$

The fluxes coming from both faces so change the vapour content in the venting air. With the cavity temperature T_{cav} assumed equal to the asymptotic $T_{cav,\infty}$, which is a simplification, the vapour pressures p_{s1} and p_{s2} turn that balance into:

$$\left[\dfrac{\dfrac{p_1}{Z_1}\left(\dfrac{1}{Z_2} + \beta\right) + \dfrac{p_2}{Z_2}\left(\dfrac{1}{Z_1} + \beta\right)}{\dfrac{1}{Z_1}\left(\dfrac{1}{Z_2} + \beta\right) + \dfrac{1}{Z_2}\left(\dfrac{1}{Z_1} + \beta\right)} - p_{cav}\right] dz = \dfrac{d_{cav}v}{RT_{cav,\infty}\left(\dfrac{1}{Z_1+\beta^{-1}} + \dfrac{1}{Z_2+\beta^{-1}}\right)} dp_{cav}$$

With the vapour pressure on face 1 known, this equation simplifies to:

$$\underbrace{\left[\dfrac{P_i\left(\dfrac{1}{Z_2} + \beta\right) + \dfrac{p_2}{Z_2}}{\dfrac{2}{Z_1} + \beta} - p_{cav}\right]}_{a} dz = \underbrace{\dfrac{d_{cav}v}{RT_{cav\infty}\left(\beta + \dfrac{1}{Z_2+\beta^{-1}}\right)}}_{b} dp_{cav}$$

with as solution:

$$z/b = -\ln[(a - p_{cav})/C]$$

In it, C is the integration constant. An infinite long cavity sees the ratio z/b turning ∞, or:

$$\ln[(a - p_{cav,\infty})/C] = -\infty$$

meaning $a - p_{cav,\infty} = 0$ or $a = p_{cav,\infty}$. The parameter a so equals the asymptotic value the vapour pressure reaches in an infinitely high cavity. Because without air moving, $v = 0$, z/b also turns ∞, a additionally represents the vapour pressure in a non-vented cavity. At $z = 0$, that vapour pressure equals $p_{cav,0}$, or:

$$\ln[(p_{cav,\infty} - p_{cav,0})/C = 0] \quad \text{giving} \quad C = p_{cav,\infty} - p_{cav,0}.$$

Hence, the overall vapour pressure build-up can be written as:

$$p_{cav} = p_{cav,\infty} - (p_{cav,\infty} - p_{cav,0})\exp(-z/b) \quad (2.119)$$

Or, the change is exponential, from the value at the inflow up to the value it should have in a non-vented cavity. The term b, dimension m, is called the venting length.

The slower the air moves along and the more vapour is picked up, the shorter this length.

Does venting so accelerate veneer drying? The vapour flow taken away approximately equals:

$$G_v = b_{cav} v \frac{p_{cav,L} - p_{cav,o}}{R T_{cav,\infty}} = \frac{b_{cav} v}{R T_{cav,\infty}} (p_{cav,\infty} - p_{cav,o}) \left[1 - \exp\left(-\frac{L}{b_1}\right)\right] \quad (2.120)$$

Or, the acceleration looks disappointing in cases where the vapour pressure in the inflowing air hardly differs from the value in a non-vented cavity. But there is more. As the difference in wind pressure between open head joints at different height may change continuously, cavity venting often turns into the cavity air moving up and down, which makes better drying even more untrue. Stack is more stable, but a partial fill with thickness adapted to the low U-values actually mandated moves the cavity temperature close to the one outside. Happily, the sun warming a wet veneer lifts the vapour release to and the stack flow in the cavity, what anyhow accelerates drying to some extent.

2.3.8.2.3 Surface Condensation and the Vapour Balance in Spaces

Without hygroscopic buffering, the vapour balance in a ventilated space suffering from surface condensation and drying becomes:

$$x_{ve} G_a + G_{vP} + G_{vc/d} = x_{vi} G_a + \frac{d(x_{vi} M_l)}{dt}$$

If surface condensation remains limited to the inner face of one envelope assembly, this balance rewrites as:

$$p_e + \frac{RT_i}{nV}[G_{vP} - \beta_i A(p_i - p_{sat,A})] = p_i + \frac{1}{n}\frac{dp_i}{dt} \quad (2.121)$$

The steady state solution ($G_{vP} = C^t$, $n = C^t$, $p_e = C^t$, $dp_i/dt = 0$) so looks:

$$p_i = \frac{p_e + \frac{RT_i}{nV} G_{vP} + \frac{RT_i}{nV} \beta_i A p_{sat,A}}{\underbrace{\left(1 + \frac{RT_i}{nV}\beta_i A\right)}_{c}} = p_e + \frac{RT_i[G_{vP} + \beta_i A(p_{sat,A} - p_e)]}{nVc}$$

$$= p_{io} - \frac{RT_i}{nVc}[G_{vP}(c-1) + \beta_i A(p_{sat,A} - p_e)] \quad (2.122)$$

For a same vapour release and a same ventilation rate, the deposit lowers the vapour pressure indoors even more if the surface involved is larger and the saturation pressure on it lower than the one convening with the dewpoint indoors. Some therefore believe that single glass could exclude mould growth elsewhere on the inside face of an envelope. However, this is hardly the case as 1 m² of single glass only gives a small drop in vapour pressure indoors. Much lower so requires several m² of single glass, bad in terms of energy efficiency and thermal comfort but also an illusion since surface drying in turn will lift the vapour pressure indoors.

Transient solutions look as those without surface condensation and drying, provided the steady state equation is used to fix the asymptote of the vapour pressure

($p_{i,\infty}$) and the product of ventilation rate n and denominator c replaces n as exponent. By the way, surface condensation and drying go hand in hand with latent heat release and uptake, per m²:

$$q = \beta(p_i - p_{sat,A})l_b \quad (l_b \text{ in J/kg})$$

2.3.9 Evaluating Interstitial Condensation in Practice

2.3.9.1 Boundary Conditions Used

2.3.9.1.1 Diffusion Only

If diffusion only is an acceptable choice, then the calculations are currently done on a monthly mean basis, using as climate conditions outdoors either the monthly mean air temperature and monthly mean vapour pressure over the last 30 years period or a fictitious monthly mean temperature, which integrates the effects of solar gains, long wave losses and the exponential relation between temperature and vapour saturation pressure and is called the equivalent temperature for condensation and drying for the location considered:

$$\theta_{ce}^* = \overline{\theta}_{ce}^* + \widehat{\theta}_{ce}^* \, C(t)$$

In it, $\overline{\theta}_{ce}^*$ is the annual mean and $\widehat{\theta}_{ce}^*$ the annual amplitude, both in °C. $C(t)$ is the month-based time function also used to quantify the vapour pressure in- and outdoors, with as values for Uccle, Belgium, those given in Table 2.6. Table 2.7 in turn gives the annual mean and the amplitude of that fictitious temperature for an outer face with short-wave absorptivity $a_K = 1$ and long wave emissivity $e_L = 0.8$. Correcting the value for short-wave absorptivities lower than 1 is done, using as formulas (the corrected between [ ]):

$$\left[\overline{\theta}_{ce}^*\right] = \alpha_S \left(\overline{\theta}_{ce}^* - \overline{\theta}_e'\right) + \overline{\theta}_e' \quad \left[\widehat{\theta}_{ce}^*\right] = \alpha_S \left(\widehat{\theta}_{ce}^* - \widehat{\theta}_e'\right) + \widehat{\theta}_e'$$

with $\overline{\theta}_e'$ the annual mean for a north-looking outer face with slope 45°, shortwave absorptivity 0 and long wave emissivity 0.9, for Uccle 8.5 °C and $\widehat{\theta}_e'$ the annual amplitude for that outer face, for Uccle 7.1 °C.

To get the corrected values for a long wave emissivity <0.9 of non-shaded, sun-radiated surfaces, a linear interpolation is applied on $\overline{\theta}_e'$ and $\widehat{\theta}_e'$, for Uccle:

$$\overline{\theta}_e' = 8.5 + 1.3 \left(\frac{0.9 - e_L}{0.9}\right) \quad \widehat{\theta}_e' = 7.1 - 0.2 \left(\frac{0.9 - e_L}{0.9}\right)$$

Table 2.6 Uccle, month-based time function.

Month	C(t)	Month	C(t)	Month	C(t)	Month	C(t)
January	−0.98	April	−0.10	July	+1.00	October	−0.10
February	−0.85	May	+0.55	August	+0.85	November	−0.55
March	−0.50	June	+0.90	September	+0.55	December	−0.90

Table 2.7 Equivalent temperature for condensation and drying ($a_K = 1$, $e_L = 0.9$).

	Annual mean $\overline{\theta}^*_{ce}$ (°C)					Annual amplitude $\hat{\theta}^*_{ce}$ (°C)				
					Or. →					
↓Slope	N	NW/NE	W/E	SW/SE	S	N	NW/NE	W/E	SW/SE	S
0	14.4	14.4	14.4	14.4	14.4	12.6	12.6	12.6	12.6	12.6
15	13.7	13.9	14.4	14.7	14.8	12.4	12.5	12.6	12.7	12.7
30	12.9	13.4	14.2	14.9	16.0	11.7	12.0	12.3	12.4	12.4
45	12.2	12.8	13.9	14.7	14.9	10.9	11.3	11.6	12.0	11.8
60	11.9	12.6	13.7	14.6	14.7	9.8	10.6	11.2	11.3	11.1
75	11.8	12.4	13.5	14.3	14.6	9.3	10.0	10.7	10.6	10.3
90	12.1	12.6	13.6	14.2	14.4	9.1	9.5	9.9	9.7	9.5

Table 2.8 Indoor temperature.

Building type	Annual mean (°C)	Annual amplitude (°C)
Dwellings, schools, office buildings	20	3
Hospitals	23	2
Natatoriums	30	2

To get the monthly mean and amplitude of this fictitious temperature for other locations, the methodology given in 'Applied Building Physics', Chapter 1 "Ambient conditions Out- and Indoors", should be applied. The same for the month-based time function. For the air temperature indoors, the values of Table 2.8 are often used, coupled to the time function of Table 2.6.

The excess vapour pressure indoors compared to the monthly mean outdoors is coupled to the indoor climate class the building considered belongs to, see Table 2.9.

2.3.9.1.2 Diffusion and Air Moving Through

If the case, often predictions are limited to evaluating one cold week, for Uccle:

Temperature (°C)	Relative humidity (%)	Radiation, hor. surface (W/m²)	Surface film coefficient, W/(m²·K)	Wind speed (free field, west), m/s
−2.5	95	−30	17	3.8

The wind speed allows to calculate the wind pressure on and so, the difference in air pressure between the in- and the outdoors for the envelope assembly considered,

Table 2.9 Indoor climate classes, pivot values.

Indoor climate class	Pivot $\Delta \bar{p}_{ie}$ (Pa)		Which buildings
	$\bar{\theta}_{e,m} < 0$	$\bar{\theta}_{e,m} \geq 0$	
1–2	270	270–13.5$\bar{\theta}_{e,m}$	Dry storage rooms, sport arenas, etc.
2–3	540	540–27.0$\bar{\theta}_{e,m}$	Offices, schools, shops, large dwellings, apartments
3–4	810	810–40.5$\bar{\theta}_{e,m}$	Small dwellings, hospitals, pubs, restaurants
4–5	1080	1080–54.0$\bar{\theta}_{e,m}$	Natatoriums, breweries, industrial complexes

while the difference in in- and outdoor temperature helps quantifying the difference in stack pressure the assembly experiences.

2.3.9.2 Calculation Sequence

2.3.9.2.1 Diffusion Only

All layers in an assembly are assumed dry. After calculating the thermal resistances, the diffusion resistances and the temperature course for January, the assembly is redrawn in a [Z, p]-axis system and the saturation pressure curve (p_{sat}) traced. For that, the values on the inner face, in the successive interfaces and on the outer face are connected in the correct order with line segments. If the vapour pressure course, a straight line from the values in- ($Z = Z_T$) to outdoors ($Z = 0$) then intersects this trapped vapour saturation course, interstitial condensation is a fact. To know where condensate is depositing, the tangents from the vapour pressures in- and outdoors to the saturation curve are drawn. Does their contact coincide in an interface, condensate deposits there. Otherwise, a tangent scan from the highest to the lowest possible contact points will fix all interfaces or zones with deposit. Per interface or zone, the difference in slope between the in- and outgoing tangent then gives the amounts condensing there in January:

$$m_c = 86\ 400 d_{mo} g_c \ (kg/m^2)$$

In it, g_c is the condensing flux in kg/(m²·s) and d_{mo} the 31 days of January. If only one interface suffers, per successive month the amounts deposited are added and, when drying starts, those evaporating subtracted. If in month x what's left drops below 0, that 0 is kept till the vapour pressure line restarts intersecting saturation. When over a year, the December deposit added to the January deposit remains constant, then condensate does not accumulate. If instead that sum increases over the years, then condensate will accumulate until a limit state is reached. More interfaces showing interstitial condensation complicate counting. Quantifying the monthly totals depositing or drying requires a redistribution per month over all condensing interfaces, according to the difference in slope between the incoming and outgoing tangents. An interface seeing the deposit dropping below zero is turning dry, which requires adapting the tangents.

2.3.9.2.2 Diffusion and Moist Air Moving Through

Although mostly limited to that one week, calculation may also consider a whole year, as with diffusion only. How to do, resembles diffusion only, be it that the exponential curves are more complex to handle and the deposit can increase substantially.

2.3.9.3 Example

A damage case, a single-storey detached house with low-sloped roof, that suffered from severe moisture problems, had to be cured. A look to the architectural drawings and control on-site gave as roof-composing layers in- to out (Figure 2.41):

- Inner plaster against gypsum boards
- Joists with section 75/280 mm, centre to centre 0.6 m apart, with on top sloped laths
- The bays in between filled with 2 × 12.5 cm thick glassfibre blankets
- Air space
- Chipboard as boarding
- Extra layer of 5 cm thick cork board thermal insulation
- Bituminous roofing membrane

Could interstitial condensation be the cause?

First, the properties of the composing layers are collected using measured data and standard lists. The outcome is:

Layer	d (m)	λ (W/(m·K))	R (m²·K/W)	ΣR (m²·K/W)	μ	μd (m)	$\Sigma \mu d$ (m)
R_i			0.10	0.10		0.0	0.0
Inner finish (1)	0.03	0.7	0.04	0.14		3.0	3.0
Glass fibre (2)	0.25	0.04	6.25	6.39	1.2	0.3	3.3
Cavity (3)	0.05		0.17	6.56	0	0.0	3.3
Boarding (4)	0.022	0.12	0.18	6.75	32	0.7	4.0
Cork (5)	0.06	0.04	1.43	8.18	22	1.3	5.3
Bituminous membrane (6)	0.01	0.20	0.05	8.23	10 000	100.0	105.3
R_e			0.04	8.27		0.00	105.3
			$U=$	0.12	W/(m² K)		

Figure 2.41 The house with low-slope roof, the multi-layer roof section.

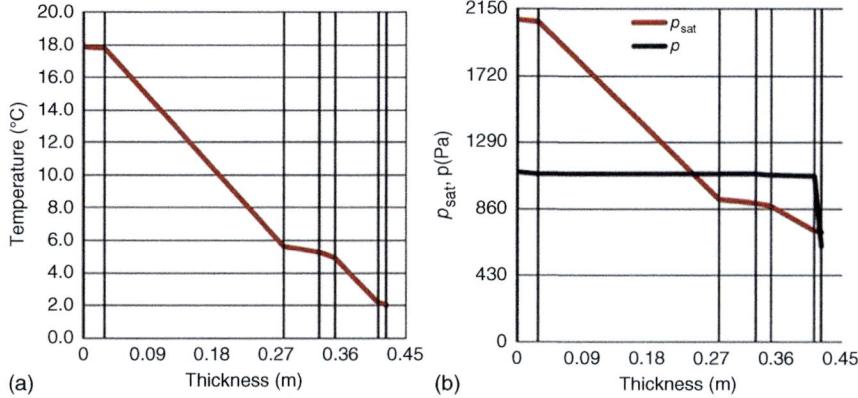

Figure 2.42 (a) January, the mean temperature, (b) the mean saturation and vapour pressure course in the low-slope roof. They intersect, or, interstitial condensation is a fact.

Next, looking to where the house is standing, ≈20 km from Uccle, the equivalent temperature for condensation and drying at Uccle for a surface with slope 0 and solar absorptivity 0.9, applies. The monthly mean temperature courses (°C) through so become (see Figure 2.42a):

	J	F	M	A	M	J	J	A	S	O	N	D
Days	31	28,25	31	30	31	30	31	31	30	31	30	31
Indoors	18.1	18.5	19.5	20.7	22.7	23.7	24.0	23.6	22.7	20.7	19.4	18.3
Inner face	17.9	18.3	19.4	20.6	22.6	23.7	24.0	23.6	22.6	20.6	19.2	18.1
(1)/(2)	17.8	18.2	19.3	20.6	22.6	23.7	24.0	23.6	22.6	20.6	19.1	18.0
(2)/(3)	5.6	6.9	10.4	14.4	20.9	24.4	25.4	23.9	20.9	14.4	9.9	6.4
(3)/(4)	5.3	6.6	10.2	14.3	20.9	24.5	25.5	23.9	20.9	14.3	9.7	6.1
(4)/(5)	5.0	6.3	9.9	14.1	20.8	24.5	25.5	24.0	20.8	14.1	9.4	5.8
(5)/(6)	2.2	3.7	7.9	12.7	20.5	24.6	25.8	24.0	20.5	12.7	7.3	3.1
Outer face	2.1	3.6	7.8	12.6	20.4	24.7	25.9	24.1	20.4	12.6	7.2	3.0
Outdoors	2.0	3.6	7.8	12.6	20.4	24.7	25.9	24.1	20.4	12.6	7.2	3.0

Then, related vapour saturation pressure courses are fixed. For that, the tables or the formulas listed when discussing vapour in the air are used (Pa):

Days	J 31	F 28,25	M 31	A 30	M 31	J 30	J 31	A 31	S 30	O 31	N 30	D 31
Indoors	2106	2159	2307	2487	2805	2991	3046	2964	2805	2487	2285	2139
Inner face	2080	2135	2287	2472	2801	2993	3050	2965	2801	2472	2264	2114
(1)/(2)	2069	2124	2278	2465	2799	2994	3052	2965	2799	2465	2255	2103
(2)/(3)	918	1005	1278	1668	2524	3128	3322	3034	2524	1668	1235	971
(3)/(4)	897	984	1257	1650	2517	3131	3329	3036	2517	1650	1215	949
(4)/(5)	874	961	1235	1630	2509	3135	3338	3038	2509	1630	1192	927
(5)/(6)	716	801	1075	1486	2450	3167	3402	3054	2450	1486	1032	768
Outer face	711	796	1070	1482	2448	3168	3405	3055	2448	1482	1026	763
Outdoors	707	792	1066	1478	2446	3169	3407	3055	2446	1478	1022	759

After, the monthly mean vapour pressure courses, assuming no interstitial condensate, are quantified. To get these, the indoor climate cmass, the house belongs to, has to be chosen. Considering the internal volume, the number of inhabitants, their daily activities and the ventilation possibilities available, an indoor climate class 2 situation looks a fair guess. This gives as monthly mean vapour pressure values indoors and monthly mean vapour pressure courses (Pa):

Days	J 31	F 28,25	M 31	A 30	M 31	J 30	J 31	A 31	S 30	O 31	N 30	D 31
Indoors	1097	1126	1203	1291	1434	1511	1533	1500	1434	1291	1192	1115
Inner face	1097	1126	1203	1291	1434	1511	1533	1500	1434	1291	1192	1115
(1)/(2)	1084	1113	1192	1283	1430	1509	1531	1497	1430	1283	1181	1102
(2)/(3)	1082	1112	1191	1282	1429	1508	1531	1497	1429	1282	1180	1101
(3)/(4)	1082	1112	1191	1282	1429	1508	1531	1497	1429	1282	1180	1101
(4)/(5)	1079	1109	1189	1280	1428	1508	1531	1496	1428	1280	1177	1097
(5)/(6)	1073	1103	1184	1276	1426	1507	1530	1495	1426	1276	1172	1092
Outer face	621	677	827	999	1279	1429	1472	1408	1279	999	806	655
Outdoors	621	677	827	999	1279	1429	1472	1408	1279	999	806	655

The result is clear: during January, the vapour saturation and vapour pressure course intersect. The interface, where the difference between the two looks largest, is (5)/(6), see Figure 2.42b.

The vapour pressure there now is set equal to the vapour saturation pressure there and the adapted vapour pressure course, given by the tangents to that interface is recalculated, giving as a result (Pa):

	J	F	M	A	M	J	J	A	S	O	N	D
Days→31	31	28,25	31	30	31	30	31	31	30	31	30	31
Indoors	1097	1126	1203	1291	1434	1511	1533	1500	1434	1291	1192	1115
Inner face	1097	1126	1203	1291	1435	1511	1533	1500	1434	1291	1192	1115
(1)/(2)	883	943	1131	1401	2007	1509	1531	1497	1430	1283	1102	919
(2)/(3)	861	925	1124	1412	2064	1508	1531	1497	1429	1282	1093	900
(3)/(4)	861	925	1124	1412	2064	1508	1531	1497	1429	1282	1093	900
(4)/(5)	811	882	1107	1438	2199	1508	1531	1496	1428	1280	1071	854
(5)/(6)	716	801	1075	1486	2450	1507	1530	1495	1426	1276	1032	768
Outer face	621	677	827	999	1279	1429	1472	1408	1279	999	806	655
Outdoors	621	677	827	999	1279	1429	1472	1408	1279	999	806	655

This shows that interface (5)/(6) is the only point of tangency, although the interface (3)/(4) between the air space left and the boarding is close, see Figure 2.43.

In January, February, March, November and December, the vapour saturation pressure in the interface (5)/(6) remains lower than the vapour pressure there without deposit. In April, May, June, July, August, September and October, it lays above, so no deposit anymore. Instead drying proceeds till all condensate is evaporated.

Figure 2.43 January, [Z, p] axis system, tangents traced, deposit at the back of the roofing.

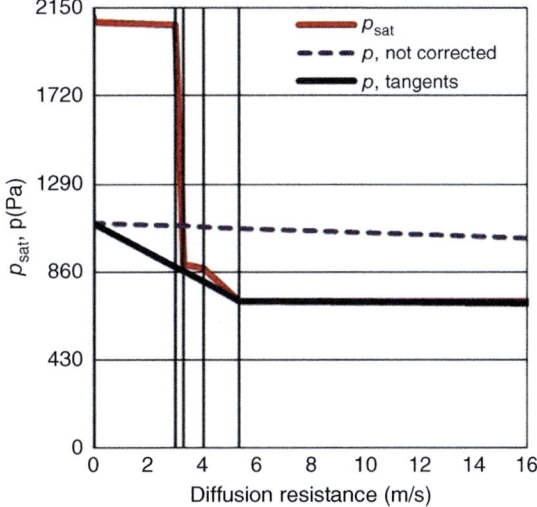

Control of if accumulation could be expected gives:

	J	F	M	A	M	J	J	A	S	O	N	D
Days	31	28.25	31	30	31	30	31	31	30	31	30	31

Monthly amount of vapour diffusing from inside to interface (5)/(6) (kg:m²)
 0.035 0.027 0.012 −0.018 −0.094 0.000 0.000 0.000 0.001 0.001 0.014 0.032

Monthly amount of vapour diffusing from interface (5)/(6) to outside (kg:m²)
 0.000 0.001 0.001 0.002 0.006 0.000 0.000 0.000 0.001 0.001 0.001 0.001

Monthly deposit in interface (5)/(6) (kg:m²)
 0.035 0.027 0.011 −0.020 −0.100 0.000 0.000 0.000 0.000 0.000 0.013 0.032

Accumulates in interface (5)/(6) (kg:m²)
First year 0.035 0.062 0.073 0.053 0.000 0.000 0.000 0.000 0.000 0.000 0.013 0.045

Second year, 0.045 kg/m² end of December, so, end of all other months:
 0.080 0.107 0.118 0.098 0.000 0.000 0.000 0.000 0.000 0.000 0.013 0.045

End of March sees the highest deposit at the back of the bituminous membrane. This annual course looks as shown in Figure 2.44.

The quantities involved are minimal, a peak of only 118 g/m² and no annual accumulation! Or, interstitial condensation does not explain the serious moisture problems encountered. Could air outflow play and give deposits, large enough to explain them? The answer is no. A bituminous roofing membrane is not only vapour- but also airtight. With other words, a control on combined diffusion and air exfiltration adds nothing. Of course, it should be if roof vents were perforating the membrane, if the boarding and cork were laid with open joints and if electric lamp sockets were

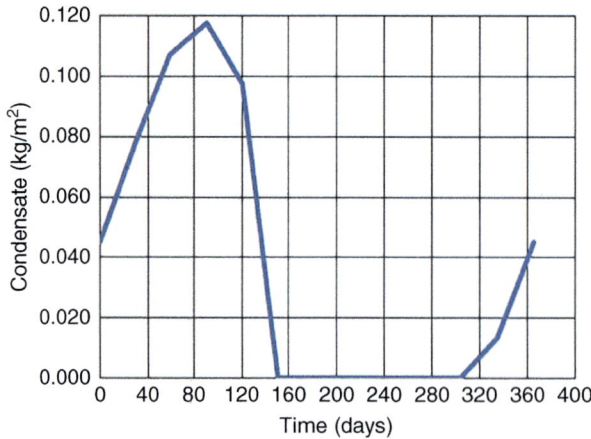

Figure 2.44 Interstitial condensation, deposit at the back of the bituminous membrane.

2.4 Moisture

perforating the ceiling. Was the real cause of the moisture problems: accidental rain leakage at the eaves.

2.4 Moisture

2.4.1 In General

As long as the RH in capillary-porous materials remains below the threshold ϕ_M, often set 95%, equivalent diffusion and convection go on acting as only 'moisture movers'. However, construction moisture, rising damp, rain absorption and interstitial condensation, all give moisture contents convening with 100% RH, whereby liquid moisture flows get capillary suction, gravity, external pressure heads and internal pressure differences as driving forces. Air and vapour flow were analysed in an phenomenological way, using macroscopic properties such as vapour resistance factor, air permeability, etc., which keep the same value independent of how large the material volume is. Contrary to that, although a pore filled with moist air, water and dissolved substances is a too simple surrogate, analysing liquid moisture flow starts there. After, the outcome is generalized to materials as a whole.

2.4.2 Water Flow in a Pore

2.4.2.1 Capillarity

Capillary action in a pore filled with water and air follows from the cohesion between the water molecules, their impairment by contact with air and their's and the air molecule's adhesion to the pore wall. Each water molecule experiences attraction from the other, most from the adjacents and ever less from those more than 1 nm (nm = 10^{-9} m) away. Water molecules so seem embedded in a cohesive sphere with radius 1 nm (Figure 2.45).

Spheres far enough below the contact plane between water and air do not experience any resulting attraction but, due to the weaker attraction to air, those nearing, more, intersecting the contact plane do. This creates a force sticking these spheres against those farther in the water. Since the potential energy of the surface layer,

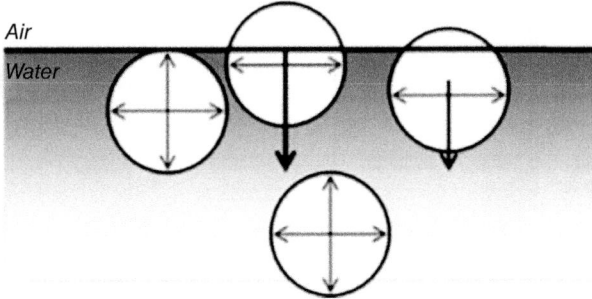

Figure 2.45 Surface tension.

called the meniscus, so formed passes that in the water, separating requires work. As 1 J/m² work equals 1 N/m, the term surface tension (σ_w) for it is commonly used. For water contacting air, its value decreases with temperature:

$$\sigma_w = (75.9 - 0.17\theta)10^{-3}$$

When a meniscus now touches a pore wall, then, provided adhesion wins from the attraction to air, it creeps up the wall under a 0°–90° contact angle ϑ. Since in nearly all pores in stony and wood-based materials, this is the case, they behave hydrophilic. If instead adhesion loses from the attraction to air, the meniscus is pushed away from the pore wall under a contact angle 90°–180°, the case in most synthetics and against water-repellent surfaces, which so behave hydrophobic (Figure 2.46).

If a sufficiently thin circular pore of a hydrophilic material so contacts a water surface, the concave meniscus formed will draw water into the pore, an action called capillary suction, symbol p_c or s, units Pa (Figure 2.46):

$$p_c = s = -4\sigma_w \cos\vartheta/d \tag{2.123}$$

with d the pore diameter. In a crack with width b, suction becomes (Figure 2.46)

$$p_c = s = -2\sigma_w \cos\vartheta/b \tag{2.124}$$

Suction so changes proportionally to the surface tension between water and air and inversely proportional to the pore radius or the crack width, while the material intervenes via the contact angle ϑ. Suction is a strong force. In a 1 μm wide, ideal hydrophilic circular pore ($\vartheta = 0$), it touches 300 kPa, which equals the pressure exerted by a 30 m high water column. To compare with, the pressure a 175 km/h strong wind gives hardly passes 1 kPa at the windward side, what anyhow may push water through a 0.15 mm wide crack. Is that a crack in a hydrophobic layer, then strong winds so can short-circuit its water repellency.

Figure 2.47 shows how suction looks in an ideal hydrophilic pore contacting water and wetted over L m: linear from 0 Pa at the water surface up to $p_{c,max} = -4\sigma_w/d$ Pa at L beyond that surface, with in between a value p_c equal to $p_{c,max}x/L$ Pa, x being the distance to the water surface. Because air exerts the same pressure on the meniscus at L as on the water surface, that result reflects the difference between the water

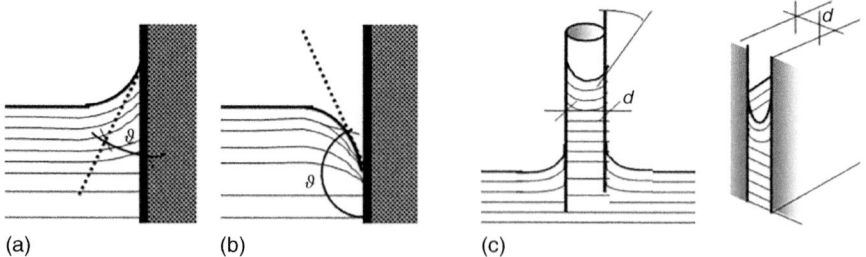

Figure 2.46 In (a) water creeps up a pore wall under a contact angle 0°–90°, while in (b) water it is pushed away under a contact angle 90°–180° and (c) shows capillary suction in a cylindric pore and in a crack.

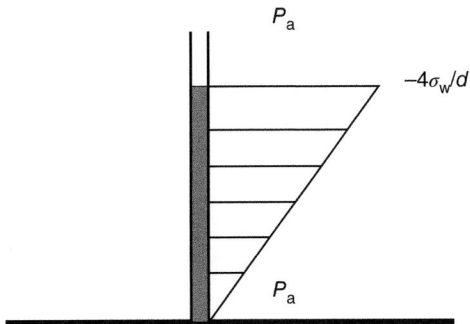

Figure 2.47 Suction in an ideally hydrophilic pore ($\cos \vartheta = 1$).

pressure just below and the air pressure just above the meniscus:

$$p_c = s = P_{w,L} - P_a$$

Capillary repellency in turn is also inversely proportional to the pore diameter or crack width and equal to the difference in the water pressure just below and the air pressure just above the meniscus.

2.4.2.2 Poiseuille's Law

The pressure differences, suction creates in capillary active pores, displace the water these contain. Whether the flow will be laminar, transitional or turbulent, depends on the Reynolds number (Re). Water at 10 °C has a kinematic viscosity of $1.25 \cdot 10^{-6}$ m/s², which gives Re = 800 000$v d_H$ with d_H the hydraulic diameter, for a circular pore the diameter, for a crack twice the width, and v the velocity. Flows now are laminar for Re $\leq$ 2000, which convenes with a velocity $\leq 0.0025/d_H$. Or, in a 1 µm wide circular pore, the flow will only turn turbulent above a supersonic velocity of 2500 m/s! Clearly, water movement in pores by capillary suction remains laminar under all circumstances, giving as flux:

$$g_w = -\frac{\rho_w d_H^2}{32\eta_w}\mathbf{grad}\, P_w \tag{2.125}$$

This equation is called Poiseuille's law. In it, η_w is the dynamic viscosity of water:

Temperature	η_w (N·s/m²)
0	0.001820
20	0.001025
100	0.000288

For a cylindric pore, the proof of this goes as follows. For laminar flow without surface slip, the velocity against the pore wall is 0. Per concentric water cylinder with length dy, the equilibrium between the pressure differential and the viscous friction along the perimeter gives (Figure 2.48):

$$\pi\, r^2 dP_w = 2\pi\, r\eta_w dy \frac{dv}{dr} \rightarrow dv = \frac{1}{2\eta_w}\frac{dP_w}{d\vec{y}}(r\, dr)$$

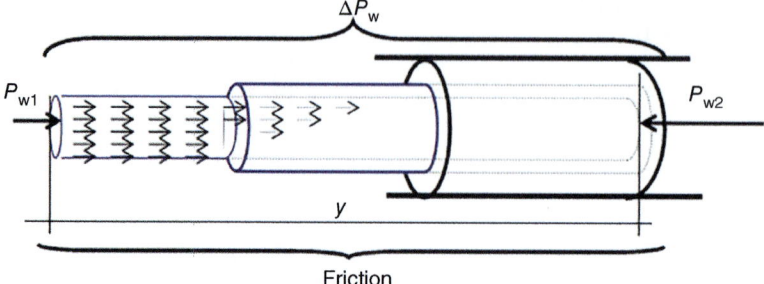

Figure 2.48 Equilibrium between pressure differential and viscous friction.

Integration results in:

$$\int_v^0 dv = \frac{1}{2\eta_w} \frac{dP_w}{dy} \int_r^R r\, dr \rightarrow v(r) = -\frac{1}{4\eta_w}(R^2 - r^2)\frac{dP_w}{dy}$$

The water flow through a cylindric pore so becomes:

$$G_w = \rho_w \int_0^R v(r) 2\pi r\, dr = -\rho_w \frac{\pi R^4}{8\eta_w} \frac{dP_w}{dy}$$

The outcome thus is a flux, on average equal to:

$$g_w = -\frac{4G_w}{\pi d^2} = -\rho_w \frac{d^2}{32\eta_w} \frac{dP_w}{dy}$$

Since for a cylindric pore, the diameter d is also the hydraulic one (d_H), Poiseuilles's law is proven. The term $32\eta_w/d_H^2$ is called the specific flow resistance of the pore, symbol W_m, units kg/(m³·s). Air flowing through has a density ρ_a, ≈ 1.2 kg/m³ versus $\rho_w = 1000$ kg/m³ for water, and a dynamic viscosity $\eta_a \approx 1.74 \cdot 10^{-5}$ versus $\eta_w \approx 1.025 \cdot 10^{-3}$ N·s/m² for water, giving a specific flow resistance 70 times lower than for water.

Making the specific flow resistance fluid-independent demands rewriting the law as:

$$g_w = -\frac{\rho}{\eta} \frac{d_H^2}{32} \frac{dP_w}{dy} = -\frac{\rho}{\eta} \left[\frac{1}{W'} \frac{dP_w}{dy}\right] \tag{2.126}$$

A reshuffle with v_m the average velocity in the pore (= dy/dt) gives as pressure differential for water:

$$dP_w = -\frac{32\eta_w dy}{d^2}\left(\frac{g_w}{\rho_w}\right) = -\frac{8\eta_w dy}{r^2} v_m$$

2.4.2.3 Isothermal Water Flow in a Pore Contacting Water
2.4.2.3.1 Balancing the Forces Interacting

Taken is a cylindric pore with radius r, length L and slope α to the vertical. One side contacts water, the other air. Applying Newton's law to the capillary sucked water gives (Figure 2.49):

$$\sum F_i = (\rho_w \pi r^2 l)\frac{d^2 l}{dt^2} \tag{2.127}$$

Figure 2.49 Suction by a pore from a water surface.

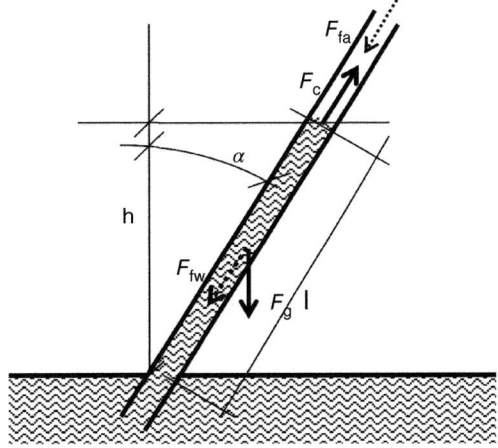

with $\rho_w \pi r^2$ the mass of the water column, l the ordinate where the meniscus is in the pore and d^2l/dt^2 the acceleration of its moving.

The most important forces (F_i) intervening are:

Gravity $\qquad F_g = -\rho g \pi r^2 l \cos \alpha$

Capillary suction $\qquad F_c = \pi r^2 \dfrac{2\sigma_w \cos \vartheta}{r} = 2\pi r \sigma_w \cos \vartheta$

Water friction $\qquad F_{fw} = -\dfrac{8\eta_w l}{r^2}\left(\dfrac{dl}{dt}\right)\pi r^2 = -8\pi \eta_w l \left(\dfrac{dl}{dt}\right)$

Air friction above the meniscus $F_{fa} \approx -8\pi \eta_a (L-l)\left(\dfrac{dl}{dt}\right)$

In the last equation, $\approx$ indicates that Poiseuille's law is an approximation for a compressible gas as air. In fact, at the start, a shortly lasting peak pressure in the air will affect the water flow for a while. Introducing the four forces in Newton's law gives:

$$\dfrac{d^2l}{dt^2} + \dfrac{8\eta_w}{\rho_w r^2}\left(1 + \dfrac{\eta_a(L-l)}{\eta_w l}\right)\dfrac{dl}{dt} - \left(\dfrac{2\sigma_w \cos \vartheta}{\rho_w r l} - g \cos \alpha\right) = 0 \qquad (2.128)$$

This second-order differential equation with variable coefficients cannot be solved analytically. However, due to friction ($\div 1/r^2!$), the meniscus moves so slowly that the acceleration barely plays. The second derivative so can be omitted, reducing the kinetics to:

$$v_m = \dfrac{dl}{dt} = \dfrac{\varrho_w r^2}{8}\left[\underbrace{\dfrac{1}{\eta_w + \eta_a \dfrac{(L-1)}{l}}}_{\eta_r}\right]\left(\dfrac{2\sigma_w \cos \vartheta}{\varrho_w r l} - g \cos \alpha\right) \qquad (2.129)$$

The term η_r now acts as a variable viscosity, infinite when suction starts ($l = 0$) and approaching the viscosity of water once it's moving. The equation underlines that every further movement stops once a maximum suction height h_{max} is reached. Indeed, the velocity turns 0 for (Figure 2.50):

$$l_{max} = 2\sigma_w \cos\vartheta/(\rho_w g \, r \cos\alpha) \quad \text{giving} \quad h_{max} = l_{max} \cos\alpha$$

The general solution now links the time to the wet length, as Figure 2.50 shows for a vertical pore:

$$t = \frac{8(\eta_w - \eta_a)}{\rho_w g \, r^2 \cos\alpha}\left[-1 - \left(\frac{(\eta_w - \eta_a)l_{max} + \eta_a L}{\eta_w - \eta_a}\right)\ln\left(1 - \frac{l}{l_{max}}\right)\right] \quad (2.130)$$

2.4.2.3.2 Horizontal Pore as a Substitute for Rain Absorption

An outside finish absorbing rain is a common phenomenon (Figure 2.51).

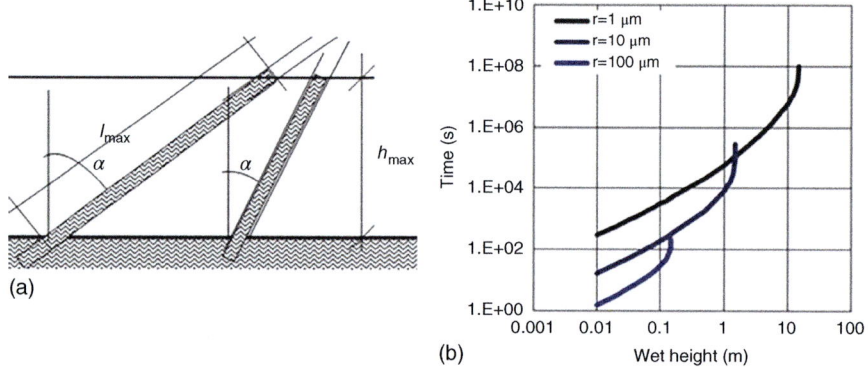

Figure 2.50 (a) The maximum suction height and (b) the wet height versus time for a 15 m long vertical pore, $r = 1\,\mu m$, $h_{max} = 14.8\,m$ / $r = 10\,\mu m$, $h_{max} = 1.48\,m$ / $r = 100\,\mu m$, $h_{max} = 0.148\,m$.

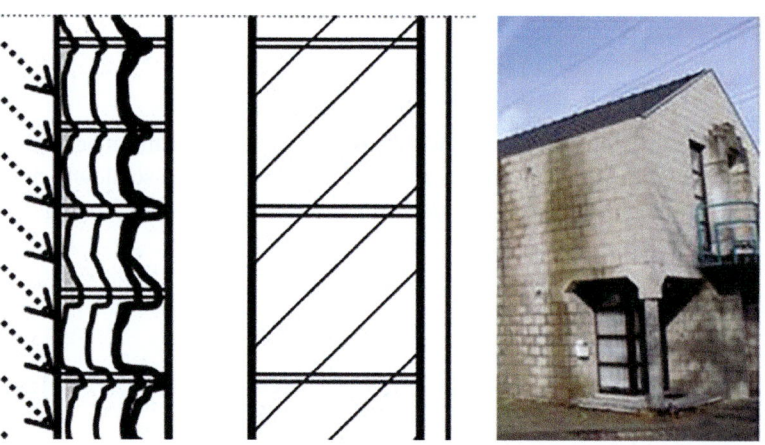

Figure 2.51 Rain absorption.

The horizontal pore, acting as substitute, now has a length L and a radius r. When sucking water, the kinetics, with x the position of the meniscus, look:

$$\frac{dl}{dt} = \frac{dx}{dt} = \frac{r\sigma_w \cos\vartheta}{4[(\eta_w - \eta_a)x + \eta_a L]}$$

Integration gives:

$$2(\eta_w - \eta_a)x^2 + 4\eta_a L x - r\sigma_w \cos\vartheta \, t = 0$$

The positive root of that quadratic equation is:

$$x = -\frac{\eta_a L}{\eta_w - \eta_a} + \sqrt{\left(\frac{\eta_a L}{\eta_w - \eta_a}\right)^2 + \frac{r\sigma_w \cos\vartheta}{2(\eta_w - \eta_a)}t}$$

The pore turning wet over a length x so not only depends on time but also on the pore radius and length. The narrower and longer it is, the slower it sucks. Otherwise said, the speed at which capillary outer finishes pick up rain decreases when thicker and their pores thinner. Neglecting now the friction in the air column in front of the sucked water has hardly any impact in wide pores but induces some early deviation in finer pores. When neglected, the root simplifies to:

$$x = \underbrace{\sqrt{\frac{r\sigma_w \cos\vartheta}{2\eta_w}}}\sqrt{t} \qquad (2.131)$$

The term above the brace gets the name 'capillary water penetration coefficient B in m/s$^{0.5}$'. Multiplying both sides with the density of water gives what the pore absorbs in kg/m^2:

$$m_w = \rho_w B \sqrt{t} = A\sqrt{t}$$

The quantity A in turn is called the capillary water absorption coefficient in kg/(m$^2\cdot$s$^{1/2}$). The lower A, the slower a horizontal pore sucks water and the longer it takes before the water will reach the other end. Both the correct and simplified $\sqrt{t}$-relations underline that, whatever the pore length is, the meniscus straightens once there, so forces suction to stop. This makes water outflow by capillarity impossible. Rain absorbed by porous materials can so give wet stains on the inner face but no water outflow. If the last anyhow happens, then the culprits must be sought in extra pressures on the outer face and gravity intervening see Figure 2.52.

The horizontal pore model with constant section is easily expanded to two horizontal pores in series, having different sections and contact angles. Plaster on a wall is an example. The plaster is modelled as having identical pores with radius r_1, a contact angle ϑ_1 and a length l_1, the wall as having only identical pores with radius r_2, a contact angle ϑ_2 and a length l_2. Each plaster pore contacts a wall pore in a way their centerlines are in line with each other. Neglecting air friction, the horizontal flow equation applies to the plaster pore with the uptake slower, the lower its capillary water absorption coefficient (A_1), thus, the finer the pore and/or the larger the contact angle. Once the meniscus touches the wall pores, then, if thinner than the plaster pores, these will start sucking, turning the plaster pores into a hydraulic

Figure 2.52 Seeping rain.

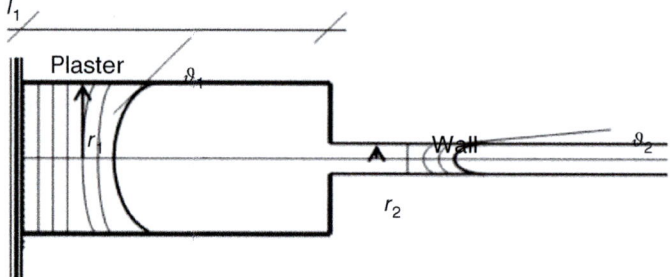

Figure 2.53 Serial two pores model in case of water contact.

resistance (see Figure 2.53). If instead, the wall pores are wider, then, in the contact, the meniscus in the plaster pores will straighten till the wall pores can start sucking, again turning the plaster pores into a, now higher hydraulic resistance. This gives as total friction:

$$F_{\text{fw}} = \left(\frac{8\pi\eta_{\text{w}} l_1 r_2^2}{r_1^2} + 8\pi\eta_{\text{w}} x \right) \frac{dx}{dt} \tag{2.132}$$

with x the wet length in the wall pores.

Solving the force equilibrium, see above, with the friction F_{fw} equal to the capillary suction F_c, gives as quadratic equation in x and as positive root:

$$x^2 + \left(\frac{2l_1 r_2^2}{r_1^2} \right) x - \left(\frac{r_2 \sigma_{\text{w}} \cos \vartheta}{2\eta_{\text{w}}} \right) t = 0 \rightarrow \quad x = \underbrace{\frac{l_1 r_2^2}{r_1^2}}_{} + \sqrt{\left(\frac{l_1 r_2^2}{r_1^2} \right)^2 + B_2^2 t}$$

Clearly, water sorption by the wall pores (x) is slowing down when the plater pores are longer and finer, see the ratio above the brace. Or, a coarse hydrophilic plaster will hardly increase the rain resistance of a wall made of fine porous blocks. Only when water repellent, it will. Included air friction, the solution of the quadratic equation giving the wet length in the plaster pore changes into:

$$X = \underbrace{\frac{\eta_a d_1}{\eta_w - \eta_a}}_{(1)} + \underbrace{\frac{\eta_a d_2 r_1}{(\eta_w - \eta_a) r_2^2}}_{(2)} \sqrt{\left(\frac{\eta_a d_1}{\eta_w - \eta_a} + \frac{\eta_a d_2 r_1}{(\eta_w - \eta_a) r_2^2}\right)^2 + \frac{r \sigma_w \cos \theta}{2(\eta_w - \eta_a)} t}$$

For a wall with radius r_2 smaller than the plaster radius r_1 and the thickness d_2 larger than the thickness d_1 of the plaster, term (2) overwhelms term (1), meaning that due to air friction in the wall pores, the meniscus in the wider plaster pores will move much slower. Or, a coarse plaster on a fine porous wall material will absorb water slower than if no wall.

Of course, the pore model used is one-dimensional, whereas pore systems in materials act as three-dimensional networks. Surface wetting also never occurs uniformly. Even the air outflow is a three-dimensional reality.

2.4.2.3.3 Vertical Pore as a Substitute for Rising Damp

Rising damp is a problem in older walls lacking a watertight insert just above grade (Figure 2.54). The vertical pores acting as its substitute have a cosine of the slope α equal to 1, converting the suction velocity for a radius r into:

$$v_m = \frac{dz}{dt} = \frac{\rho_w r^2}{8} \frac{1}{\eta_w + \eta_a (L - z)/z} \left(\frac{2\sigma_w \cos \vartheta}{\rho_w r z} - g\right)$$

with L the height. That velocity turns 0 at the maximum suction height:

$$2\sigma_w \cos \vartheta / (\rho_w r z) = g \text{ or } z = h_{max} = \frac{2\sigma_w \cos \vartheta}{\rho_w g r} \qquad (2.133)$$

Smaller pores and lower contact angles lift that maximum. Or, the finer the pores and the more hydrophilic the wall material, the higher the damp will rise. Instead, for wall materials having very wide pores with contact angle nearing 90° the suction height nears zero.

Preventing rising damp in walls so requires either a watertight layer inserted just above grade, a layer built-in with really coarse pores or a water-repellent substance injected that lifts the contact angle above 90°. In the two last cases, the

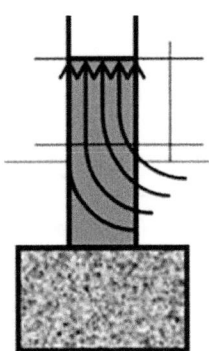

Figure 2.54 Rising damp.

suction left should lay below $\rho_w gh$, with h the height of the coarse layer or the water-repellent zone.

2.4.2.4 Isothermal Water Flow in a Pore After Water Contact

Known is that after a water contact, a water isle with concave meniscuses at both sides remains in a pore. If horizontal and having a constant section (Figure 2.55), both meniscuses will pull equally, excluding the isle to move. If inclined and having a constant section, gravity will make the lower meniscus less, the upper more concave, so inducing a flow if the water isle is heavy enough to straighten the lower, a fact in case the isle has a length equal to the maximum rise:

$$l_{max} = h_{max} / \cos \alpha$$

Or, the wider a pore, the easier gravity and external pressures displace the suked water isle. If having a variable section (Figure 2.55) the meniscus with smallest diameter d_1 (the other is d_2) will pul the most, moving the water isle in the direction of largest suction:

$$4\sigma_w \cos \vartheta / d_1 > 4\sigma_w \cos \vartheta / d_2$$

The flux will increase when the isle is shorter and the difference in suction larger, thus, when the suction gradient enlarges:

$$\mathbf{g_w} = -\left(\frac{\rho_w d^2}{32\eta_w}\right) \operatorname{grad} p_c = -k_w \operatorname{grad} s \quad (2.134)$$

The term k_w in the equation stands for the water permeability of the pore, units s.

The flux in the pore will last until elsewhere in it both meniscuses turn equal again, making the suction gradient zero. A water isle partly in a narrow, partly in a wide pore reacts the same way, with the narrow draining the wide. In case the contact angle is variable but the section is constant, the water isle will move direction the smallest angle as this is most sucking. When also the section varies, the isle will move direction highest ratio between the cosine of the contact angle and the pore diameter ($\cos\vartheta/d$), i.e. the largest suction.

2.4.2.5 Non-isothermal Water Transfer in a Pore After Water Contact

Back to the constant contact angle and section pore. Since the surface tension of water and, with it, the suction increases when colder, a temperature difference will push the water isle in the direction of lower temperature. With s_0 the surface tension

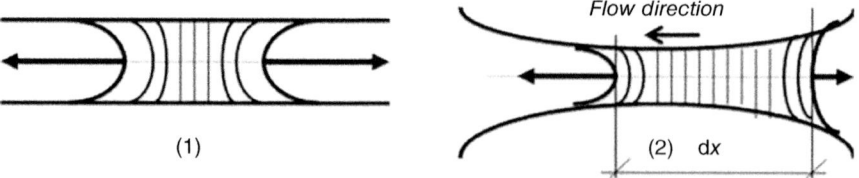

Figure 2.55 Isothermal water transport in a pore after water contact is interrupted, constant contact angle, (1) constant, (2) variable section.

at reference temperature, take 20 °C, in general suction at any temperature can be written as:

$$s = s_o \sigma_w / \sigma_{wo} = s_o \frac{75.9 - 0.17\,\theta}{72.5} \quad (2.135)$$

The gradient so becomes:

$$\mathbf{grad}\, s = \frac{75.9 - 0.17\,\theta}{72.5}\mathbf{grad}\, s_o + 0.17 \cdot 10^{-3} \frac{s_o}{72.5 \cdot 10^{-3}}\mathbf{grad}\, \theta \quad (2.136)$$

As in a pore with constant section and constant contact angle, $\mathbf{grad}\, s_o = 0$ then the flux turns into:

$$\mathbf{g_w} = -\left[0.17\frac{s_o}{72.5}\left(\frac{\rho_w d^2}{32\eta_w}\right)\right]\mathbf{grad}\, \theta = -K_{\theta,w}\mathbf{grad}\, \theta \quad (2.137)$$

In it, $K_{\theta,w}$ is called the thermal water permeability of a pore, units kg/(m·s·K). In pores with variable section and variable contact angle, both suction and temperature induce moving, giving as flux:

$$\mathbf{g_w} = -k_w\,\mathbf{grad}\, s - K_{\theta,w}\mathbf{grad}\,\theta \quad (2.138)$$

2.4.2.6 Remark

Besides temperature and suction, two other forces may move water in pores: gravity and external pressures. In wider pores, both play. There, the driving force in isothermal conditions becomes:

$$\mathbf{grad}\, s + \mathbf{grad}(\rho_w g\, z) + \mathbf{grad}\, P.$$

To remind, suction reflects the difference in water and air pressure over a meniscus. When a water isle starts moving, it expels air, causing a rise in air pressure in front of the water moving and a drop in the air pressure behind. This has an effect on how water flow in a pore starts.

2.4.3 Vapour Flow in a Pore Containing Water Isles with Air Inclusions in Between

2.4.3.1 A Short Description

When in a pore a series of water isles move, then the water they contain is flowing. At the same time, from isle to isle, vapour is exchanged through the air inclusions in between. To get such situation, the contact with water should be repeatedly broken and resumed. The vapour transfer in between has two driving forces: gradients in pore diameter and gradients in temperature.

2.4.3.2 Isothermal

When a pore's contact angle and section remain constant, the vapour saturation pressure in it also does (see Thompsons law):

$$p'_{sat} = p_{sat}(\theta)\exp[s/(\rho_w RT)] = C^t$$

This excludes vapour transfer between the water isles (Figure 2.56), be it that the two isles at to the pore's ends may exchange vapour with both ambients each time

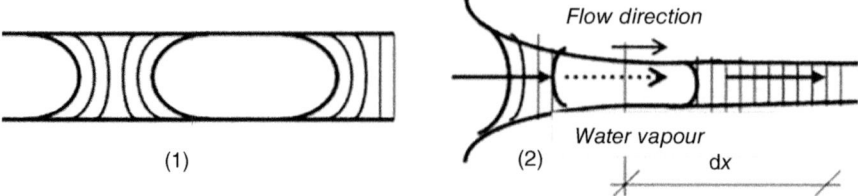

Figure 2.56 Vapour transport in the air inclusions between water isles, isothermal conditions, constant contact angle, (1) constant, (2) variable section.

the RH on their meniscuses at the ambient side differs from the ratio between the vapour saturation pressure in the pore (p'_{satt}) and in the ambient (p_{sat}).

In pores with variable section but constant contact angle, suction differs between the meniscuses of the successive water isles (Figure 2.51), resulting in a difference in saturation pressure at their meniscuses with as a result vapour fluxes in the air inclusions in between:

$$\mathbf{g_v} = -\delta_a \mathbf{grad}\, p'_{sat} = -\delta_a \frac{\partial p'_{sat}}{\partial s}\mathbf{grad}\, s = -\delta_a \frac{\rho_v}{\rho_w}\mathbf{grad}\, s \qquad (2.139)$$

This last equality is based on Thompson's law, stating that for a stable meniscus having a constant contact angle, vapour pressure and suction are interchangeable. In principle, vapour transfer disturbs the water isle's stability, but for small fluxes, this effect is too small to consider. If the contact angle, not the section varies, or, if vice versa, differences in suction between meniscuses exist, then the vapour flux induced in the air inclusions will obey the equation given.

2.4.3.3 Non-isothermal

With the contact angle and section constant, differences in temperature give different vapour saturation pressures (p'_{sat}) between the meniscuses delimiting the air inclusions, giving as vapour flux:

$$\mathbf{g_v} = -\delta_a\,\mathbf{grad}\, p_{sat} = -\delta_a \phi \left[\frac{dp_{sat}}{d\theta} + \underbrace{\frac{p_{sat}}{\rho_v RT}\left(\frac{\partial s}{\partial \theta} - \frac{s}{T}\right)} \right]\mathbf{grad}\, \theta$$

Since the term above the brace has hardly any impact except in really small pores, the equation can be simplified to:

$$\mathbf{g_v} = -\delta_a \phi \frac{dp_{sat}}{d\theta}\mathbf{grad}\, \theta \qquad (2.140)$$

When also the contact angle and the section vary, suction and temperature then act together, turning the equation for the vapour flux between the ambient and the first meniscus, the vapour flux in the air inclusions and the vapour flux between the last meniscus and the ambient into:

$$\mathbf{g_v} = -\delta_a \left[\frac{\rho_v}{\rho_w}\mathbf{grad}\, s + \phi \frac{dp_{sat}}{d\theta}\mathbf{grad}\, \theta\right] = -\delta_a \left(p_{sat}\,\mathbf{grad}\,\phi + \phi\frac{dp_{sat}}{d\theta}\mathbf{grad}\,\theta\right)$$

$$(2.141)$$

Otherwise said, the moisture flow consists of a serial transfer of liquid and vapour. As a result, the water isles with colder and smaller meniscuses grow, and those with warmer and larger meniscuses contract. Flowing will stop once all isles are in balance again.

2.4.4 Moisture Flow in and Through Materials and Assemblies

2.4.4.1 Transport Equations

How moisture flow in capillary porous materials develops, differs between the humidity intervals the moisture content belongs to: from dry to hygroscopic at RH ϕ_M, from hygroscopic to capillary and from capillary to saturation. Materials in turn are modelled as consisting of a fixed matrix traversed by a labyrinth of pores that are neither circular nor straight, change section, deviate, come together, split, etc. Each representative material volume contains that complete labyrinth, in which the same driving forces as in one pore cause moisture flow. However, differentiating between vapour and liquid and if in series or in parallel is no longer doable. Any plane crossing a representative volume in fact cuts water isles in one and air inclusions in other pores. Or, suction and temperature gradients apparently force water and vapour to flow in parallel. Replacing the water permeability and thermal permeability in the water flux equation analysed for one pore by analogous properties at material level and exchanging the vapour permeability of air for the vapour permeability of the material in the vapour flux equation gives as moisture transfer equation:

$$\begin{aligned} \mathbf{g_w} &= -k_w\,\mathbf{grad}\,s - K_{\theta,w}\mathbf{grad}\,\theta \\ +\mathbf{g_v} &= -\delta\frac{\rho_v}{\rho_w}\mathbf{grad}\,s - \delta\phi\frac{dp_{sat}}{dT}\mathbf{grad}\,\theta \\ \hline \mathbf{g_m} &= -(k_m\mathbf{grad}\,s + K_\theta \mathbf{grad}\,\theta) \end{aligned} \qquad (2.142)$$

with:

$$k_m = \left(k_w + \delta\frac{\rho_v}{\rho_w}\right) \quad K_\theta = K_{\theta,w} + \delta\phi\frac{dp_{sat}}{d\theta}$$

Unsaturated water flow in a material so is fixed by a property noted as k_m with units s and called the moisture permeability. Vapour transfer, in turn, depends on K_θ, a property with units kg/(m·s·K), called the thermal moisture diffusion coefficient. For most capillary-porous materials, both are tensors. Just like temperature, in case of ideal contact between materials, the suction (s) is a real potential unambiguously determined in all interfaces. Anyhow, once above the capillary moisture content, the two potentials still govern drying but wetting then is limited to vapour diffusion from water isle to water isle, coupled to a slow dissolution of the air left in some pores in the water filling the other.

Besides hygroscopic type 1 and non-hygroscopic capillary type 2 materials, they can also be non-hygroscopic, non-capillary, type 3 materials, and hygroscopic, non-capillary, type 4 materials. Type 3's have such wide pores that capillarity hardly plays and gravity plus external pressures only intervene when soaked This limits the moisture flow in and through to vapour diffusion and convection. Type 4's have such fine pores that the flow resistance is too large to see vapour move by convection and water by capillarity, gravity and external pressures. Remains as possible: vapour diffusion. Or, in type 1 and type 2 materials, capillary suction remains absent below a limit relative humidity ϕ_M = 95–98%, in type 3 and 4 materials capillary never intervenes, RH there is the only driver, giving as equations and transfer coefficients:

Diffusion only
$$\mathbf{g_m} = -(k_{\phi,m}\mathbf{grad}\,\phi + K_\theta \mathbf{grad}\,\theta) \qquad (2.143)$$

Diffusion and convection
$$\mathbf{g_m} = -(k_{\phi,m}\mathbf{grad}\,\phi + K_\theta \mathbf{grad}\,\theta) + 6.21 \cdot 10^{-6}\mathbf{g_a}\phi p_{sat} \qquad (2.144)$$

So:

Material	$k_{\phi,m}$	K_θ
Type 1, 2: capillary porous	$k_m \dfrac{ds}{d\phi}$	$\delta\phi \dfrac{dp_{sat}}{d\theta}$
Type 3: non-hygroscopic, non-capillary	δp_{sat}	$\delta\phi \dfrac{dp_{sat}}{d\theta}$
Type 4: hygroscopic, non-capillary	δp_{sat}	$\delta\phi \dfrac{dp_{sat}}{d\theta}$

Also, external water heads (P_w), if large enough, may act as driver. They simultaneously push the moisture content in open porous materials to saturation, while imposing as water flux:

$$\mathbf{g_w} = -k_w \mathbf{grad}\,P_w$$

with k_w the water permeability at saturation (also called the Darcy coefficient). Fluxes in moisture-saturated porous materials lack inertia. Across a d m thick single-layer assembly, they obey:

$$g_w = k_w \frac{\Delta P_w}{d} = \frac{\Delta P_w}{d/k_w} = \frac{\Delta P_w}{W_w}$$

with W_w the water resistance in m/s. In a multi-layer assembly, the flux and water pressure curve become:

$$g_w = \frac{\Delta P_w}{\sum_{i=1}^{n} W_{w,i}} \qquad P_{w,x} = P_{w1} + g_w W_{w,1}^x \quad (P_{w1} < P_{w2})$$

Waterproofing an assembly requires protecting it by a layer with infinite water resistance. Of course, preventing water from reaching the other face of an assembly can be dome by activating drying to a level that the saturation front is halted in the assembly itself.

2.4.4.2 Moisture Permeability

For capillary materials, moisture permeability is function of suction, replaceable by RH. The higher suction is, meaning the lower the RH is, i.e. the finer the pores and the lower the contact angles, the more the moisture permeability is dropping. For water in a pore, the formula was:

$$k_w = \frac{\rho_w d^2}{32\eta_w} = \frac{\rho_w \sigma_w^2}{2\eta_w}\left[\frac{\cos\vartheta}{s}\right]^2$$

For materials, the property is determined experimentally as the product of the moisture diffusivity D_w (see below) with the derivative of the relation between moisture content and suction, see Figure 2.57. An alternative to get the moisture permeability could be by using a pore model.

The water part in the thermal moisture diffusion coefficient is often set zero, although the temperature dependencies of both suction and moisture permeability must still be considered.

2.4.4.3 Mass Conservation

The mass balance per representative material volume (V_{rep}) is:

$$(\text{div }\mathbf{g_m} \pm G')dV_{rep} = \left(-\frac{\partial w}{\partial t}\right)dV_{rep}$$

In it, w is moisture content and G' any moisture source or sink.

In type 1, 2 and 4 materials, vapour displacement linked to air moving in and through hardly intervenes. For a representative volume infinitely small, combining this mass balance with the general transfer equation gives:

$$\text{div}(k_{\phi,m}\,\mathbf{grad}\,\phi + K_\theta\,\mathbf{grad}\,\theta) \pm G' = \frac{\partial w}{\partial t} = \rho_w \xi_\phi \frac{\partial \phi}{\partial t} \qquad (2.145)$$

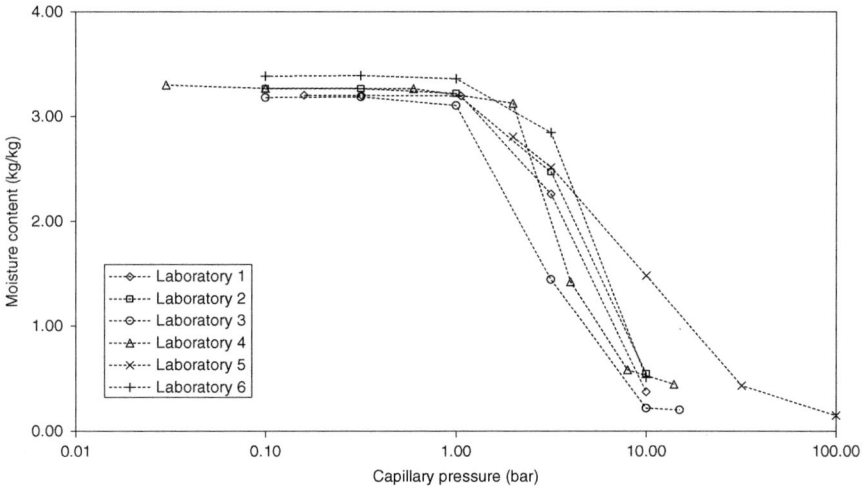

Figure 2.57 Calcium-silicate stone, suction characteristic measured in six laboratories. The capillary pressure on the horizontal axis is suction (s).

In type 3 materials vapour convection could add, or:

$$\text{div}[(k_{\phi,m}\textbf{grad}\,\phi + K_\theta \textbf{grad}\,\theta) + 6.21 \cdot 10^{-6}\textbf{g}_a \phi p_{sat}] \pm G' = \rho_w \xi_\phi \frac{\partial \phi}{\partial t} \quad (2.146)$$

In both formulas, $\rho_w \xi_\phi$ is the specific moisture content with ξ_ϕ the derivative of the suction characteristic multiplied with the derivative of suction to the RH (ϕ):

$$\xi_\phi = \xi_s\, \partial s/\partial \phi$$

The outcome convenes with the derivative between 0 and in theory 100% RH of the hygroscopic curve, except for type 3 materials where the specific moisture content becomes:

$$\rho_w \xi_\phi = \frac{\Psi_o}{\rho_w R} \frac{\partial(\phi p_{sat}/T)}{\partial \phi}$$

In type 1 and type 4 materials, the source or sink term (G') is part of the hygroscopic moisture content, becoming condensation when somewhere in the material/assembly the RH touches 100% or becoming evaporation where that RH is 100% and the temperature there high enough to lift the vapour saturation pressure above the vapour pressure at the warm end face.

2.4.4.4 Starting, Boundary and Contact Conditions

In most cases, construction moisture ensures that assemblies are wetter than hygroscopic at the start. During drying, desorption, sorption, possibly interstitial condensation, all may mix up, while the boundary conditions give vapour fluxes, equal to:

$$g_v = \beta(p - p_s)$$

with p the vapour pressure in the ambient considered, p_s the vapour pressure at the assembly's end face there and β the related surface film coefficient for diffusion. With RH as driving force, the vapour pressure at an end face writes as $p_s = p_{sat,s}\phi$ with $p_{sat,s}$ the vapour saturation pressure on it. Other possible boundary conditions are the moisture flux at an end face known, take wind-driven rain impinging on it, or an end face humidified by surface condensation or rain run-off.

A first possible contact condition is suction ensuring continuity of the water and vapour flow. in interfaces. This seldom happens. Most of the time, suction encounters a contact resistance such as in mortar-to-brick and brick-to-plaster interfaces. A second possible contact condition is small air gaps left between layers, so allowing diffusion but not suction. A third possible contact condition is interstitial condensation and rain penetration building-up a water film in an air gap separating two faces, which gravity forces to run off and capillarity to stay. In real contacts, these three contact conditions may occur in parallel.

2.4.4.5 Remarks

The balance equations work as long as capillary porous materials do not contact water. If doing, water will not only move from the wider to the narrower pores but the wider pores will also suck water from the narrower, as shown when discussing isothermal water transfer in a serial two-pore system contacting water. This difference in moisture movement is accounted for by valuing the moisture permeability and the specific moisture content between dry and capillary. For non-capillary materials, a water contact means 100% RH in the contact plane, while for capillary materials it matches with suction 0, 100% RH and the moisture content capillary.

2.4.5 Simple Moisture Flow Model

2.4.5.1 How to Do?

The properties embedded in the balance equations are all function of suction (or RH) and of temperature. Given an assembly, its initial situation and boundary conditions in terms of temperatures and moisture loads, these suction (or RH) and temperature dependencies exclude any analytical solution. Using CVM can but requires software. Favoured to get some understanding therefore are models where suction is replaced by moisture content as an 'improper' driving force. The water flux in a material then writes as:

$$\mathbf{g_w} = -D_w \mathbf{grad}\ w \qquad (2.147)$$

with D_w its water diffusivity in m^2/s, a property derived from the material's water permeability:

$$D_w = k_w ds/dw = k_w/(\rho_w \xi_s) \qquad (2.148)$$

A preset function advanced for the relation between D_w and moisture content is:

$$D_w = a(A/w_c)^2 \exp(bw/w_c) \qquad (2.149)$$

In it, A is the capillary water absorption coefficient and w_c the capillary moisture content of the material, while the numbers a and b are material-dependent constants. In the simple model handled here, D_w is further simplified to being a step function of moisture content, zero below the critical moisture content assumed to convene with RH = ϕ_M = 95–98%, and constant above with a sudden step from zero to that constant value at the critical moisture content($w = w_{cr}$) (Figure 2.58):

$$w \leq w_{cr}: D_w = 0 \quad w > w_{cr}: D_w = a(A/w_c)^2 \exp(bw_{cr}/w_c) \qquad (2.150)$$

Below critical, only vapour moves, or:

$$g_w = -\delta \mathbf{grad}\ p$$

Above, unsaturated water flow dominates, with a vapour flow added in case of non-isothermal boundary conditions:

	$w < w_{cr}$	$w \geq w_{cr}$
Isothermal	$g_v = -\delta \mathbf{grad}\ p$	$\mathbf{g_w} = -D_w \mathbf{grad}\ w$
Non-isothermal	$g_v = -\delta \mathbf{grad}\ p$	$\mathbf{g_w} = -D_w \mathbf{grad}\ w - \delta \dfrac{dp_{sat}}{d\theta} \mathbf{grad}\ \theta$

Typical for the simple model is a sharp front between air-dry and wet. In fact, in an isothermal monolayer, hygric in steady state, the moisture content at one end face

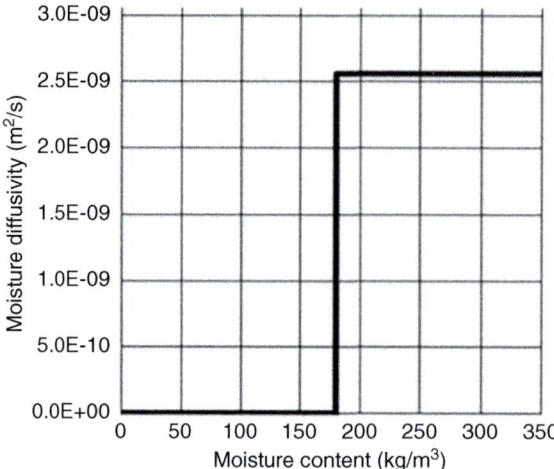

Figure 2.58 Moisture diffusivity as simplified function of moisture content.

below and at the other end face above critical, each infinitesimal small sublayer will see the same moisture flux (g_m) passing:

$$\text{Below: } g_m = -\delta(\mathbf{grad}\ p)_{\text{below}} = -\delta \frac{dp}{dw}(\mathbf{grad}\ w)_{\text{below}} = -D_{w,\text{below}}(\mathbf{grad}\ w)_{\text{below}}$$

$$\text{Above : } g_m = -D_{w,\text{above}}(\mathbf{grad}\ w)_{\text{above}}$$

with $D_{w,\text{below}}$ really small compared to $D_{w,\text{above}}$. Where the moisture content turns critical, the driving term (**grad** w) so is suddenly jumping up, indicating that the moisture content must directly increase to critical there, a fact observations confirm. An ink stain on blotting paper spreads out for a while and then stops spreading, with the ink content in the stain critical and outside the stain hardly visible. Evenly sharp is the front between dry and wet in the outflow zone at sandy beaches. The same is seen on brick walls, see Figure 2.59.

So, although the flux equations look similar, the existence of a moisture front distinguishes unsaturated water flow from heat conduction, air passage and vapour diffusion.

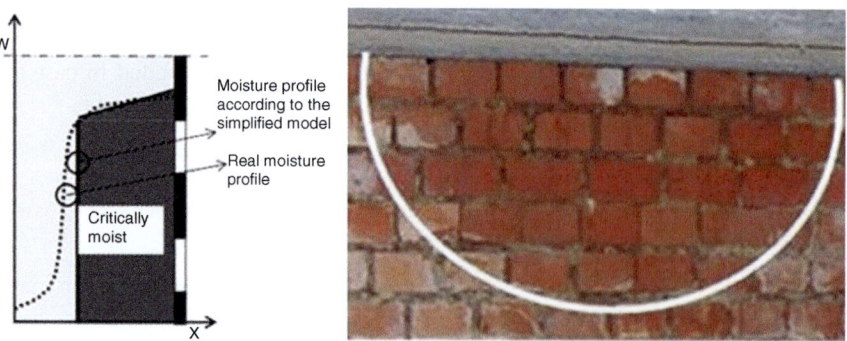

Figure 2.59 Masonry, moisture front between critically wet and air-dry.

2.4.5.2 Applying the Simple Model
2.4.5.2.1 Remark
The analytical solutions based on the simple model advanced and discussed in what follows, all presume that any final moisture distribution of the moisture in an assembly is known in advance.

2.4.5.2.2 Capillary Suction
When a sample of a hydrophilic, homogeneous, capillary material, sealed along its perimeter, is contacting water, then regular weighting makes a linear relation emerge between the moisture sucked per m² of contact surface and the square root of time (Figure 2.60):

$$m_c = A\sqrt{t} \tag{2.151}$$

In it, A is the capillary water sorption coefficient of the material, units $kg/(m^2 \cdot s^{1/2})$.

Monitoring the moving moisture front in turn gives as relation between its position and time:

$$h = B\sqrt{t} \tag{2.152}$$

with B the water penetration coefficient of the material, units $m/s^{1/2}$. Once that front reaches the other end of the sample, suction switches to a second $[\sqrt{t},m]$-line, be it with a much lower slope called the coefficient of secondary water sorption A' of the material. Related water uptake then is mainly linked to a slow dissolution in the sucked water of the air inclusions left in the pores. The intersection of that second line with the $A\sqrt{t}$-line fixes the capillary moisture content (w_c).

Assuming the whole height of the sample turned capillary wet, the relation between water sorption and water penetration coefficient should be:

$$A \approx B w_c \tag{2.153}$$

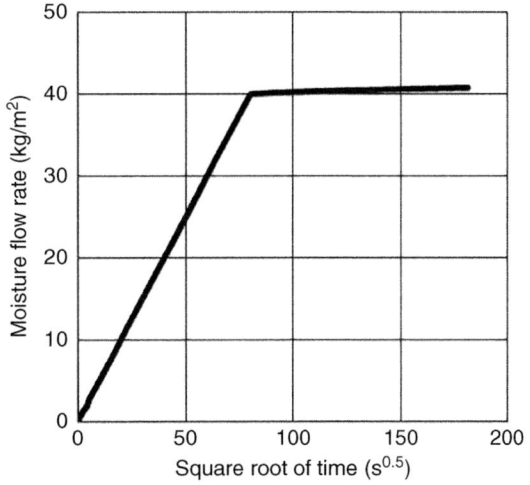

Figure 2.60 Capillary suction, $m_w = A\sqrt{t}$.

A, B and w_c figure as material properties. The larger the water sorption and water penetration coefficient, the more capillary a material, the larger the capillary moisture content, the more moisture it can buffer. The capillary water sorption coefficient is comparable to the contact coefficient in heat transfer. It instructs how easily a material sucks water while the contact coefficient tels how easily a material picks up heat. Anyhow, the linearity between $\sqrt{t}$ and m_c is overlooking the anomalies an unavoidable air egress during sucking may induce. These appear when high samples of fine-porous materials or samples sealed at the top are tested. High ones see the curve bending towards the time axis, while a sealed top retards water sorption because the air bubbles present have to leave the sample at the water side. Also in non-homogeneous materials, the relation deviates from linear, while capillary moisture uptake by hydrophobic materials hardly exists, and if it should, will show no linearity with $\sqrt{t}$.

The question now is whether the simplified model is predicting that linearity with $\sqrt{t}$. Assuming a one-dimensional flow, mass conservation gives:

$$D_w \frac{\partial^2 w}{\partial x^2} = \frac{\partial w}{\partial t} \quad (2.154)$$

an equation, analogous to heat conduction in a flat layer containing sources nor sinks. Capillary water sorption yet is comparable to a sudden temperature increase at the surface of a semi-infinite medium. Of course, the existence of a critical moisture content creating a moisture front complicates the solution of this hygric equal. In addition to a starting and boundary condition, this imposes two additional limitations. First, the moisture content there must remain critical during suction. Second, related water flux must supply the moisture needed to move that front over a distance dx_{fr} per time unit, or:

$$t \leq 0, 0 = x \leq \infty : w = w_H \quad t \geq 0, x = 0 : w = w_c \quad t > 0, x = x_{fr} : w = w_{cr}$$

$$\text{and} -D_w \left(\frac{\partial w}{\partial x}\right)_{x=x_{fr}} = w_{cr}\left(\frac{dx_{fr}}{dt}\right)_{x=x_{fr}}$$

Mathematically, that front at $x = x_{fr}$ is a singular point, excluding an analytical solution except for (i) a hygroscopic and critical moisture content ≈ 0 and (ii) a critical moisture content hardly different from capillary. For (i), when sucking (w_c), the moisture content in an infinitely long sample at a distance x from the water contact will become:

$$w = w_c \left[\underbrace{\frac{2}{\sqrt{\pi}} \int_{\frac{x}{2\sqrt{D_w t}}}^{\infty} \exp(-q^2) dq} \right] \quad (2.155)$$

with the term above the brace equal to the inverse error function, see Figure 2.61. In the wet zone, the moisture content is zero at the moisture front and capillary at the water contact. The quantity of water absorbed equals:

$$m_c = 2w_c \sqrt{D_w/\pi} \sqrt{t}$$

Figure 2.61 Capillary suction, moisture content in the material for a critical moisture content close to zero ($w_{cr} \ll w_c$).

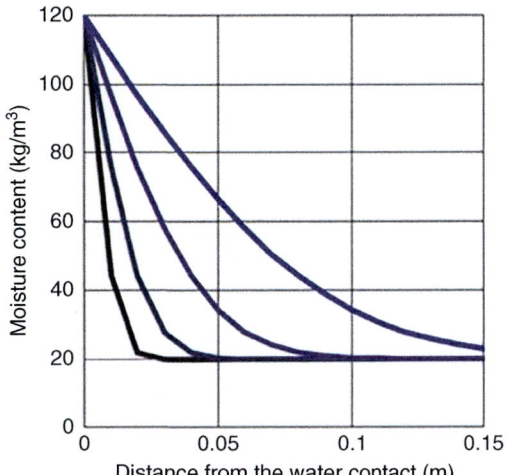

Or, the relation between the capillary water uptake and $\sqrt{t}$ looks linear with the water sorption coefficient and the moisture diffusivity equal to:

$$A = 2w_c \sqrt{D_w/\pi} \quad D_w = \frac{\pi}{4}\left(\frac{A}{w_c}\right)^2 \tag{2.156}$$

The water content equation in turn allows to fix the position of the ≈ 0 moisture front. The water penetration coefficient becomes:

$$B = 2\,\mathrm{erfc}^{-1}(w_{cr}/w_c)\sqrt{D_w}$$

Materials that have a critical moisture content ≈ 0 are rare. For (ii), assumed is that the moisture profile between the sucking face and the critically wet moisture front is linear (Figure 2.62).

Figure 2.62 Capillary suction, critical and capillary moisture content hardly different ($w_{cr} \approx w_c$).

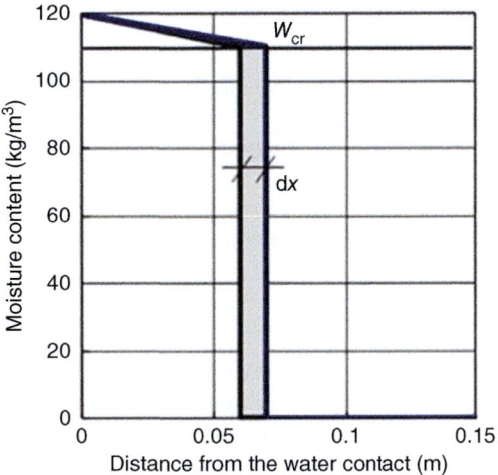

The mass balance then looks:
$$\frac{w_{cr}+w_c}{2}dx = D_w \frac{w_c - w_{cr}}{x} dt$$
Solving for x gives the linear relation sought:
$$x = 2\sqrt{D_w \frac{w_c - w_{cr}}{w_c + w_{cr}}} \sqrt{t}$$
The water penetration coefficient so becomes:
$$B = 2\sqrt{D_w \frac{w_c - w_{cr}}{w_c + w_{cr}}} \tag{2.157}$$
The amount of water absorbed between $t = 0$ and $t = t$ touches (m_c, kg/m^2):
$$m_c = x \frac{w_c + w_{cr}}{2} = \sqrt{D_w(w_c + w_{cr})(w_c - w_{cr})}\sqrt{t}$$
giving as water sorption coefficient and moisture diffusivity:
$$A = \sqrt{D_w\left(w_c^2 - w_{cr}^2\right)} \quad D_w = \frac{A^2}{\left(w_c^2 - w_{cr}^2\right)} \tag{2.158}$$

In reality, moisture diffusivity is a function of the moisture content, which turns mass conservation into:
$$\frac{\partial}{\partial x}\left(D_w \frac{\partial w}{\partial x}\right) = \frac{\partial w}{\partial t} \tag{2.159}$$

With the relation as advanced, with a material volume close to semi-infinite, with the moisture content in the water contact equal to capillary (w_c) at $t = 0$ and remaining so for $t \geq 0$, a Boltzmann transformation converts this partial differential in a second order differential equation:
$$\lambda = x/\sqrt{t} - \frac{\lambda}{2}\frac{dw}{d\lambda} = \frac{d}{d\lambda}\left[D_w(w)\frac{dw}{d\lambda}\right] \tag{2.160}$$
with moisture content depending on the variable λ. In a [λ,w] axis system, all time-related curves now melt together, with as surface between that one left and the λ-axis:
$$\int_0^{\lambda_m} w(\lambda)\,d\lambda = C^{te} \tag{2.161}$$
Moisture absorbed beyond $t = 0$ now is:
$$m_w(t) = \left[\int_0^{x_m} w(x)dx\right]_t \text{ with } x_m = \lambda_m \sqrt{t}.$$

As the integration for λ is time independent, a combination with the surface equation gives as water sorption coefficient:
$$A = C^{te} = \int_0^{\lambda_m} w(\lambda)\,d\lambda \tag{2.162}$$

The relation $m_c = A\sqrt{t}$ so looks generally correct, provided air egress has no impact and $D_w(w)$ is univocal. When assuming a profile $w(\lambda)$, the last equation allows to calculate the moisture diffusivity as function of the water sorption coefficient and moisture content.

Figure 2.63 Rain sorbed before a water film forms ($w_{cr} \ll w_c$, $w_s < w_c$ until $w_s = w_c$).

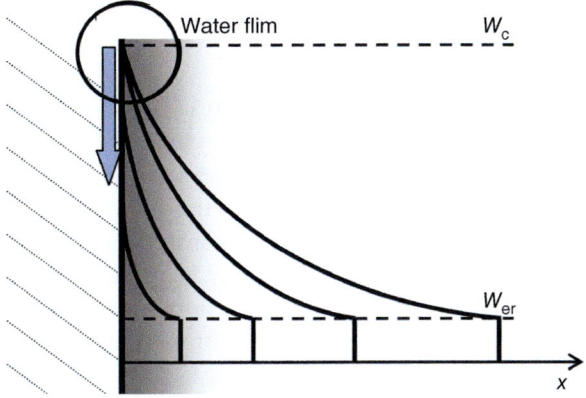

2.4.5.2.3 Wind Driven Rain

Rain hitting a vertical wall creates two successive boundary conditions. First, as long as the moisture content of the outer face is below capillary, the rain intensity impinging creates a water flux absorbed (g_{ws}). Second, once capillary wet, a water film forms that is partly sucked by the wall (Figure 2.63).

In case the critical moisture content is ≈ 0, then with wetness at the outer face (w_s) below capillary, the boundary conditions are: $t = 0$, $0 \leq x \leq \infty$: $w = 0$ / $t > 0$, $x = 0$: $-D_w(dw/dx)_{x=0} = g_w$, g_w being the constant wind-driven rain intensity in kg/(m²·s) hitting the wall's outer face. Solving of the mass conservation equation then gives:

$$w = \frac{2 g_{ws}}{D_w} \left[\sqrt{\frac{D_w t}{\pi}} \exp\left(-\frac{x^2}{4 D_w t}\right) - \frac{x}{2} \mathrm{erfc}\left(\frac{x}{2\sqrt{D_w t}}\right) \right]$$

At the wet outer face ($x = 0$), the moisture content becomes:

$$w_s = \frac{2 g_{ws}}{D_w} \sqrt{\frac{D_w t}{\pi}} \qquad (2.163)$$

Once a water film is formed, the water uptake becomes horizontal suction. This happens when:

$$t_f = \frac{\pi D_w w_c^2}{4 g_{ws}^2} = \frac{\pi^2 A^2}{16 g_{ws}^2} \qquad (2.164)$$

Or, the smaller the water sorption coefficient and the higher the wind-driven rain intensity, the faster run-off starts, see brick veneers and stuccoed walls. Bricks have quite a high water sorption coefficient ($0.2 \leq A \leq 0.8$ kg/(m²·s$^{1/2}$)), making run off the exception. Water-repellent stuccos instead hardly absorb water ($A < 0.005$ kg/(m²·s$^{1/2}$)), so give almost instantly run off.

Just before film formation, the water absorbed is $m_w = g_{ws} t_f$ while the flux still equals the impinging rain intensity, or:

$$g_{ws} = w_c \sqrt{\frac{\pi D_w}{4}} \frac{1}{\sqrt{t_f}} = 0.89 w_c \sqrt{\frac{D_w}{t_f}} = 0.79 \frac{A}{\sqrt{t_f}}$$

With the solution for capillary sorption, once a water film formed, the flux sucked becomes:

$$g_{ws2} = w_c \sqrt{\frac{D_w}{\pi} \frac{1}{\sqrt{t_f}}} = 0.56 w_c \sqrt{\frac{D_w}{t_f}} = \frac{A}{2\sqrt{t}} = g_{ws} \frac{2}{\pi} < g_{ws}$$

or, once a water film forms, the material suddenly is picking up less rain, which is not logic. Therefore, with this switch, a transition regime linked to the moisture profile in the material must link 'hit by a constant rain intensity' to 'part of the water film sucked'. To circumvent that anomaly, further discussion presumes a critical moisture content near capillary and the moisture content changing linearly between the wet outer face and a front at critical moisture content x metres deep in the wall. As long as the moisture content at the outer face (w_s) stays below capillary (w_c), the water flux in the material is:

$$g_{ws} = D_w \frac{w_s - w_{cr}}{x}$$

From the mass balance $g_{ws} dt = w_s dx$ follows for the position of the front at critical moisture content and the moisture content at the outer face:

$$x = \frac{D_w w_{cr}}{g_{ws}} \left(\sqrt{1 + \frac{2 g_{ws}^2 t}{D_w w_{cr}^2}} - 1 \right) \quad w_s = w_{cr} + g_{ws} x / D_w.$$

At the moment the outer face turns capillary wet ($w_s = w_c$), the moisture front in the material reached a depth (x_{fr}):

$$x_{fr} = [D_w(w_c - w_{cr})]/g_{ws} = \frac{A^2}{(w_c + w_{cr})g_{ws}}$$

Or, film formation starts at:

$$t_f = A^2 / (2 g_{ws}^2)$$

2.4.5.2.4 Drying

Drying happens each time the RH in the ambient drops below the RH convening with the moisture content in a material. To test how a material, saturated with water, dries, a beam-like sample with one face unsealed is hung in a climate chamber at constant temperature and RH. Regularly weighting allows quantifying the drying flux as function of the average moisture content left in the sample, giving a curve as shown in Figure 2.64. Two stages appear: first a constant drying flux at higher average moisture content, second a fast drop at lower average moisture content.

Can the simple moisture transfer model advanced explain this? In a saturated sample, the RH everywhere is 100%. With the vapour pressure on the drying face the saturated one at the surface temperature θ_s ($p_{sat,s}(\theta_s)$), drying to the chamber values:

$$g_{vd} = \beta (p_{sat,s} - \phi p_{sat}) \tag{2.165}$$

In it, g_{vd} is the drying flux in kg/(m²·s), β the surface film coefficient for diffusion, ϕ the RH and p_{sat} the vapour saturation pressure for the temperature in the chamber.

Figure 2.64 Isothermal drying, the two stages.

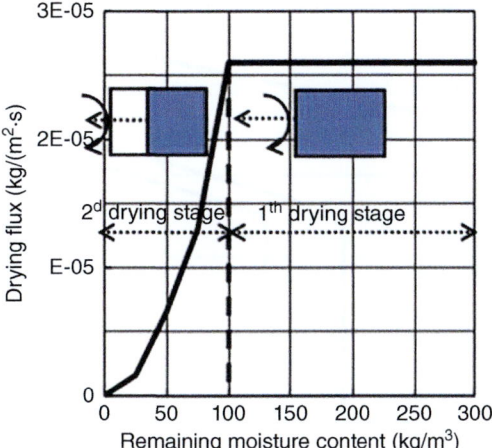

As drying means evaporation, related heat of evaporation will simultaneously lower the temperature and the vapour saturation pressure on the drying face. Assumed now is that the wet sample has a thermal conductivity ∞, while at the drying face, no heat transfer by radiation occurs. The face's vapour saturation pressure then must be the one in the chamber, or, the drying flux can be rewritten as:

$$g_{vd} = \beta \, p_{sat} (1 - \phi) \tag{2.166}$$

For any material, drying apparently only depends on the air temperature, the RH, and, via the surface film coefficient for diffusion, the air speed in the chamber. The warmer, the drier and the more intense the air moves, the faster the sample dries. People use this. To dry, laundry is hung outdoors when sunny (=high temperature) and dry (=low RH). If also windy, drying goes still faster. Anyhow, as a result, the moisture content at the drying face is dropping and a moisture gradient originates in the sample, which initiates flowing. At any moment, the following then applies:

$$x = 0: \; -D_w (dw/dx)_{x=0} = \beta \, p_{sat}(1 - \phi)$$

Or, a capillary water flux to the drying face develops, where it gives evaporation while the moisture content in the sample slowly drops. All this fixes a 'first drying stage'. Anyhow, once the moisture content at the drying face turns critical, the moisture diffusivity there becomes ≈0 and the moisture front starts a slow retreat into the sample, while the moisture flow to the drying face turns into a vapour only flow, the start of the 'second drying stage' (Figure 2.65):

$$g_{vd} = \frac{p_{sat}(1 - \phi)}{1/\beta + \mu N x} \tag{2.167}$$

In it, μ is the vapour resistance factor of the 'dry' material and x the distance between the retreating moisture front and the drying face.

As the distance between moisture front and drying face increases, the diffusion flux drops and drying is slowing down. The second stage lasts until the whole sample regains hygroscopic equilibrium with the ambient. However, after some time, the

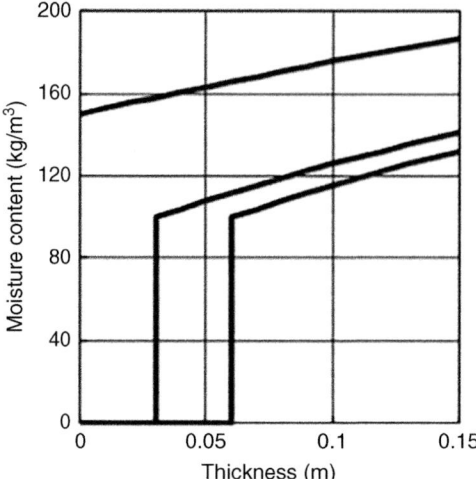

Figure 2.65 Moisture content in the sample.

material will look superficially dry, which is often perceived as 'it's dry', although it may still contain quite some moisture.

The first drying stage can be described using mass equilibrium with as starting and boundary conditions: face $x = 0$ the drying one, the perimeter and face $x = d$ tight, at $t = 0$ a constant moisture content w_o between $x = 0$ and $x = d$, for $t > 0$ a gradient in moisture content at $x = 0$ equal to $-g_{vd}/D_w$ and at $x = d$ a zero gradient in moisture content. The solution of the mass balance then is:

$$w = w_o - \frac{g_{vd}d}{D_w} \underbrace{\left\{ \frac{3x^2 - 6dx + 2d^2}{6d^2} + \frac{2}{\pi^2} \sum \left[\frac{(-1^n)}{n^2} \exp\left(-D_w \frac{n^2\pi^2 t}{d^2}\right) \cos\left(\frac{n\pi(d-x)}{d}\right) \right] \right\}}$$

The term above the brace describes how the change in moisture profile slowly moves from the drying face $x = 0$ to the tight face $x = d$. Once there, the term loses impact, leaving a dropping parabolic moisture profile with lowest value at the drying face:

$$w = w_o - \frac{g_{vd}t}{d} - \frac{g_{vd}d}{D_w}\left(\frac{3x^2 - 6dx + 2d^2}{6d^2}\right) \rightarrow \quad (2.168)$$

$$\text{drying face: } w = w_o - \frac{g_{vd}t}{d} - \frac{g_{vd}d}{3D_w} \quad (2.169)$$

At any moment, the average moisture content so touches:

$$w = w_o\left(1 - \frac{g_{vd}t}{w_o d}\right) \quad (2.170)$$

Compared to the start of the first drying stage (t_{start}), the second (t_{second}) does when the moisture content at the drying face turns critical, or:

$$t_{second} - t_{start} = d\left[\frac{w_o - w_{cr}}{g_{vd}} - \frac{d}{3D_w}\right] \quad (2.171)$$

At that moment, the moisture left in the sample fixes what is called the transition moisture content (w_{tr}):

$$w_{tr} = w_{cr} + g_{vd}d/3D_w \qquad (2.172)$$

The larger the drying flux, the more the difference with the critical moisture content increases. Or, faster drying during the first stage does not guarantee a sooner dryness. The second stage simply starts at a higher transition moisture content. At equal flux, due to their higher moisture diffusivity compared to less capillary ones, the difference with the critical moisture content will be lower in really capillary materials. More generally, a low diffusion resistance factor μ, a high moisture diffusivity D_w and a low critical moisture content w_{cr}, take bricks, will accelerate drying. Instead, less capillary materials with the critical moisture content close to the capillary one, take concrete, will show a short first and long second drying stage.

A simple model of the second stage assumes a transition and critical moisture content so close that during the first stage, the flux remains really low. During the second stage, mass equilibrium between the retreat of the critically wet moisture front and the drying flux writes as:

$$\frac{p_{sat}(1-\phi)}{1/\beta + \mu N x}dt = (w_{cr} - w_H)dx \qquad (2.173)$$

with w_H the hygroscopic equilibrium with the ambient air. In steady-state and with the vapour resistance factor μ constant, solving gives as position of the moisture front:

$$x = \frac{1}{\mu N \beta}\left(\sqrt{1 + \frac{2p_{sat}(1-\phi)\mu N \beta^2}{w_{cr} - w_H}\sqrt{t}} - 1\right) \qquad (2.174)$$

Once the vapour resistance μNx outranges the surface vapour resistance $1/\beta$, this equation simplifies to:

$$x = \sqrt{\frac{2p_{sat}(1-\phi)}{\mu N(w_{cr} - w_H)}}\sqrt{t} \qquad (2.175)$$

Or, also drying gives a square root relation between the position of the moisture front and time. Multiplying left and right with $w_{cr} - w_H$ gives as dried moisture quantity:

$$m_{vd} = \sqrt{\frac{2p_{sat}(1-\phi)(w_{cr} - w_H)}{\mu N}}\sqrt{t} = A'\sqrt{t} \; (kg/m^2) \qquad (2.176)$$

With the critical and hygroscopic moisture content known, this equation allows calculating the vapour diffusion factor μ from a drying test on a, except the drying face, vapour tight wrapped sample. The slower drying proceeds, the higher the μ-value found will be.

The simplified model as explained is usable at low drying rates and easy heat supply. High rates do not proceed isothermally, what bends the drying curve to the $\sqrt{t}$-axis. The heat of evaporation in fact not only lowers the temperature at the drying front but till the start of the second drying stage also the drying flux. In the second stage, as the front moves deeper in the sample, the decreasing drying flux requires ever less heat of evaporation, which moves the curve back to linear with $\sqrt{t}$.

The drying model as discussed is extendable to multi-layer assemblies with a wet interface (*j*) or a more than critically moist layer between non-capillary layers. Is <equivalent> diffusion the only driver, then the steady state vapour fluxes between the wet interface *j* or the interfaces *k* and *l* bounding the wet layer and both ambients look (also see Figure 2.66a):

$$\text{Wet interface} \quad g_{md} = -\left(\frac{p_{sat,j} - p_1}{Z_1} + \frac{p_{sat,j} - p_2}{Z_2}\right) = \frac{p_2 - p_{sat,j}}{Z_2} - \frac{p_{sat,j} - p_1}{Z_1} \tag{2.177}$$

$$\text{Wet layer} \quad g_{md} = -\left(\frac{p_{sat,k} - p_1}{Z_1} + \frac{p_{sat,l} - p_2}{Z_2}\right) = \frac{p_2 - p_{sat,l}}{Z_2} - \frac{p_{sat,k} - p_1}{Z_1} \tag{2.178}$$

In the wet interface or wet zone, the RH is 100%, meaning $p_{sat,j}$, $p_{sat,k}$ and $p_{sat,l}$ are the vapour saturation pressures in the interfaces *j*, *k* and *l*, while Z_1 and Z_2 are the diffusion resistances between *j* and the ambients at both sides or between *k* and the cold ambient 1, vapour pressure p_1, and *l* and the warm ambient 2, vapour pressure p_2. In the [Z, p] axis system, the slope of the lines connecting the saturation pressure in the wet interface *j* with the vapour pressure p_1 in the cold ($Z = 0$) and p_2 in the warm ambient ($Z = Z_T$) or the slope of the line connecting the saturation pressure in *k* with the vapour pressure p_1 in the cold and the saturation pressure in *l* with the vapour pressure p_2 in the warm ambient, give the fluxes. The line to the cold can of course intersect saturation. To see where part of the vapour diffusing from *j* or *l* will condense, demands tracing the tangents from $p_{sat,j}$ or $p_{sat,l}$ and p_1 in the cold ambient to the saturation curve. The difference in slope gives what deposits (Figure 2.66b).

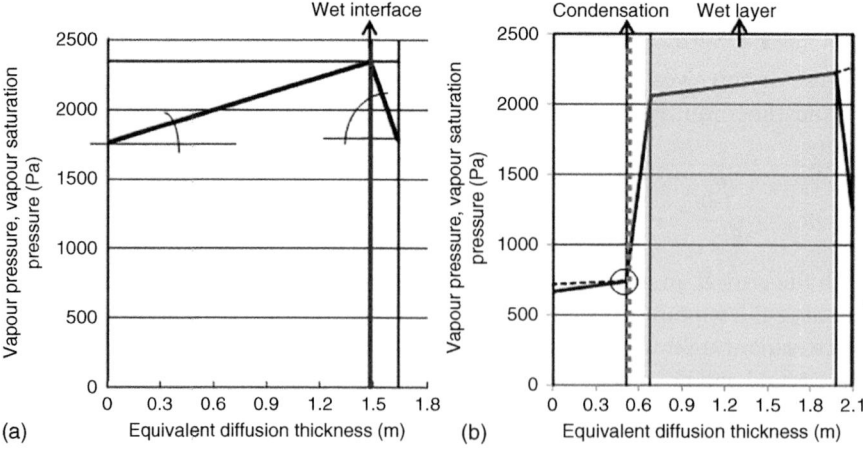

Figure 2.66 (a) Drying of a moist interface and (b) drying of a moist layer in an assembly giving condensation in an interface at the cold side elsewhere as a result.

Interstitial condensation elsewhere in a multi-layer assembly could be a problem when containing construction moisture, having one or more layers wetted by rain or in which vapour from indoors condensed during the cold months, whereby in cases with air outflow, drying will go faster, but the condensate present could be a multiple of only diffusion.

2.4.5.2.5 Interstitial Condensation, Limit State

With interstitial condensation accumulating year after year, the question is if it ever stops? As in case 1, a flat multi-layer assembly is taken, composed of a capillary, vapour permeable, better-insulating material with diffusion resistance Z_2 at the warm side and a vapour retarding, airtight layer with diffusion resistance Z_1 at the cold side. Assumed is that the yearly mean temperature and vapour pressure at the warm side ($\theta_2, p_2, Z = Z_1 + Z_2$) is causing condensation year round on the back of the cold side layer ($Z = Z_1$). With constant boundary conditions the deposit initially equals (g_{wc}):

$$g_{wc} = g_{v2} - g_{v1} = \frac{p_2 - p_{sat,Z_1}}{Z_2} - \frac{p_{sat,Z_1} - p_1}{Z_1}$$

In it, p_1 the vapour pressure at the cold side and $p_{sat,Z1}$ the saturation pressure at the back of the cold side layer.

The capillary material at the warm side now will start sucking back that deposit, so inducing a liquid flux to the warm side:

$$g_w = D_w \frac{w - w_{cr}}{x} = \frac{p_2 - p_{sat,Z_1}}{Z_2} - \frac{p_{sat,Z_1} - p_1}{Z_1}$$

Slowly a wet zone will start extending direction warm side, in which the vapour pressure turns saturated and moisture content passes the critical somewhat, except at the moisture front in the further still dry capillary material. There, it remains critical. Simultaneously, the ingoing vapour pressure line will rotate from tangent to the saturation pressure at the back of the cold side layer to contacting the saturation pressure at the moisture front ($p_{sat,front}$) (Figure 2.67).

Condensate will go on accumulating until the decreasing ingoing vapour flux g_{v2} equals what's diffusing through the cold side layer (g_{v1}). The limit state then is reached. In the wet zone of the capillary, vapour permeable, better insulating material at the warm side, a capillary water flux from cold to warm will persist in equilibrium with a vapour flux direction cold with as in- and as outflow:

$$g_{v2\infty} = g_{v1} = \frac{p_2 - p_{sat,f}}{Z_2 - \mu N d_w} = \frac{p_{sat,f} - p_{sat,Z_1}}{\mu_w - N d_w} - D_w \frac{w_{Z_1} - w_{cr}}{d_w} \quad (2.179)$$

In it, w_{Z1} is the moisture content in interface Z_1 between both material layers, $p_{sat,f}$ the vapour saturation pressure at the moisture front, μ and μ_w the vapour resistance factors of the dry and critically wet part of the material layer at the warm side and d_w the thickness of the wet zone in it. Solving this equation for w_{Z1} gives:

$$w_{Z1} = w_{cr} + \frac{1}{D_w}\left(\frac{p_{sat,f} - p_{sat,Z_1}}{\mu_w N} - g_{v1} d_w\right) \quad (2.180)$$

The term above the brace is mostly too small to consider, or $w_{Z1} \approx w_{cr}$. The limit state of the two-layer assembly so consists of a ≈critically wet zone in the layer at the warm side contacting the layer at the cold side and extending over a thickness d_w in it, a thickness extending when the difference in vapour pressure between the cold and warm ambient increases.

A fair guess of the final position of the moisture front is given by the intersection between the original saturation curve and a line parallel to the outgoing tangent through the vapour pressure at the warm side (see Figure 2.67). The time to reach that limit state follows from:

$$w_{cr}\, dx_w = \left(\frac{p_2 - p_{sat,f}}{Z_2 - \mu N x} - \frac{p_{sat,Z_1} - p_1}{Z_1} \right) dt \qquad (2.181)$$

with x the thickness of the wet zone. Solving this equation requires a preset relation between the vapour saturation pressure ($p_{sat,f}$) and the diffusion resistance at the position (x) of the moisture front. Assuming a linear relation between vapour saturation pressure and diffusion resistance:

$$p_{sat,f} = aZ + p_{sat,Z_1} \quad p_{sat,Z_1} = C^t \text{ and } Z = \mu N x,$$

gives as approximate solution:

$$(\mu N / w_{cr}) t = a_1 Z - a_2 \ln(1 - a_3 Z) \qquad (2.182)$$

$$a_1 = \frac{1}{a - g_{v1}} \quad a_2 = \frac{a - g_{v1}}{Z_2 (g_{v2} - g_{v1})} \quad a_3 = \frac{Z_2}{(a - g_{v1})}$$

$$g_{v1} = \frac{p_{sat,Z_1} - p_1}{Z_1} \quad g_{v2} = \frac{p_2 - p_{sat,Z_1}}{Z_2}$$

With the diffusion resistance (Z_w) and thus the thickness of the wet zone (x_w) known as implicit function of time, the deposit in kg/m² after t days of constant boundary conditions will equal:

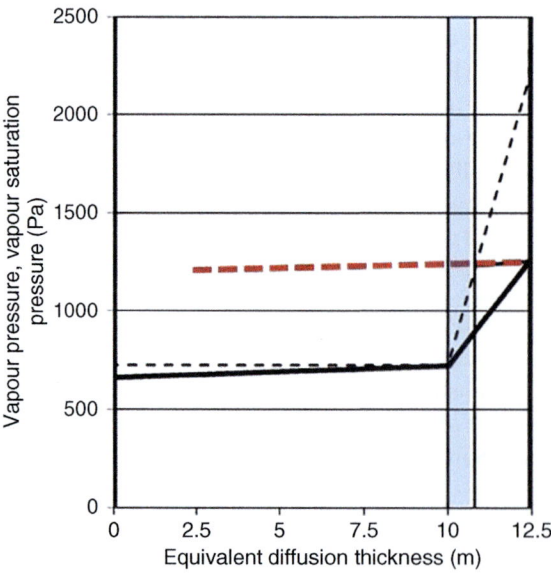

Figure 2.67 Annually accumulating condensate, limit state in the case a capillary, vapour permeable, better insulating material at the warm side has as finish at the cold side a non-capillary, vapour retarding, non-insulating material.

$$m_c = 86\ 400\ w_{cr} Z(t)/(\mu N).$$

Table 2.10 lists the limit states, named 2 to 5, for other layer combinations.

Whether interstitial condensation will be acceptable so depends not only on the nature of the material in or against which condensation deposits and on the quantities involved but also on the limit state. Case 4 looks most negative. It occurs each time a capillary layer is sandwiched between a non-capillary thermal insulation at its warm and a vapour retarding finish at its cold side. An example in cold and temperate climates is the deck of a low-slope roof with the insulation inside. An example in warm and humid climates where colder is inside an warmer outside, is a mineral wool insulated timber-frame wall with inside a gypsum board finish covered with a vapour-tight vinyl paper and outside a vapour-permeable, non-capillary finish.

Table 2.10 Limit state in case of annual accumulation of condensate.

Cold side layer	Warm side layer	Limit state
Case 2 Vapour retarding, non-capillary	Better insulating, vapour permeable, non-capillary	Layer at the warm side turning saturated from the interface in-be-tween to the warm side over a thickness $(d_w) \approx$ equal to the distance between this interface and the intersection of a line through the vapour pressure at the warm side parallel to the outgoing tangent with the original saturation course.
Case 3 Vapour permeable, capillary	Insulating, vapour permeable, non-capillary	Layer at the cold side turning critically wet from the interface between the two towards the cold side over a thickness (d_w) approximately equal to the distance between this interface and the intersection of a line through the vapour pressure at the cold side parallel to the ingoing tangent at warm side with the original saturation course
Case 4 Vapour permeable, capillary, finished vapour retarding	Insulating, vapour permeable, non-capillary	Interstitial condensation starts in the cold side capillary layer and its interfaces with the vapour retarding outer finish and with the non-capillary layer at the warm side. The limit state consists of the capillary layer ending above capillary wet, with water run-off in the interface with the non-capillary layer at the warm side.
Case 5 Vapour permeable, capillary	Insulating, vapour permeable, capillary	Both layers turning critically wet by sucking condensate from the interface in between. At the limit state, thickness of the critically moist part in both is so that the in- and outgoing vapour pressure lines in the $[Z, p]$-plane have the same slope. Limit state cannot be determined graphically, but the assumption is that the wet thicknesses in both layers are equal: $$d_1 = \frac{1}{w_{kr,1}}\left(\frac{A_1^2}{A_1^2 + A_2^2}\right), d_2 = \frac{1}{w_{kr,2}}\left(\frac{A_2^2}{A_1^2 + A_2^2}\right)$$

Problems and Solutions

Problem 17 The monthly mean temperature outdoors is $0\,°C$ and the dewpoint $-2\,°C$, the temperature indoors $22\,°C$. The lowest temperature factor somewhere on the envelope is 0.4, while the daily mean vapour release indoors touches 20 kg. How much outside air ventilation in m^3/h is needed to avoid mould, which means to exclude a surface relative humidity touching 80%?

Solution 17 The vapour pressure outdoors is:

$$p_e = 611 \exp\left(\frac{22.44 \theta_{d,e}}{272.44 + \theta_{d,e}}\right) = 518\,Pa$$

with $\theta_{d,e}$ the dew point. The lowest inside surface temperature so equals $\theta_s = 0 + 22 \cdot 0.4 = 8.8\,°C$. The allowable vapour pressure indoors becomes:

$$p_i = 0.8 \cdot 611 \exp\left(\frac{17.08 \theta_s}{234.18 + \theta_s}\right) = 907\,Pa$$

The outside air ventilation flow needed then is:

$$\dot{V}_a = \frac{R_v T_i G_{v,P}}{p_i - p_e} = \frac{462 \cdot 295.15 \cdot 20/24}{907 - 518} = 292\,m^3/h$$

To avoid surface condensation, which means 100% relative humidity against the surface, $184\,m^3/h$ suffices, 63% of the flow excluding mould.

Problem 18 An extract fan delivers the ventilation flow calculated in problem (17) needed to avoid mould. However, the designer forgot to provide the necessary air inlets. As a result, air leakage through the envelope has to do the job. The air permeance of the envelope is $0.083 \Delta P_a^{-0.37}\,kg/(s \cdot Pa)$. Which air pressure excess indoors will generate the inflow needed?

Solution 18 The relation between airflow in kg/s and air pressure difference in Pa is:

$$G_a = K_a \Delta P_a = 0.083 \Delta P_a^{0.63}$$

A volumetric flow of $292\,m^3/h$ means a weight-related flow of $345.3\,kg/h$ or $0.095922\,kg/s$. The air pressure excess indoors needed to get this amount supplied by air leakage so equals:

$$\Delta P_a = \left(\frac{0.095922}{0.083}\right)^{1/0.67} = 1.24\,Pa$$

a result, which learns the envelope must be highly air permeable for that. In fact, at a pressure difference of 50 Pa, a volumetric airflow of $3469\,m^3/h$ should be measured, which is a lot.

Problem 19 Repeat problem (18) in case the air tightness of the envelope is 10 times higher, albeit with the same air pressure difference exponent. What is the conclusion?

Solution 19 Pressure excess needed now touches 38.6 Pa. Or, quite an airtight envelope demands unrealistic high pressure differences before infiltration could deliver the ventilation needed.

Problem 20 Outdoors, the temperature touches 34 °C for 80% relative humidity. Indoors, air conditioning provides 24 °C at 60% relative humidity. The building has a volume of 600 m³. The ventilation rate is 0.6/h and the daily mean vapour release indoors 15 kg. How much vapour must be removed by the system (in g/h) to keep this 60%? Could this quantity be influenced by selecting appropriate glazing?

Solution 20 The formula describing the indoor/outdoor vapour pressure difference is:

$$\Delta p_{i,e} = RT_i X / nV$$

with X the vapour flow, which must be removed out of ventilation air coming from outdoors. In this case being, the difference equals:

$$611 \left[0.6 \cdot \exp\left(\frac{17.08 \cdot 24}{234.18 + 24}\right) - 0.8 \cdot \exp\left(\frac{17.08 \cdot 34}{234.18 + 34}\right) \right] = -2468 \text{ Pa.}$$

giving $X = 6471$ g/h, added the 625 g/h released indoors! The glazing chosen has no impact. In fact, as the temperature outdoors passes the one indoors, the inside surface temperature of the glass is higher than the air temperature indoors, prohibiting surface condensation.

Problem 21 Given is a dwelling where the bathroom, having a volume of 15 m³, is directly accessible from the 50 m³ large parent's bedroom. The daily vapour release in the bathroom is 2 kg, in the bedroom 1 kg. Which average vapour pressure will be measured in the bedroom when the ventilation air entering the bathroom at a rate of 1.7 ach compared to its volume moves into that bedroom to mix in it with an outside airflow of 30 m³/h. Outdoors, it's −10 °C for 90% RH. The temperature in the bathroom is 24 °C, in the bedroom 18 °C.

Solution 21 The formula to calculate vapour pressure in the bathroom ($p_{i,bath}$) is:

$$p_{i,bath} = p_e + RT_i \left(\frac{G_{v,P}}{(nV)}\right)_{bathroom} \text{ with}$$

$$p_e = 0.9 \cdot 611 \exp\left(\frac{22.44 \cdot -10\theta_{d,e}}{272.44 - 10}\right) = 234 \text{ Pa.}$$

With $2/24 = 0.127$ kg/h vapour produced in it, related value becomes:

$$p_{i,bath} = 234 + 462 \cdot 297.15 \left(\frac{2/24}{1.7 \cdot 15}\right) = 683 \text{ Pa}$$

The vapour balance in the bedroom so writes as:

$$0.127 + \frac{30 \cdot 234}{462 \cdot 291.15} + 1/24 - \frac{p_{i,bed}(30 + 1.7 \cdot 15)}{462 \cdot 291.15} = 0$$

giving as value:

$$p_{i,bed} = 234 \left(\frac{30}{30 + 1.7 \cdot 15} \right) + 462 \cdot 291.15 \left(\frac{0.127 + 1/24}{30 + 1.7 \cdot 15} \right) = 535 \text{ Pa}$$

Or, thanks to the extra ventilation, vapour pressure in the bedroom drops below the one in the bathroom.

Problem 22 Repeat problem (21) for the case the ventilation in the bedroom comes from the airflow leaving the bathroom. Conclusions?

Solution 22 The vapour pressure in the bedroom now is 902 Pa, which is higher yet than the vapour pressure in the bathroom.

Problem 23 Given is a cavity wall with as section:

Layer	Thickness, m	λ-value. W/(m · K)	Air permeance K_a, m³/(m²·s·Pa)
Inside leaf (no fines blocks)	0.14	1.00	$33 \cdot 10^{-4} \Delta P_a^{-0.42}$
Cavity fill (mineral fibre)	0.10	0.04	0.00081
Air layer	0.01	0.067	∞
Brick veneer	0.09	1.00	$0.32 \cdot 10^{-4} \Delta P_a^{-0.19}$

Calculate the air flux across the wall, knowing wind induces a 6 Pa higher average air pressure outdoors than indoors and the flux develops perpendicular to the wall.

Solution 23 The air flux equals:

$$g_a = \frac{6}{g_a^{\frac{1}{0.58}-1}/(33 \cdot 10^{-4})^{\frac{1}{0.58}} + 1/0.00081 + g_a^{\frac{1}{0.81}-1}/(0.32 \cdot 10^{-4})^{\frac{1}{0.81}}}$$

This formula requires iteration, starting from assuming a flux at the right-hand side, here the one if the wall consisted of the fill only:

$$g_a = 6 \cdot 0.00081 = 0.00486 \text{ m}^3/\text{s}$$

Iterating gives as final result:

$$g_a = 0.000133 \text{ m}^3/(\text{m}^2 \cdot \text{s})$$

As the graph on top in the figure below shows, few iterations already give a fair approximation. So, 0.48 m³/h of air passes across each m² of cavity wall! The air pressures curve through the wall is found by introducing the flux in the equation per layer ($g_a = K_a \Delta P_a$), calculating the air pressures in the interfaces and coupling the successive values with line segments:

$$P_a = P_{a,0} - \sum_{i=1}^{5} \Delta P_a$$

See the graph down left in the figure below.

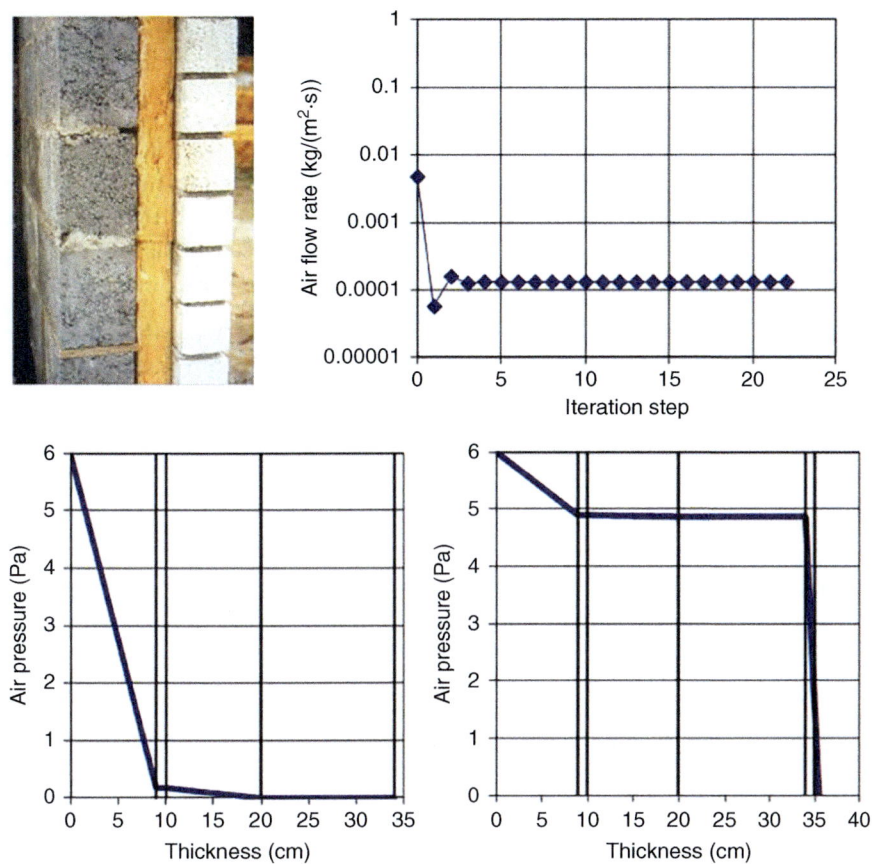

Apparently, the veneer bears the largest pressure difference, which means excessive wind load may threaten its stability. Rain will also easily seep across its head joints and run off at its back. In other words, try to avoid! A better choice consists of having the largest pressure difference over the inside leaf. This is realized by plastering its inside, turning the section into:

Layer	Thickness, m	λ-value, W/(m·K)	Air permeance, m³/(m²·s·Pa)
Plaster	0.01	0.3	$0.1 \cdot 10^{-4} \Delta P_a^{-0.23}$
Inside leaf (no fines blocks)	0.14	1.00	
Cavity fill (mineral fibre)	0.10	0.04	0.00081
Air layer	0.01	0.067	∞
Brick veneer	0.09	1.00	$0.32 \cdot 10^{-4} \Delta P_a^{-0.19}$

The result is a 75% decrease in infiltration rate, down to $0.12\,\mathrm{m^3/(m^2 \cdot h)}$. Also, the air pressure in the interfaces changes substantially as the graph down right in the figure above shows. It's the plaster that faces the largest difference now, while the veneer only has to withstand some 20% of the total, meaning less wind load and less rain seeping across the head joints.

Problem 24 Return to the cavity wall, problem (23) started with. Outdoors it's $-10\,°\mathrm{C}$, indoors $21\,°\mathrm{C}$. Calculate the temperatures in the wall and evaluate the heat loss by conduction at the inner face in case a $4\,\mathrm{m/s}$ wind gives $6\,\mathrm{Pa}$ over-pressure on its outer face. The surface film coefficient inside is $8\,\mathrm{W/(m^2 \cdot K)}$, outside $22.2\,\mathrm{W/(m^2 \cdot K)}$. To simplify things, stack must not be considered.

Solution 24 The temperature difference over the wall in this windy weather induces a combined heat and enthalpy flow, without stack equally distributed over the wall's surface and giving as temperature curve in the wall:

$$\theta = \theta_e + (\theta_i - \theta_e)F_1(R) \text{ with } F_1(R) = \frac{1 - \exp(c_a g_a R)}{1 - \exp(c_a g_a R_T)}$$

The wall experiences an infiltration rate of $0.000133\,\mathrm{m^3/(m^2 \cdot s)}$ or $0.00016\,\mathrm{kg/(m^2 \cdot s)}$, inducing an enthalpy flow of $0.1602\,\mathrm{W/(m^2 \cdot K)}$ ($c_a = 1008\,\mathrm{J/(kg \cdot K)}$). The temperatures in all interfaces become:

Layer	Thickness, (m)	λ-value, (W/(m·K))	R, (m²·K/W)	ΣR (m²·K/W)	$F_1(R)$, (−)	Temp., (°C)
						21.0
Surface film resistance (R_i)			0.125	0.125	0.0514	19.4
Inside leaf (no fines blocks)	0.14	1.00	0.14	0.265	0.1077	17.7
Cavity fill (mineral fibre)	0.10	0.04	2.5	2.765	0.9260	−7.7
Air layer	0.01	0.067	0.15	2.915	0.9654	−8.9
Brick veneer	0.09	1.00	0.09	3.005	0.9885	−9.6
Surface film resistance (R_e)			0.045	3.05	1	−10

The full line gives the course with infiltration, the dashed without. Infiltration cools the wall a bit. Heat flux by conduction at the inside surface is

$q = F_2(\theta_i - \theta_e)$ with: $F_2 = \frac{c_a g_a \exp(c_a g_a R_T)}{1-\exp(c_a g_a R_T)}$, $R_T = 3.05 \, m^2 \cdot K/W$, $c_a g_a = 0.1602 \, W/(m^2 \cdot K)$, giving $q = 12.9 \, W/m^2$. Without infiltration, the result should be $10.2 \, W/m^2$. Or, infiltration lifts the conductive losses at the inside face a little. Simultaneously, an enthalpy flow of $3.36 \, W/m^2$ passes through the wall.

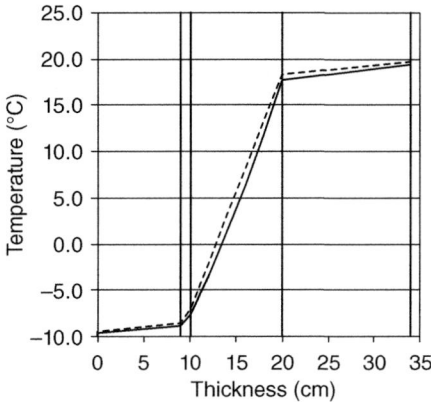

Problem 25 Return to the cavity wall plastered inside of problem (23). Calculate the temperatures across and heat flux by conduction at the inside face for the same in- and outdoor conditions as in problem (24). What are the conclusions?

Solution 25

Layer	Thickness (m)	λ-value (W/(m·K))	R (m²·K/W)	ΣR (m²·K/W)	F_1 (R)	Temp (°C)
					0.000	21.0
Surface film resistance (R_i)			0.125	0.125	0.043	19.7
Plaster	0.01	0.3	0.035	0.16	0.055	19.3
Inside leaf (no fines blocks)	0.14	1.00	0.14	0.30	0.103	17.8
Cavity fill (mineral fibre)	0.10	0.04	2.50	2.80	0.913	−7.3
Air layer	0.01	0.067	0.15	2.95	0.959	−8.7
Brick veneer	0.09	1.00	0.09	3.04	0.986	−9.6
Surface film resistance (R_e)			0.045	3.09	1.000	−10.0

The conductive heat flux at the inside surface of the plastered cavity wall equals 10.7 W/m², whereas pure transmission should have given 10.0 W/m². Or, the plaster is quite effective in assuring acceptable airtightness.

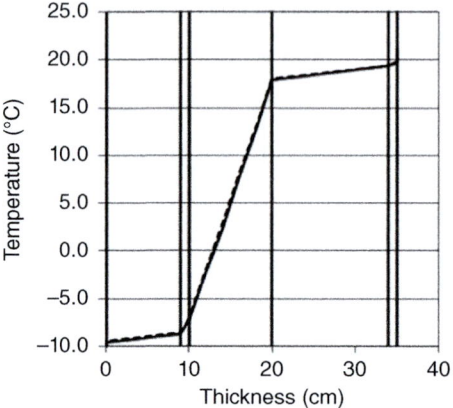

Problem 26 The cavity wall, problem (23) started with, is 2.7 m high. Outdoors it's −10 °C, indoors 21 °C. Given the wall is air permeable, thermal stack flow develops. How do stack and related heat fluxes look for such window-less one-storey high wall?

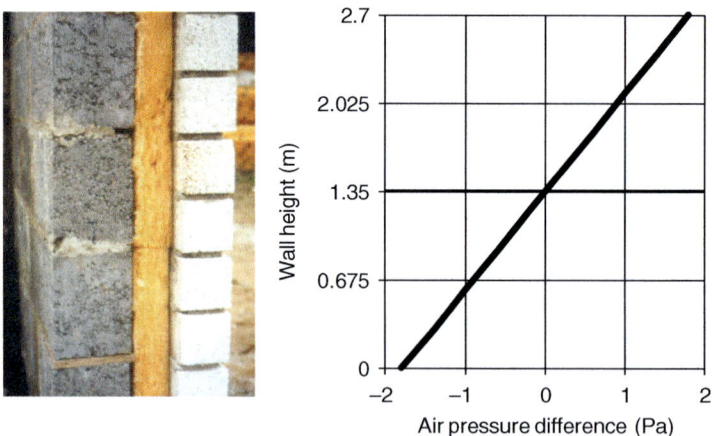

Solution 26 In such cases, the neutral plane sits at mid-height, giving a triangular pressure profile. Just above the floor 1.8 Pa underpressure forces outside air to infiltrate. Above mid-height inside air exfiltrates, with the maximum just below the ceiling, where overpressure touches 1.8 Pa. The air flux at different heights is calculated the same way as in problem (23).

For the results, see the table and figure below.

Height m	Air flow rate m³/(m²·h)
2.7	0.180
2.4	0.147
2.1	0.112
1.8	0.074
1.5	0.030
1.2	−0.030
0.9	−0.074
0.6	−0.112
0.3	−0.147
0	−0.180

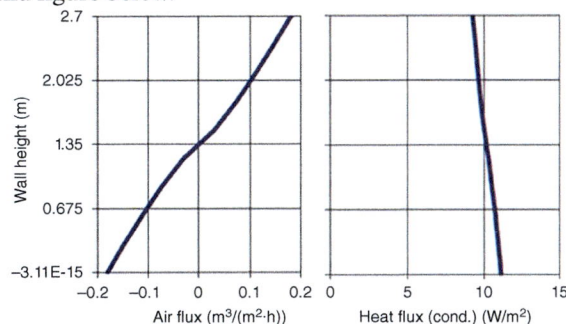

The profile stack generates is close to linear. In- and exfiltrion touch in toral a moderate 0.13 m³/(h·m). As a consequence, heat flux by conduction at the inner face only varies a little: highest above the floor, pure conduction at mid-height and lowest below the ceiling. See the figure above right. Calculating those fluxes does not differ from problem (24).

Problem 27 Given is a timber-frame wall with section:

Layer	Thickness (m)	Density (kg/m³)	λ-value (W/(m·K))	μ-value (−)	Air permeability (m³/(m²·s·Pa))
Gypsum board lining	0.012	—	0.1	12	$3.1 \cdot 10^{-5} \Delta P_a^{-0.19}$
Insulation (mineral fibre)	0.15	20	0.04	1.2	0.00081
OSB-sheathing	0.01	400	0.13	15	$3.5 \cdot 10^{-4} \Delta P_a^{-0.31}$
Building paper	0.0005	—	0.2	200	$4.2 \cdot 10^{-4} \Delta P_a^{-0.4}$
Cavity	0.03	—	0.18	0	∞
Brick veneer	0.09	1600	0.9	5	$0.32 \cdot 10^{-4} \Delta P_a^{-0.19}$

Will this design suffer from unacceptable interstitial condensation? If so, which vapour retarder should be added where? As the wall looks north–east, the boundary conditions are:

	J	F	M	A	M	J	J	A	S	O	N	D
Outdoors												
Air temp. (°C)	2.3	3.3	5.9	8.8	13.5	16.1	16.8	15.7	13.5	8.8	5.5	2.9
Eq. temp. (°C)	3.1	4.2	7.2	10.6	16.1	19.1	19.9	18.6	16.1	10.6	6.7	3.8
Vapour press. (Pa)	619	675	825	997	1277	1427	1470	1406	1277	997	804	653

	J	F	M	A	M	J	J	A	S	O	N	D
Indoors												
Air temp. (°C)	20.0	20.3	21	21.8	23.1	23.8	24	23.7	23.1	21.8	20.9	20.2
Vapour press. (Pa)	1224	1253	1330	1418	1561	1638	1660	1627	1561	1418	1319	1242

The surface film resistance inside is 0.13 m²·K/W, outside 0.04 m²·K/W. Calculation uses the equivalent temperature (Eq. temp), a value accounting for solar gains, under-cooling and the exponential relationship between temperature and vapour saturation pressure.

Solution 27 'Equivalent' diffusion is assumed to be the only driving force. Whether interstitial condensation will happen, follows from a tangent analysis for the coldest month, January.

Step 1 Temperatures in all interfaces

Thickness (m)	R (m²·K/W)	ΣR (m²·K/W)	Temp. (°C)
0	0	0	20.0
0	0.13	0.13	19.5
0.012	0.12	0.25	19.0
0.162	3.75	4.00	3.9
0.172	0.08	4.08	3.6
0.172	0	4.08	3.6
0.2025	0.17	4.25	2.9
0.2925	0.1	4.35	2.5
0.2925	0.04	4.39	2.3

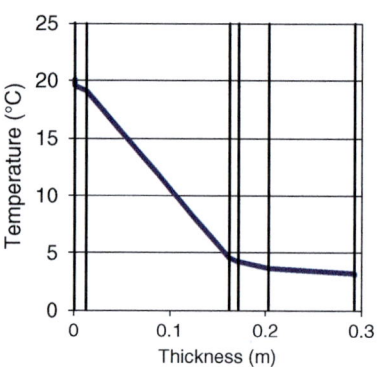

Step 2 Vapour saturation pressure in all interfaces

Temp. (°C)	p_{sat} (Pa)
20.0	2348
19.5	2273
19.0	2206
3.9	808
3.6	791
3.6	791
2.9	754
2.5	732
2.3	724

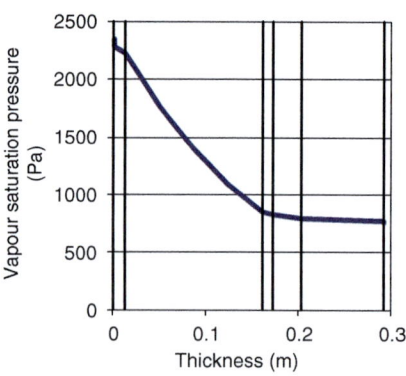

Step 3 Vapour pressure in all interfaces

Thickness (m)	μd (m)	Σμd (m)	Vapour pressure (Pa)
0	0	0	1224
0	0.007	0.007	1220
0.012	0.144	0.151	1136
0.162	0.180	0.331	1030
0.172	0.150	0.481	942
0.172	0.100	0.581	884
0.2025	0.000	0.581	884
0.2925	0.450	1.031	620
0.2925	0.001	1.032	619

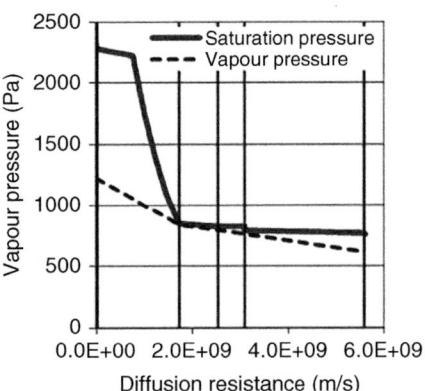

Step 4 Vapour saturation and vapour pressure intersect, or, condensate will deposit. To find the interface involved, the corrected vapour pressure curve and the amount depositing, the wall is redrawn in the $[Z, p]$ axis system and the tangents to the saturation line traced. For the result, see the last figure above. Condensate deposits at the back of the OSB, that sucks it up, so turns slowly wet. The deposit in January totals:

$$g_c = \left(\frac{1224 - 842}{1.72 \cdot 10^9} - \frac{842 - 619}{5.62 \cdot 10^9 - 1.72 \cdot 10^9} \right) \cdot 3600 \cdot 24 \cdot 31 = 0.41 \text{ kg/m}^2$$

As the 1 cm thick OSB sheathing has a density of 400 kg/m³, the result is a 10% increase in moisture ratio by weight! The condensation/drying cycle over a whole year looks (kg/m²):

	J	F	M	A	M	J	J	A	S	O	N	D
Deposit	0.41	0.33	0.16	−0.16	−0.95	−1.48	−1.72	−1.43	−0.92	−0.17	0.18	0.38
Cumulated	0.98	1.31	1.46	1.30	0.35	0	0	0	0	0	0.18	0.56

In winter, the deposit accumulating in the OSB gives as maximum moisture ratio in it 36.5% kg/kg, which is unacceptable for a timber-based material. Drying in springtime goes fast. In summer, nothing happens. As the accumulated deposit is far too high, a vapour retarder at the back of the thermal insulation is needed to return to acceptability. A rule of the thumb says that the maximum deposit in timber-based materials should not pass a 3% kg/kg, a limit convening with an end of the winter maximum of 0.12 kg/m². An iterative calculation shows a vapour retarder with vapour permeance below $3.3 \cdot 10^{-10}$ s/m suffices, which is not really a severe requirement. Expressed in terms of μd_{eq}-value, 0.56 m suffices, a value that a leak-free vapour-retarding paint could guarantee.

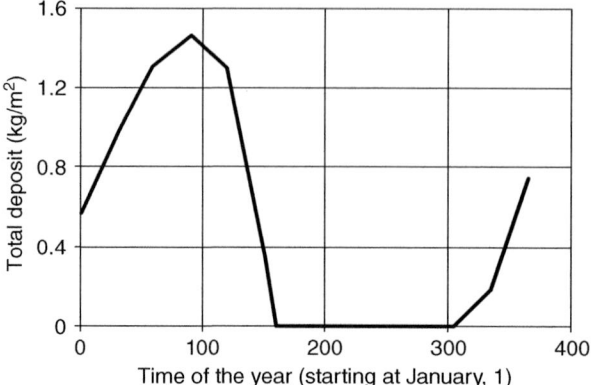

But, does this result reflect reality? Three facts, demanding a correction, were overlooked: air ex- and infiltration, hygric inertia and wind-driven rain sucked by the capillary veneer.

Correction 1. Assumed is that the difference in monthly mean in- and outdoor air temperature is the main cause for a permanent air ex- or infiltration through the upper part of the 2.7 m high wall. For reasons of simplicity, the neutral plane is supposed at mid-height. The air pressure difference under the ceiling then is:

	J	F	M	A	M	J	J	A	S	O	N	D
ΔP_a (Pa)	0.41	0.33	0.16	−0.16	−0.95	−1.48	−1.72	−1.43	−0.92	−0.17	0.18	0.38

Steady-state 'equivalent' diffusion plus convection through the timber-frame wall now gives as condensation deposit and cumulation per month (in kg/m²):

	J	F	M	A	M	J	J	A	S	O	N	D
Deposit	0.54	0.43	0.25	−0.10	−0.92	−1.47	−1.71	−1.42	−0.89	−0.10	0.28	0.50
Cumulated	1.32	1.75	2.00	1.90	0.98	0	0	0	0	0	0.28	0.78

Also, see the figure below.

Although a gypsum board typically lifts the air tightness, exfiltration remains strong enough to increase the accumulated maximum above in the OSB-sheathing with 21%, giving a moisture ratio of 44% kg/kg, which is worse than 'equivalent' diffusion only gave. Caring for air tightness is clearly a prime requirement, this, in combination with a correct vapour resistance at the back of the thermal insulation, see above. At mid-height diffusion remains the only driving force, while just above floor level, infiltration lowers the amounts deposited.

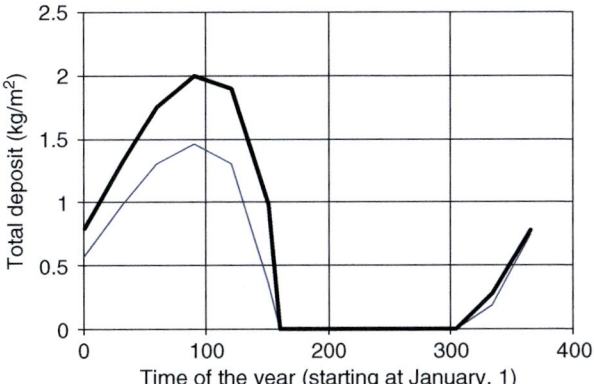

Correction 2 Calculating the impact of sorption/desorption as correctly as possible demands appropriate software. A first approximation anyhow is doable, assuming gypsum board and OSB have constant material properties, included a constant specific moisture content, while the wall is modelled as a serial network of two capacitances, coupled by resistances, whereby the cavity between building paper and brick veneer is assumed so well vented that the vapour pressure in it is the same as outdoors. Related equations so are (forward differences):

Capacitance 1: gypsum board

$$\frac{p_i^t - p_{sat,1}^t \phi_1^t}{\mu_{gyps} N d_{gyps}/2 + 1/\beta_i} + \frac{p_{sat,2}^t \phi_2^t - p_{sat,1}^t \phi_1^t}{\mu_{gyps} N d_{gyps}/2 + \mu_{insul} N d_{insul} + \mu_{OSB} N d_{OSB}/2}$$
$$= \rho_{gyps} \xi_{gyps} d_{gyps} \frac{\phi_1^{t+\Delta t} - \phi_1^t}{\Delta t}$$

Capacitance 2: OSB:

$$\frac{p_{sat,1}^t \phi_1^t - p_{sat,2}^t \phi_2^t}{\mu_{gyps} N d_{gyps}/2 + \mu_{insul} N d_{insul} + \mu_{OSB} N d_{OSB}/2}$$
$$+ \frac{p_{sat,e}^t \phi_e^t - p_{sat,2}^t \phi_2^t}{\mu_{OSB} N d_{OSB}/2 + \mu_{bp} N d_{bp} + 1/\beta_{cavity}}$$
$$= \rho_{OSB} \xi_{OSB} d_{OSB} \frac{\phi_2^{t+\Delta t} - \phi_2^t}{\Delta t}$$

Solving this system of two equations demands knowledge of the boundary conditions. For that, the climate data given above are used. At the beginning of every month, the temperature course, related vapour saturation pressure course and the vapour pressures at both sides change suddenly to remain invariable till the next month, To calculate the change in vapour pressure course during the month considered, the time step used is one day. If steady state is reached and condensate is depositing before the end of the month, calculation started, the steady state diffusion calculation for the months to come will reflect the condensate added or lost quite well.

The extra material properties needed are:

Layer	$\rho\xi d$, kg/m²
Gypsum board lining	0.46
Insulation (mineral fibre)	0
OSB-sheathing	0.93, 2.83 beyond 90% RH
Building paper	0

Starting conditions on January 1, are 80% RH in the OSB and 50% RH in the gypsum board.

The transient result learns that hygroscopic uptake during January, 0.43 kg/m² in total, progresses till January 29, see the figure below, when the RH against the backside of the OSB touches 100%, meaning condensation there starts, depositing 0.048 kg/m² during the two January days left. The droplets formed thereby start running off to the sill plate. Also in February and March, the deposit runs off, 0.485 kg/m² in total. In April, drying starts.

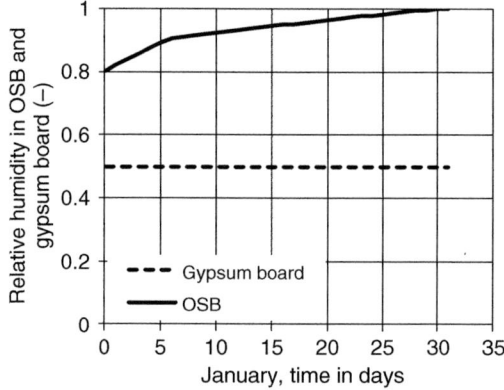

Correction 3 Inserting an air and vapour retarder at the backside of the thermal insulation is a valid solution as long as wind-driven rain does not wet the brick veneer. Assumed is that the timber-frame wall now looks south-west, the rain orientation in Western Europe, while the building paper covering the OSB, $\mu d_{eq} = 0.1$ m, is replaced by a high permeance vapour-open wrap, $\mu d_{eq} = 0.01$ m. In the cool, humid climate of Western Europe, drying of a brick veneer, wetted by successive rain events, proceeds so slowly that its RH remains 100% from November till March, often even year round.

The boundary conditions are:

	J	F	M	A	M	J	J	A	S	O	N	D
					Outdoors							
Air temp. (°C)	2.3	3.3	5.9	8.8	13.5	16.1	16.8	15.7	13.5	8.8	5.5	2.9
Eq. temp. (°C)	5.5	7.0	9.2	12.3	17.3	20.0	20.8	19.7	17.3	12.3	8.8	6.1
Vap. pres. (Pa)	619	675	825	997	1277	1427	1470	1406	1277	997	804	653
					Indoors							
Air temp. (°C)	20.0	20.3	21	21.8	23.1	23.8	24	23.7	23.1	21.8	20.9	20.2
Vap. pres. (Pa)	1224	1253	1330	1418	1561	1638	1660	1627	1561	1418	1319	1242

Although the brick veneer is kept at vapour saturation pressure from November till March, in winter, things hardly change. A tangents calculation gives some condensation at the backside of the OSB with an accumulated maximum of 0.47 kg/m², see the figure below on top. In reality, this amount mainly represents hygroscopic wetting, followed by a little deposit. In summer, each time the sun heats the veneer after being wetted by rain, a solar-driven vapour flow inwards originates with as a result condensate deposited on the wrap. On a sunny day, the amount could reach 0.16 kg/m², close to the vapour adsorbed by the OSB during the whole month of December! Without wrap, the condensate will wet the OSB! That a wet veneer gives very high vapour saturation pressures is underlined by the figures below down.

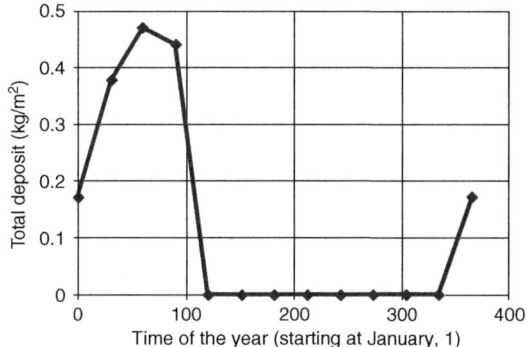

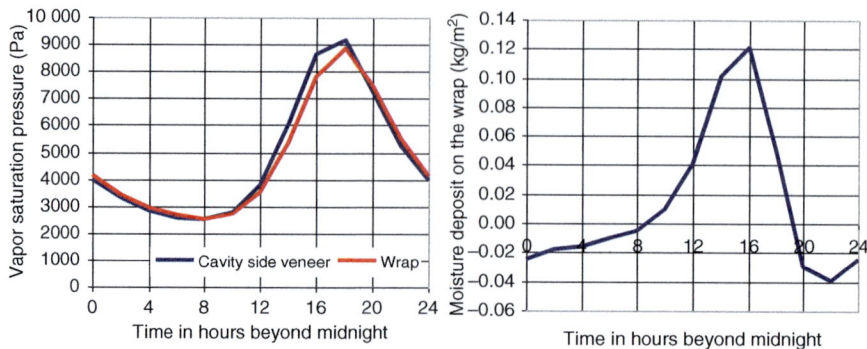

Problem 28 Given is a pitched roof assembly with as section (see figure below):

Layer	Thickness (m)	Density (kg/m³)	λ-value (W/(m·K))	μ-value (−)	Air permeability (m³/(m²·s·Pa))
Lathed ceiling	0.01	450	0.14	85	$2 \cdot 10^{-4} \Delta P_a^{-0.32}$
Insulation (mineral fibre)	0.15	20	0.04	1.2	0.00081
Capillary underlay (fibre cement)	0.0032	—	0.2	45	$4.2 \cdot 10^{-4} \Delta P_a^{-0.4}$
Cavity	0.06	—	$\lambda_{eq} = 0.33$	0	∞
Tiles	0.012	1800	1	20	$1.6 \cdot 10^{-2} \Delta P_a^{-0.5}$

Will this roof suffer from unacceptable interstitial condensation? If so, which vapour retarder should be added where?

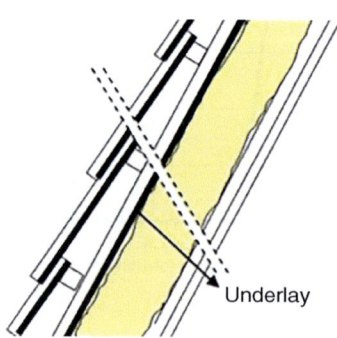

The boundary conditions to be considered are (slope: 30°, SW):

	J	F	M	A	M	J	J	A	S	O	N	D
					Outdoors							
Air temp. (°C)	2.3	3.3	5.9	8.8	13.5	16.1	16.8	15.7	13.5	8.8	5.5	2.9
Eq. temp. (°C)	3.9	5.2	8.7	12.6	19.1	22.5	23.5	22.1	19.1	12.6	8.2	4.7
Vap. pres. (Pa)	619	675	825	997	1277	1427	1470	1406	1277	997	804	653
					Indoors							
Air temp. (°C)	20.0	20.3	21	21.8	23.1	23.8	24	23.7	23.1	21.8	20.9	20.2
Vap. pres. (Pa)	1224	1253	1330	1418	1561	1638	1660	1627	1561	1418	1319	1242

The surface resistances are 0.13 m²·K/W indoors and 0.04 m²·K/W outdoors. Stack height above the neutral plane is 3 m. The underlay is capillary active. Although the tiled roof cover is wind washed, the <equivalent> diffusion mode neglects this. Tiles and cavity are considered to be separate layers with as properties the ones listed in the table. Instead, when looking to the impact of stack, wind-washing is accounted for by replacing the cavity and tiled cover by a surface resistance 0.1 m² · K/W and a vapour resistance 0.

Calculations are done using the equivalent temperature.

Solution 28 It is up to the reader to solve and comment on the case.

If <equivalent> diffusion was the only driving force, no interstitial condensation should be noted against the backside of the underlay at the end of January, see the first figure below. Anyhow, with the air outflow equal to $1.49 \cdot 10^{-4} \Delta P_a^{0.68}$ m³/(m²·s), convection adds, causing condensation at the backside of the underlay. The amounts touch:

End of	J	F	M	A	M	J	J	A	S	O	N	D
Stack (Pa)	2.28	2.18	1.91	1.62	1.17	0.93	0.87	0.97	1.17	1.62	1.95	2.22
Airflow (kg/(m²·h))	1.13	1.10	1.00	0.90	0.72	0.61	0.58	0.63	0.72	0.90	1.02	1.11
Cond (kg/m²)	3.26	4.39	4.55	3.35	0.00	0.00	0.00	0.00	0.00	0.00	0.32	1.69

Interface	Temp. (°C)	p_{sat} (Pa)	p (Pa)
	20	2343	1224
$1/h_i$	19.5	2271	1221
Lathed ceiling	19.2	2233	859
Insulation (mineral fibre)	4.9	865	783
Underlay	4.8	861	722
Cavity	4.1	820	722
Tiles	4.1	817	620
$1/h_e$	3.9	808	619

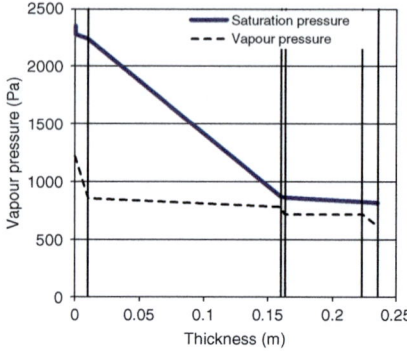

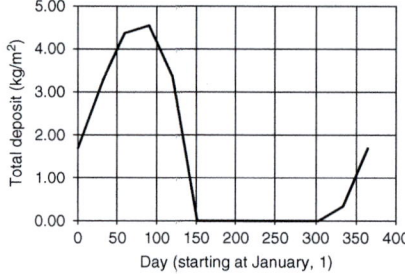

A maximum of 4.55 lkg per m² at the end of March looks really problematic. So, not a vapour but an air retarder at the warm side of the insulation is needed! In reality, however, these amounts will not cause problems. In fact, the capillary underlay will turn wet over a thickness such that the condensate sucked will evaporate and diffuse to the wind-washed air layer under the tiled deck at a rate equal to the vapour flux coming from indoors. Anyhow, an air retarder still remains positive as it will add airtightness to the building enclosure.

Problem 29 Back to the timber-frame wall of problem (27). It now looks north. Assumed is that it was designed for use in a cold climate with as air and vapour retarder a gypsum board lining, finished with vinyl wallpaper. But, as designed, it is applied now in a hot and humid climate, where cooling and dehumidification are needed to keep the building comfortable. The indoors are cooled to an air temperature not passing 20 °C, which may be quoted as too low from a thermal comfort point of view, but is not the exception as experience learns. What problems could be expected in case the veneer stays either dry or wet, due to wind-driven rain each day.

The layer properties are:

Layer	Thickness (m)	Density (kg/m³)	λ-value (W/(m·K))	μ-value (−)	μd-value (m)
Vinyl paper	—	—	—	—	1
Gypsum board lining	0.012	—	0.1	12	
Insulation (mineral fibre)	0.15	20	0.04	1.2	
OSB-sheathing	0.01	400	0.13	15	
Building paper	0.0005	—	0.2	200	
Cavity	0.03	—	0.18		0
Brick veneer	0.09	1600	0.9	5	

The maximum amount of moisture that can stick to the building paper without running-off is $100\,\text{g/m}^2$, while the capillary moisture content of gypsum board is $150\,\text{kg/m}^3$. The surface film resistances are $0.13\,\text{m}^2\cdot\text{K/W}$ indoors and $0.04\,\text{m}^2\cdot\text{K/W}$ outdoors. As the wall looks north, the boundary conditions to be considered are:

	J	F	M	A	M	J	J	A	S	O	N	D
					Outdoors							
Air temp. (°C)	22.5	23.2	25.3	28.1	30.9	32.9	33.6	32.9	30.9	28.1	25.3	23.2
Vap. pres. (Pa)	2185	2279	2585	3047	3579	4008	4168	4008	3579	3047	2585	2279
					Indoors							
Air temp. (°C)	20	20	20	20	20	20	20	20	20	20	20	20
Vap. pres. (Pa)	1910	1913	1895	1808	1656	1512	1455	1512	1656	1808	1895	1913

Solution 29 In case of <equivalent> diffusion only without rain, interstitial condensate will accumulate year round at the backside of the gypsum board with following quantities at the end of each month during the first year:

End of cond (kg/m²)	J	F	M	A	M	J	J	A	S	O	N	D
	0.000	0.000	0.000	0.071	0.217	0.460	0.749	1.000	1.141	1.146	1.016	0.773

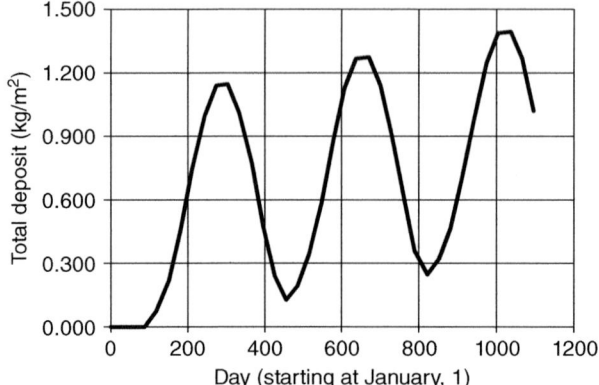

When additionally abundant wind-driven rain keeps the veneer wet, 100% relative humidity will be maintained at its backside with quite devastating consequences for the wall as a whole, see the figure below (Glaser analysis)

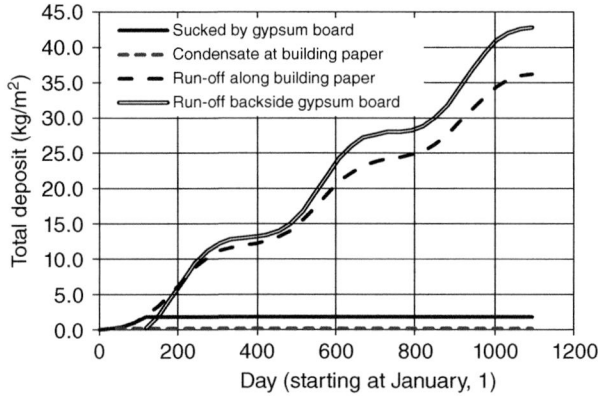

Problem 30 Given is a 3 m high masonry cavity wall with as layer properties:

Layer	Thickness (m)	Density (kg/m³)	λ-value (W/(m·K))	μ-value (–)	μd-value (m)
Gypsum plaster	0.01	980	0.3	7	
Inside leaf (fast bricks)	0.14	1400	0.5	5	
Insulation (mineral fibre)	0.12	30	0.04	1.2	
Cavity	0.03	400	0.18		0
Brick veneer	0.09	1600	0.9	5	

The cavity is vented by two open head joints per running metre at the bottom and top, the four with section $1 \times 6.5 \text{ cm}^2$ and 9 cm deep. Either the insulation is correctly mounted, or it sits centrally in the cavity with a 1.5 cm wide air layer left at both sides, or it touches the brick veneer. The tray at the bottom guarantees an open contact between the cavity or the remaining air layers and the open head joints, while up, the insulation stops just below the open head joints. This gives some thermal bridging, which is not considered. To what extend in the three cases does the venting flow affects the thermal transmittance when on average a 4 Pa air pressure difference exists between the open head joints on top (highest air pressure) and at the bottom (lowest air pressure). The surface film coefficients are 25 W/(m²·K) outdoors and 8 W/(m²·K) indoors. Indoors, it's 20 °C, outdoors −10 °C.

Solution 30 Only the case with the insulation centrally in the cavity is considered, see the figure.

The two open head joints per meter run at the top and the bottom fix the airflow. Their hydraulic resistance in fact outweighs the flow friction exerted by the two 1.5 cm thick air layers at both sides of the insulation. Related pressure equilibrium so is:

$$\Delta P_a = 2\left[1.5\frac{\rho_a}{2}\left(\frac{G_a}{\rho_a 2A_{\text{head joint}}}\right)^2 + 0.42\frac{96\nu\rho_a 2A_{\text{head joint}}}{G_a d_H}\right.$$

$$\left.\cdot\frac{L}{d_H}\rho_a\left(\frac{G_a}{\rho_a 2A_{\text{head joint}}}\right)^2\right]$$

$$\approx 0.32\left(\frac{G_a}{0.00065}\right)^2 + 0.08\left(\frac{G_a}{0.00065}\right) = 757400 G_a^2 + 12.3 G_a$$

For $\Delta P_a = 4$ Pa, the outcome is an airflow top down of 0.0023 kg/s. As the air layers at both sides of the thermal insulation have equal thickness, this flow splits in two, one washing the air layer in front, the other the air layer at the back of the insulation. As its thermal resistance touches 3 m²·K/W, the temperature in the air layer behind the brick veneer can, without much deviation, be set equal to the one outdoors. This way the problem does not differ from calculating the equivalent thermal transmittance of a wall with vented cavity between the insulation with thermal resistance $R_1 = 3.07$ m²·K/W and the plastered inside leaf with, included the surface resistance inside, thermal resistance $R_2 = 0.44$ m²·K/W. For all constants advanced, see 'Vented cavities' in this chapter. The constant D for example equals:

$$D = (3.33 + 4.45 + 1/3.07) \cdot (3.33 + 4.45 + 1/0.438) - 4.45^2 = 61.9$$

In this sum, 3.33 W/(m²·K) is the convective surface film coefficient at the bounding faces of the cavity and 4.45 W/(m²·K) its radiant surface film coefficient. To remind, for a 1.5 cm wide air layer the Nusselt number is 1 and the convective surface film coefficient: $0.025/0.015 \cdot 2 = 3.33$ W/(m²·K). The other constants intervening are $A_1 = 0.0531$, $A_2 = 0.0235$, $B_1 = 0.1643$, $B_2 = 0.2993$, $C_1 = 0.7826$, $C_2 = 0.6773$. The temperatures at the faces bounding the cavity so are:

$$\theta_{s1} = 2.8 + 0.783\theta_{cav} \quad \theta_{s1} = 5.7 + 0.677\theta_{cav}$$

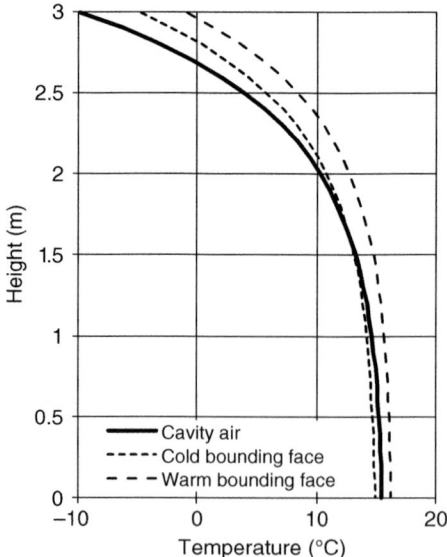

The figure gives the temperatures along the cavity. It's interesting to see how radiation impacts these temperatures. In fact, just below the air inlet, the warm and cold faces are warmer than the air in the cavity.

The equivalent thermal transmittance (U_{eq}) becomes:

$$U = U_0 \left\{ 1 + \frac{C_2 b_1 R_1}{L R_2} \left[1 - \exp\left(-\frac{L}{b_1}\right) \right] \right\}$$

with

$$b_1 = \frac{1008 G_a}{h_c(2 - C_1 - C_2)}$$

Calculation gives 0.55 W/(m²·K). Without the wind-washed 1.5 cm thick air layer behind the thermal insulation, the clear wall thermal transmittance should be 0.27 W/(m²·K) or, the increase by wind-washing passes 100%, which is unacceptable.

For the other two cases, it is up to the reader to solve.

Further Reading

Abuku, M. (2009). Moisture stress of wind-driven rain on building enclosures. Doctoral thesis, KU-Leuven.

Arfvidsson, J. (1998). Moisture transport in porous media, modelling based on Kirchhoffs potentials. Doctoral thesis. Lund University.

ASHRAE (2009). *Handbook of Fundamentals*. Atlanta, GA: ASHRAE.

ASHRAE (2021). *Handbook of Fundamentals*, SIe. Tullie Circle, Atlanta, GA: ASHRAE.

Blocken, B. (2004). Wind-driven rain on buildings. Doctoral thesis. KU-Leuven.

Blocken, B., Hens, H., and Carmeliet, J. (2002). Methods for the quantification of driving rain on buildings. *ASHRAE Transactions* 108 (Part 2): 338–350.

Bomberg, H. (1971). Waterflow Through Porous Materials, Part 1, Methods of Water Transport Measurements. Report 19, Lund Institute of Technology, Section of Building Technology.

Bomberg, H. (1971). Waterflow Through Porous Materials, Part 2, Relative Suction Model. Report 20, Lund Institute of Technology, Section of Building Technology.

Bouwresearch, S. (1969). *Vochttransport in en droging van bouwmaterialen: fundamentele grondslagen*, 82. N. Samsom nv Alphen aan den Rijn.

Bouwresearch, S. (1971). *Inwendige condensatie*, 84. N. Samsom nv Alphen aan den Rijn.

Brocken, H. (1998). Moisture transport in brick masonry: the grey area between bricks. Doctoral thesis. TU/e, Eindhoven.

Brunauer, S., Emmett, P.H., and Teller, E. (1938). Adsorption of gases in multimolecular layers. *Journal of the American Chemical Society* 60 (2): 309–319.

Cammerer, J.S. (1962). *Wärme- und Kälteschutz in der Industrie*. Berlin/Heidelberg/New York (in German): Springer Verlag.

Carmeliet, J., Hens, H., and Vermeir, G. (ed.) (2003). *Research in Building Physics*, 1020. Lisse, Abingdon, Exton, Tokyo: Balkema Publishers.

Chaddock, J.B. and Todorovic, B. (1991). *Heat and Mass Transfer in Buiding Materials and Structures*. New York: Hemisphere Publishing Corporation.

Cranck, J. (1956). *The Mathematics of Diffusion*. Oxford: Clarendon Press.

De Grave, A. (1957). *Bouwfysica 1*. Brussel (in Dutch): Uitgeverij SIC.

de Wit, M. (1995). *Warmte en vocht in constructies*. Eindhoven: diktaat TUE (textbook in Dutch).

Devries, D.A. (1958). Simultaneous transfer of heat and moisture in porous media. *Transactions of the American Geophysical Union* 39: 909–916.

Feynman, R., Leighton, R., and Sands, M. (1977). *Lectures on Physics*, vol. 1. Reading, MA: Addison-Wesley Publishing Company.

Garrecht, H. (1992). Porenstrukturmodelle für den Feuchtehaushalt von Baustoffen mit und ohne Salzbefrachtung und rechnerische Anwendung auf Mauerwerk. Doktors Abhandlung. Universität Karlsruhe, 267 p. (doctoral thesis in German).

Glaser, H. (1958). Temperatur- und Dampfdruckverlauf in einer homogenen Wand bei Feuchtigkeitsausscheidung. Kältetechnik, n° 6 (in German).

Glaser, H. (1958). Vereinfachte Berechnung der Dampfdiffusion durch geschichtete Wände bei Ausscheiden von Wasser und Eis. Kältetechnik, n° 11 & n° 12 (in German).

Glaser, H. (1958). Wärmeleitung und Feuchtigkeitsdurchgang durch Kühlraumisolierungen. Kältetechnik, n° 3 (in German).

Glaser, H. (1959). Grafisches Verfahren zur Untersuchung von Diffusionsvorgängen. Kältetechnik, n° 10 (in German).

Gösele, K. and Schüle, W. (1973). *Schall, Wärme, Feuchtigkeit*, 3e. Wiesbaden-Berlin (in German): Bauverlag GmbH.

Hagentoft, C.E. (2001). *Introduction to Building Physics*. Lund: Studentlitteratur.

Hall, C. and D'Hoff, W. (2002). *Water Transport in Brick, Stone and concrete*, 318. London and New York: Spon Press.

Harmathy, T.Z. (1967). Moisture sorption of building materials. Technical Paper nr 242, NRC, Division of Building Research, Ottawa.

Häupl, P. and Roloff, J. (1999). *Proceedings of the 10*. Bauklimatisches Symposium, Band 1 & 2, TU Dresden, Institut für Bauklimatik.

Häupl, P. and Roloff, J. (2002). *Proceedings of the 11*. Bauklimatisches Symposium, Band 1 & 2, TU Dresden, Institut für Bauklimatik.

Hens, H. (1975). Theoretische en experimentele studie van het hygrothermisch gedrag van bouw- en isolatiematerialen bij inwendige condensation en droging, met toepassing op de platte daken, doctoraal proefschrift, KULeuven. Doctoral thesis in Dutch.

Hens, H. (1978, 1981). Bouwfysica, Warmte en Vocht, Theoretische grondslagen, 1^e en 2^e uitgave. ACCO, Leuven (in Dutch).

Hens, H. (1992, 1997, 2000). Bouwfysica 1, Warmte en Massatransport, 3^e, 4^e en 5^e uitgave, ACCO, Leuven (in Dutch).

Hens, H. (1996). Modelling. Vol 1 of the Final Report Task 1, IEA-Annex 24, ACCO, Leuven, 90 p.

Holm, A. (2001). Ermittlung der Genauigkeit von instationären hygrothermischen Bauteilberechnungen mittels eines stochastischen Konzeptes. Doktors Abhandlung. Universität Stuttgart (doctoral thesis in German).

IEA- Annex 14 (1990). *Condensation and Energy: Guidelines and Practice*. Leuven: ACCO.

Janssen, H. (2002). The influence of soil moisture transfer on building heat loss via the ground. Doctoral thesis. KU-Leuven.

Janssens, A. (1998). Reliable control of interstitial condensation in lightweight roof systems. Doctoral thesis. KU-Leuven.

Kalagasidis, A. (2004). HAM-tools, an integrated simulation tool for heat, air and moisture transfer analysis in building physics. Doctoral thesis. Chalmers University of Technology.

Kiessl, K. (1983). Kapillarer und dampfförmiger Feuchtetransport in mehrschichtigen Bauteilen. Dissertation. Essen (doctoral thesis in German).

Klopfer, H. (1974). *Wassertransport durch Diffusion in Feststoffen*, 1e. Wiesbaden-Berlin (in German): Bauverlag GmbH.

Kohonen, R. and Ojanen, S. (1985). Coupled convection and conduction in two dimensional building structures. *4th Conference on Numerical Methods in Thermal Problems*, Swansea.

Kohonen, R. and Ojanen, T. (1987). Coupled diffusion and convection heat and mass transfer in building structures. *Building Physics Symposium*, Lund.

Kreith, F. (1976). *Principles of Heat Transfer*. New York: Harper & Row Publishers.

Krischer, O. (1963). *Die wissenschaftlichen Grundlagen der Trocknungstechnik*, 2e. Berlin (in German): Springer Verlag.

Krus, M. (1995). Feuchtetransport- und Speicherkoeffizienten poröser mineralischer Baustoffe. Theoretische Grundlagen und neue Messtechniken, Doktors Abhandlung, Universität Stuttgart (in German).

Künzel, H.M. (1994). Verfahren zur ein- und zweidimensionalen Berechnung des gekoppelten Wärme- und Feuchtetransports in Bauteilen mit einfachen Kennwerten. Doktors Abhandlung. Universität Stuttgart (doctoral thesis in German).

Langmans, J (2013). Feasibility of exterior air barriers in timber frame construction. Doctoral thesis. KU-Leuven.

Langmans, J., Klein, R., and Roels, S. (2012). Hygrothermal risks of using exterior air barrier systems for highly insulated light weight walls: a laboratory investigation. *Buildings and Environment* 58: 209–218.

Langmuir, I. (1918). The adsorption of gases on plane surfaces of glass, mica and platinum. *Journal of the American Chemical Society* 40 (9): 1361–1403.

Liddament, M.W. (1996). A Guide to Energy Efficient Ventilation. Report AIVC Annex 5, 254 pp.

Luikov, A. (1966). *Heat and Mass Transfer in Capillary Porous Bodies*. Oxford: Pergamon Press.

Lutz, P., Jenisch, R., Klopfer, H. et al. (1989). *Lehrbuch der Bauphysik*. Stuttgart (in German): B.G. Teubner Verlag.

Nevander, E.L. and Elmarsson, B. (1981). *Fukt-handbok*. Stockholm (in Swedisch): Svensk Byggtjänst.

Pato, Sectie Bouwkunde. (1986). Syllabus van de leergang 'Leidt energiebesparing tot vochtproblemen'. Delft 30 sept-1 okt. (a series of lectures in Dutch).

Pedersen, C.R. (1990). Combined heat and moisture transfer in building constructions. Ph.D. thesis. Technical University of Denmark.

Roels, S. (2000). Modelling unsaturated moisture transport in heterogeneous limestone. Doctoral thesis. KU-Leuven.

Rose, W.B. (2005). *Water in Buildings*, 270. Wiley.

Sedlbauer, K. (2001). Vorhersage von Schimmelpilzbildung auf und in Bauteilen. Doktors Abhandlung. Universität Stuttgart (doctoral thesis in German).

Taveirne, W. (1990). *Eenhedenstelsels en groothedenvergelijkingen: overgang naar het SI*. Wageningen (in Dutch): Pudoc.

TI-KVIV, Kursus Thermische Isolatie en Vochtproblemen in Gebouwen, 1976-1979 -1980-1981-1983-1985 (a series of lectures in Dutch).

Time, B. (1998). Hygroscopic moisture transport in wood. Doctoral thesis. NUST Trondheim.

Trechsel, H.R. (ed.) (1994). Moisture Control in Buildings. ASTM Manual Series, MNL 18.

TU-Delft, Faculteit Civiele Techniek, Vakgroep Utiliteitsbouw-Bouwfysica, Bouwfysica, naar de colleges van Prof A.C.Verhoeven, 1975–1985 (a series of text books in Dutch).

Van der Kooy, J. (1971). *Moisture Transport in Cellular Concrete Roofs*. Delft: Uitgeverij Waltman.

Van Mook, J.R. (2003). Driving rain on building envelopes. *TU/e Bouwstenen* 69: 198.

Vos, B.H. and Coelman E.J.W. (1967). Condensation in Structures. Report Nr. BI-67-33/23 TNO-IBBC, Rijswijk.

Vos, B.H. and Tammes, E. (1969). Moisture and Moisture Transfer in Porous Materials. Report Nr. B1-69-96/03.1.001, TNO-IBBC, Rijswijk.

Welty, J.R., Wicks, C.E., and Wilson, R.E. (1969). *Fundamentals of Momentum, Heat and Mass Transfer*. New York: Wiley.

Woloszyn, M. and Rode, C. (2008). Modelling Principles and Common Exercises. Final Report IEA ECBCS Annex 41, ACCO, Leuven.

Zillig, W. (2009). Moisture transport in wood using a multiscale approach. Doctoral thesis. KU-Leuven.

3

Heat, Air and Moisture Combined

3.1 Why?

In the previous two chapters, heat, air and moisture were mainly treated separately and, if combined, this was mainly limited to heat/air, heat/vapour and heat/vapour/air only. In reality, the three strongly interfere. Indeed, mass transfer means enthalpy flow, thus, heat transfer. Temperature differences in turn induce vapour and air displacement, take natural convection, but also liquid movement in and between the pores. The volumetric specific heat capacity and the thermal conductivity are both function of the moisture content, while the moisture transfer characteristics in turn depend on temperature. These couplings are the reason why a separate or a heat/air and vapour/air analysis may simplify things to a level a loss of reality value looms. That is why in this third chapter, heat, air and moisture combined is the subject of discussion.

3.2 Material and Assembly Level

3.2.1 Assumptions

Treating combined heat, air and moisture transport mathematically exact is not doable. Soluble substances precipitating, salts crystallizing and hydrating and possibly microscopic small particles depositing, all transported in and through a material with the moving water, alter the material matrix, modify the pores and change the specific pore surface. As these effects are quasi-unpredictable, therefore the material matrix has to be assumed invariable. As a consequence, material characteristics such as the dry density, the pore distribution and the specific pore surface are given. Any porous material is further seen as composed of a really large number of infinitely small representative volumes (V_{rep}), which all have the same properties as the material as a whole. A continuum approach so mirrors to some extent the calculation per representative volume of the evolving average values of the combined transfer modes analysed.

3.2.2 Solution

A combined heat, air and moisture problem is considered solved when the temperature, the air pressure, the moisture content and the related potentials and fluxes are known in each point of the solution space, here a material layer or an assembly. As potentials are scalar and fluxes vectorial, to get their values a system of three scalar and three vector equations, added the necessary equations of state, has to be solved. Equations of state describe the thermodynamic equilibrium between potentials and their relations with certain material characteristics. In addition, to solve, also the geometry, the initial, the boundary and the contact conditions have to be known. Basic to any prediction then are conservation of mass, energy and momentum. For the mass fluxes, empirical equations typically replace the conservation of momentum, albeit convection in air and mass movement in highly permeable materials can be captured using Computerized Fluid Dynamics (CFD) to solve the Navier-Stokes's equations and their backing by a turbulence model.

3.2.3 Conservation of Mass

Below $0\,^{\circ}C$, the pores in a moist material may contain dry air, liquid water, water vapour, ice, soluble substances, dissolved salts and microscopic small particles. Above $0\,^{\circ}C$, ice melting adds liquid water.

Soluble substances, dissolved salts and microscopic small particles are not discussed. For liquid water, water vapour, ice and air, the mass per unit material volume is used to describe the quantities present, whereby "moisture content (w)" bundles the three water states in one with the vapour part so unimportant compared to the other two that it looks as if negligible, or:

	$\theta < 0$	$\theta \geq 0$
Water	$w_l = \dfrac{m_l}{V}$	$w_l = \dfrac{m_l}{V}$
Vapour	$w_v = \dfrac{m_v}{V}$	$w_v = \dfrac{m_v}{V}$
Ice	$w_i = \dfrac{m_i}{V}$	
Air	$w_a = \dfrac{m_a}{V}$	
Moisture	$w = w_l + w_i$	$w = w_l$

The air content can also be written as $w_a = \Psi_{of}\rho_a$ with Ψ_{of} the open pore volume not containing liquid or ice and ρ_a the air density. This gives for the vapour present in a pore filled with air $w_v = x_v w_a$ with x_v the vapour ratio in kg/kg.

Applying mass conservation to the three components fixing the moisture content gives for liquid water:

$$\operatorname{div} \mathbf{g_w} \pm G'_1 = -\frac{\partial w_l}{\partial t} \tag{3.1}$$

with $\mathbf{g_w}$ the water flux and G'_1 a water source due to vapour condensing or a water sink due to liquid evaporating. For vapour, the mass balance looks:

$$\text{div}\, \mathbf{g_v} \pm G'_2 = -\frac{\partial w_v}{\partial t} \tag{3.2}$$

with $\mathbf{g_v}$ the vapour flux and G'_2 a vapour source due to evaporation of liquid or a vapour sink due to condensation. For hygroscopic materials, the (de)sorption moisture content w_H includes the vapour content while buffering is given by the derivative of w_H to time. Above freezing, summing up the water sources and vapour sinks let them disappear as condensation is a water source but a vapour sink, while evaporation is vapour source but a water sink. For ice, the mass balance becomes:

$$\text{div}\, \mathbf{g_i} \pm G'_3 = -\frac{\partial w_i}{\partial t}$$

with $\mathbf{g_i}$ the ice flux and G'_3 an ice source due to freezing of liquid or an ice sink due to melting or sublimating. In porous materials, ice formation in the largest pores starts at 0 °C. Below 0 °C, stepwise, water in the smaller pores starts freezing. Because being a solid, ice can hardly move, giving a flux zero, which changes the mass balance into:

$$\pm G'_3 = -\frac{\partial w_i}{\partial t} \tag{3.3}$$

As together, the liquid, vapour and ice sources and sinks obey $G'_1 + G'_2 + G'_3 = 0$, combining the three mass conservation equations into one, using moisture content in the material (w) as storage variable, gives:

$$\text{div}(\mathbf{g_w} + \mathbf{g_v}) = -\frac{\partial w}{\partial t} \tag{3.4}$$

Moving to air, since in the pores air sources nor sinks have to be considered, mass conservation gives:

$$\text{div}\, \mathbf{g_a} = -\frac{\partial w_a}{\partial t} = -\Psi_r \frac{\partial \rho_a}{\partial t}$$

with $\mathbf{g_a}$ the air flux. Compared to moisture and heat, the air content currently changes so fast that the inertia term can be set ≈ 0, or:

$$\text{div}\, \mathbf{g_a} = 0 \tag{3.5}$$

3.2.4 Conservation of Energy

Heat conduction saw conservation of energy applied to materials in which no mass flow occurred. Adding air in- and outflow could give quite an important concomitant enthalpy flow. Now, the law is extended to porous materials, who see air, liquid and vapour moving through, which is adding an even more important enthalpy flow:

$$\mathbf{q_j} = \mathbf{g_j}(h_j - h_{o,j}) \tag{3.6}$$

In it, h_j is the specific enthalpy of mass component j, units' J/kg, and $h_{o,j}$ the reference value at 0 °C, typically set zero. As the kinetic and frictional energies are too small to have impact, energy conservation becomes:

$$\text{div}\left[\mathbf{q} + \sum_{j=1}^{3}(\mathbf{g}_j h_j)\right] \pm \Phi' = -\frac{\partial}{\partial t}\left[\rho_0 c_0 \theta + \sum_{j=1}^{4}(w_j h_j)\right] \tag{3.7}$$

with $\rho_0 c_0$ the volumetric specific heat capacity of the material matrix. The term Φ' represents the heat dissipated by processes other than the heat of evaporation h_{bo} and melting h_{mo} released or absorbed and linked to changes of state of the water or ice present. These are part of the sum $\Sigma(w_j h_j)$ and have as value at 0 °C: 2 500 000 and 330 000 J/kg. The enthalpies h_j in turn look:

Water (1) $h_1 = c_1(T - T_0) = c_1\theta$
Vapour (2) $h_v = c_v(T - T_0) + h_{bo} = c_v\theta + h_{bo}$
Ice (3) $h_i = c_i(T - T_0) + h_{mo} = c_i\theta + h_{mo}$ ($T_0 = 273.15$ K or $\theta_0 = 0°$C)
Dry air (4) $h_a = c_a(T - T_0) = c_a\theta$

When the material matrix and all mass components have the same temperature in each representative volume, an assumption that holds in pores and for low flow velocities, not for air passing through apertures, cavities, cracks and fissures, then enthalpy changes can also be written as:

$$dh_j = c_j d\theta \qquad \text{grad } h_j = c_j \text{grad } \theta$$

Because of that, energy conservation converts to:

$$-\text{div } \mathbf{q} = \sum_{j=1}^{3}(h_j \text{ div } \mathbf{g_j} + \mathbf{g_j} \text{ grad } h_j) + \frac{\partial}{\partial t}\left\{\left[\rho_0 c_0 + \sum_{j=1}^{4}(c_j w_j)\right]\theta\right\}$$

$$+ \sum_{j=1}^{4}\left(h_j \frac{\partial w_j}{\partial t}\right) \pm \Phi' \tag{3.8}$$

a vector equation, whereby reshuffling gives:

$$-\text{divq} - \sum_{j=1}^{3}(\mathbf{g_j} c_j \text{grad } \theta)$$

$$= \underbrace{\sum_{j=1}^{3}(h_j \text{divg}_j) + \sum_{j=1}^{4}\left(h_j \frac{\partial w_j}{\partial t}\right)}_{(2)} + \underbrace{\frac{\partial}{\partial t}\left\{\left[\rho_0 c_0 + \sum_{j=1}^{4}(c_j w_j)\right]\theta\right\}}_{(1)} \pm \Phi'$$

Term (1) contains the volumetric heat capacity of the matrix and of the three or four mass components intervening. As the values for vapour and air are negligible, this term can be simplified to:

$$\left[\rho_0 c_0 + \sum_{j=1}^{4}(c_j w_j)\right] = (\rho_0 c_0 + c_1 w_1 + c_i w_i) = \rho_0 c' \text{ with } c' = c_0 + \frac{c_1 w_1 + c_i w_i}{\rho_0}$$

$$\tag{3.9}$$

Because for ice the flux (g_i) is zero, term (2) can be reset as:

$$\sum \left[h_j \underbrace{\left(\mathrm{div} \mathbf{g}_j + \frac{\partial w_j}{\partial t} \right)} \right]$$

The part above the brace represents mass conservation per component if no source or sink intervene, thus equals:

$$\sum_{j=1}^{3} h_j G'_j \tag{3.10}$$

Sum 1–3 indicates that changes of state are excluded for number 4, dry air. For ice, number 3, the source term (G'_3) is zero above 0 °C, turning $G'_1 = -G'_2$ and the sum in:

$$\sum_{j=1}^{3} h_j G'_j = G'_1 (c_1 \theta - l_{b0} - c_v \theta) = -G'_1 l_b(\theta) \text{ with } l_b(\theta) = l_b(0) + (c_v - c_1)\theta$$

When freezing, $G'_1 + G'_2 + G'_3 = 0$, which changes that sum to:

$$\sum_{j=1}^{3} h_j G'_j = -G'_1 l_b(\theta) - G'_3 l_m(\theta)$$

Inserting both relations in the reshuffled conservation equation gives:

Above 0 °C: $\quad \underbrace{-\mathrm{div}\mathbf{q}}_{(1)} - \underbrace{\sum_{j=1}^{3} (g_j c_j \mathbf{grad}\theta)}_{(2)} = \underbrace{G'_1 l_b}_{(3)} + \underbrace{\frac{\partial}{\partial t}(\rho_0 c' \theta)}_{(4)} \pm \underbrace{\Phi'}_{(5)}$

(3.11)

Below and at 0 °C $\underbrace{-\mathrm{div}\mathbf{q}}_{(1)} - \underbrace{\sum_{j=1}^{3} (g_j c_j \mathbf{grad}\theta)}_{(2)} = \underbrace{G'_1 l_b - G'_3 l_m}_{(3)} + \underbrace{\frac{\partial}{\partial t}(\rho_0 c' \theta)}_{(4)} \pm \underbrace{\Phi'}_{(5)}$

(3.12)

Term (1) in both gives the heat transferred by "equivalent" conduction, which, as explained in Chapter 1, includes conduction along the matrix, conduction and convection in the pore gas, radiation between the pore walls, conduction in all absorbed water layers and latent heat transfer caused by evaporation and condensation in pores containing moisture. All this turns the material property "thermal conductivity" into a function of temperature, moisture content, thickness, etc. Term (2) quantifies the sensible enthalpy transfer linked to all mass fluxes with moving air having the largest impact. Term (3) stands for the latent enthalpy transfer by changes of state in the pores between liquid, vapour and ice with evaporation and condensation dominating. Term (4) concerns the heat stored, mainly in the material matrix and the liquid present. Term (5) finally considers other heat sources and sinks, take due to chemical reactions.

3.2.5 Flux Equations

3.2.5.1 Heat
For heat conduction, Fourier's law counts:

$$\mathbf{q} = -\lambda \operatorname{grad} \theta$$

with λ the (equivalent) thermal conductivity in W/(m.K), a number for isotropic, a tensor for anisotropic materials. Air layers demand an own approach, see Chapter 1.

3.2.5.2 Mass, Air
In open porous materials, Poiseuille's law for air moving equals:

$$\mathbf{g_a} = -k_a \operatorname{grad} P_a$$

with P_a the air pressure included stack (Pa), see Chapter 2, and k_a the air permeability (s):

$$k_a = \frac{1}{W' v_a} = \frac{B}{v_a}$$

where v_a is the kinematic viscosity of air and B the penetration coefficient of the material in m²/s², a characteristic of any porous system not impacted by the kind of fluid or gas flowing through. Again, the air permeability is a number for isotropic and a tensor for anisotropic materials, while its value nears 0 for capillary wet materials. For air permeable layers, apertures, joints, cracks, leaks and cavities, the flux or flow equations introduced in Chapter 2 are:

Air permeable layers (per m²) $\quad g_a = -K_a \Delta P_a \quad G_a$ in kg/(m².s) and K_a in s/m
Joints, cracks, cavities (per m) $\quad G_a = -K_a^\psi \Delta P_a \quad G_a$ in kg/(m.s) and in s
Leaks, voids, apertures (each) $\quad G_a = -K_a^\chi \Delta P_a \quad G_a$ in kg/s and in m.s

with K_a^x the air permeance and ΔP_a the air pressure difference, included stack, along the flow path. The air permeance typically depends on the pressure differential intervening:

$$K_a^x = a \, (\Delta P_a - \rho_a \mathbf{g} z)^{b-1}$$

3.2.5.3 Mass, Moisture
Giving vapour flow is equivalent' diffusion and convection, as air moving in and through materials and assemblies also displaces vapour. Fick's law governs diffusion:

$$\mathbf{g_a} = -\delta \operatorname{grad} p$$

with δ the vapour permeability (units: s), a number for isotropic, and a tensor for anisotropic materials. The property reflects the size and deviousness of the pore system. Its value changes with moisture content and temperature. The addition 'equivalent' indicates that in reality the fluxes may not be purely Fickian. The air-driven vapour transfer equals:

$$\mathbf{g_v} = \mathbf{g_a} x_v = \frac{0.622 \, \mathbf{g_a} p}{P_a - p} \approx \frac{0.622 \, \mathbf{g_a} p}{P_a} \approx 6.21 \cdot 10^{-6} \, \mathbf{g_a} p$$

with $\mathbf{g_a}$ the air flux (kg/(m².s)) and P_a the air pressure (Pa). The sign $\approx$ mentions that the simplifications adapted only apply for temperatures <50°C and small air pressure differences.

For unsaturated water flow, an equation similar to Darcy's law for saturated water flow applies:

$$\mathbf{g_w} = -k_w \text{ grad } s$$

with s suction and k_w the unsaturated water permeability (units: s), a number for isotropic, and a tensor for anisotropic materials. The property has a very low value up to a given moisture content, but once there follows a sharp rise.

3.2.5.4 Remark
Well known is that joints, accidental cracks, voids, leaks and unplanned air layers assist in shaping the real geometry of an assembly. They may have an overwhelming impact on the air and liquid flow moving through. However, the uncertainty about where and how makes simulations troubling, with results often far from what really happens.

3.2.6 Equations of State

3.2.6.1 Enthalpy and Vapour Saturation Pressure in Relation to Temperature
See Chapter 2.

3.2.6.2 Relative Humidity in Relation to Moisture Content
As explained in Chapter 2, this relationship is known as the sorption/desorption isotherm of a material. Its derivative to RH gives the specific moisture content.

3.2.6.3 Suction in Relation to Moisture Content
This equation of state is known as the moisture characteristic. In contact with water, the moisture contents that really matter lay between dry and capillary moist. The derivative, called the suction linked specific moisture content, supersedes the RH linked specific moisture content once the RH passes ϕ_M, typically 95–98%. Above, capillary condensation starts creating continuous water paths in materials. Of course, coupling to RH remains a possibility.

3.2.7 Start, Boundary and Contact Conditions

The start conditions indicate what the heat, air and moisture situation is at time 0. Either they are assumed, follow from testing or from observation in practice. Most often, construction moisture, whether measured or guessed, figures as such. Besides, steady state and harmonic approaches do not require start conditions.

The boundary conditions include temperatures and heat fluxes, vapour pressures and vapour fluxes, suction, RH, sometimes moisture content and liquid water fluxes, plus air pressures and fluxes acting on each side of an assembly during the time period considered. Their values can remain constant, a situation called steady state, be constant or variable along the height or/and length of an assembly or vary with time. During the period considered, how varying may change, take wind-driven rain impinging on an outer wall once a water film forms. No heat, air and moisture problem can be solved without knowledge of the boundary conditions, but in most cases, assumptions replace a hardly known reality.

The contact conditions finally describe the situation in the interfaces between layers and mate-rial volumes. Most computer tools presume ideal contact with the driving forces keeping continuity and the fluxes equality when passing an interface, a choice often conflicting with reality. Thermally, for materials with high thermal conductivity, take metals, the contact resistances are, although mostly unknown, impacting. Across free contacts, characterized by thin air gaps between materials, only vapour fluxes remain the same. Liquid transfer might fill the gap with water, which then is either absorbed by the next layer if capillary active or retained in the gap till gravity overcoming friction gives run-off. Airflows in turn could move along free interfaces. Glued or chemically active interfaces add resistance, while real interfaces may combine areas of ideal contact with air inclusions and spots of extra resistance. Where, is mostly unknown.

3.2.8 Two Examples of Simplified Models

3.2.8.1 Assemblies Composed of Non-Hygroscopic, Non-Capillary Materials

Insertion of the air flux in the air balance gives:

$$\text{div}(k_a \, \mathbf{grad} \, P_a) = 0$$

The overall air pressure P_a is including stack ($=P_{a,o} - \rho_a \mathbf{g} z$). If the air permeability remains constant, then that balance simplifies to:

$$\nabla^2 (P_{a,o} - \rho_a \mathbf{g} z) = 0 \tag{3.13}$$

In isothermal conditions, thermal stack ($-\rho_a \, \mathbf{g} z$) does not play, while in non-isothermal conditions, it differs from zero in all directions except horizontally. Inserting both the air and the equivalent diffusive vapour flux in the vapour balance in turn yields:

$$\text{div}\left(\delta \, \mathbf{grad} \, p - \frac{0.622 \mathbf{g}_a}{P_a} p\right) \pm G'_2 = \frac{\partial w_v}{\partial t}$$

As long as the RH stays below 100%, the source term G'_2 remains 0, while the inertia term is limited to the vapour concentration change in the pore air:

$$\frac{\partial w_v}{\partial t} = \frac{\partial}{\partial t}\left(\frac{\Psi_o p}{RT}\right)$$

This simplifies the vapour balance to:

$$\text{div}\left(\delta \, \mathbf{grad} \, p - \frac{0.622 \mathbf{g}_a}{P_a} p\right) = \frac{\Psi_o}{R} \frac{\partial}{\partial t}\left(\frac{p}{T}\right)$$

Once RH = 100%, condensation makes G'_2 different from 0, giving:

$$\text{div}\left(\delta \, \mathbf{grad} \, p_{sat} - \frac{0.622 \mathbf{g}_a}{P_a} p_{sat}\right) \pm G'_2 = \frac{1}{R} \frac{\partial}{\partial t}\left(\frac{\Psi_o p_{sat}}{T}\right)$$

Compared to the deposit G'_2, the right-hand storage is negligible or:

$$\text{div}(\delta \,\textbf{grad}\, p_{\text{sat}}) - \frac{0.622 g_a}{P_a}\textbf{grad}\, p_{\text{sat}} = \pm G'_2 \tag{3.14}$$

The same holds when drying takes over, but only till the deposit has evaporated. Energy conservation gives:

$$\text{div}(\lambda\, \textbf{grad}\, \theta) - \sum_{j=1}^{3}(g_j c_j\, \textbf{grad}\, \theta) = -G'_2 l_b + \frac{\partial}{\partial t}(\rho_0 c' \theta) \pm \Phi'$$

In the enthalpy transfer term, only the air has impact. Without condensation or drying, the heat of evaporation does not play, while in the volumetric heat capacity, the materials dominate. This reduces the equation to:

$$\text{div}(\lambda\, \textbf{grad}\, \theta) - g_a c_a\, \textbf{grad}\, \theta = \rho_0 c_0 \frac{\partial \theta}{\partial t} \tag{3.15}$$

When condensation or drying intervene, it changes into:

$$\text{div}(\lambda\, \textbf{grad}\, \theta) - g_a c_a \textbf{grad}\, \theta + l_b \text{div}(\delta\, \textbf{grad}\, p_{\text{sat}}) - l_b \frac{0.622 g_a}{P_a}\textbf{grad}\, p_{\text{sat}} = \rho_0 c' \frac{\partial \theta}{\partial t}$$

or:

$$\text{div}\left[\left(\lambda + l_b \delta \frac{\text{d}p_{\text{sat}}}{\text{d}\theta}\right) \textbf{grad}\, \theta\right] - g_a \left(c_a + l_b \frac{0.622}{P_a}\frac{\text{d}p_{\text{sat}}}{\text{d}\theta}\right)\textbf{grad}\, \theta = \rho_0 c' \frac{\partial \theta}{\partial t} \tag{3.16}$$

The equation of state between vapour saturation pressure and temperature $p_{\text{sat}}(\theta)$, the mass balances for air and vapour and, the energy balance form together a system of four equations with as unknowns the temperature θ, the vapour pressure p, the condensation/evaporation rate G'_2 and the air pressure P'_a. Calculation starts with the assumption of a temperature curve. Then, again and again, the airflows are quantified, the temperatures recalculated, and the airflows rechecked till a pre-set accuracy between second last and last recalculation is reached. The fluxes then ensue from the flow equations.

Without airflow intervening, the model is much simpler. Does neither condensation nor drying play, the mass and heat balances become:

$$\text{div}(\delta\, \textbf{grad}\, p) = \frac{\partial}{\partial t}\left(\frac{\Psi_o p}{RT}\right) \qquad \text{div}(\lambda\, \textbf{grad}\, \theta) = \rho_0 c_0 \frac{\partial \theta}{\partial t}$$

Only when the material properties depend on temperature, then both equations are connected. Of course, a solution assuming constant properties may already give valuable information. In case condensation or drying intervene, both balances change to:

$$\text{div}\left(\delta \frac{\text{d}p_{\text{sat}}}{\text{d}\theta}\textbf{grad}\, \theta\right) = \pm G'_c \qquad \text{div}\left[\left(\lambda + l_b \delta \frac{\text{d}p_{\text{sat}}}{\text{d}\theta}\right)\textbf{grad}\, \theta\right] = \rho_0 c_0 \frac{\partial \theta}{\partial t}$$

with the temperature as only driving force to calculate. Of course, also the condensation and drying rates per m³ (G'_2) demand quantification.

3.2.8.2 Assemblies Composed of Fine Porous, Hygroscopic Materials

In case the RH in the assembly remains below the pivot value ϕ_m, materials will only show (de)sorption, turning the vapour flows into the only intervening, giving as mass conservation:

$$\text{div}\left(\delta\,\text{grad}\,p - \frac{0.622\,\mathbf{g_a}}{P_a}p\right) = \frac{\partial w_H}{\partial t}$$

As airflow is negligible in fine-porous materials, this equation simplifies to:

$$\text{div}(\delta\,\text{grad}\,p) = \frac{\partial w_H}{\partial t}$$

Introducing the temperature and RH as driving forces transforms that balance into:

$$\text{div}\left(\delta\,p_{sat}\,\text{grad}\,\phi + \delta\phi\frac{dp_{sat}}{d\theta}\text{grad}\,\theta\right) = \rho\xi_f\frac{\partial \phi}{\partial t} \tag{3.17}$$

Conservation of energy in turn yields:

$$\text{div}(\lambda\,\text{grad}\,\theta) - \sum_{j=1}^{3}(\mathbf{g_j}c_j\,\text{grad}\,\theta) = -G'_1 l_b + \frac{\partial}{\partial t}(\rho_0 c'\theta) \pm \Phi'$$

The source or sink term Φ' remains zero, except if condensate will deposit or what is deposited evaporates somewhere in the assembly, which the assumption about the RH excludes. Diffusion alone now gives such small vapour fluxes that related enthalpy displacement hardly has an impact, while phase changes are restricted to the moisture adsorbed. All this allows writing the source term G'_1 as:

$$G'_1 = \rho\xi_\phi\frac{\partial \phi}{\partial t} = \text{div}\left(\delta\,p_{sat}\,\text{grad}\,\phi + \delta\phi\frac{dp_{sat}}{d\theta}\text{grad}\,\theta\right)$$

Insertion in the energy balance so gives:

$$\text{div}\left[l_b\delta p_{sat}\,\text{grad}\,\phi + \left(\lambda + l_b\delta\phi\frac{dp_{sat}}{d\theta}\right)\text{grad}\,\theta\right] = \rho_0 c'\frac{\partial \theta}{\partial t} \tag{3.18}$$

Both the mass and energy balance, having the RH (ϕ) and the temperature (θ) as driving forces, allow quantifying the combined heat and moisture transfer in assemblies made of fine porous, hygroscopic materials for RHs below the pivot ϕ_m. With the material properties variable, solving the system anyhow demands iteration and the use of a control volume methodology (CVM).

3.3 Whole Building Level

3.3.1 In General

Heat, air and moisture are impacting whole buildings. Indoors, the temperature fixes thermal comfort, ventilation helps upgrading the indoor air quality and the moisture produced and entering with the ventilation air is impacting the RH, whereby too high values can activate mould growth on parts of the envelope's inside face, which troubles the mental, sometimes the physical health of inhabitants and building users. The harmonic thermal balance at zone level was analysed in Chapter 1. In Chapter 2, inter-zone airflow passed the review. Here, what matters is the vapour balance at zone and whole building level.

3.3.2 Balance Equations

3.3.2.1 Vapour

Adding vapour storage and release by absorbing surfaces to the vapour entering a zone with the air inflow, vapour ratio x_j, the vapour released by sources and surface drying, the vapour removed by sinks and surface condensation, the vapour leaving that zone with the air outflow, vapour ratio x_e, and the vapour stored in the zone air changes the vapour balance to (Figure 3.1):

$$\sum (x_j G_{a,ji}) + \sum (x_j G_{a,ij}) + G_{a,e} x_e + G_{vP,i} + \underbrace{\sum (G_{vc/d,ik})}_{(1)} + \underbrace{\sum (G_{vH,il})}_{(2)} = \varrho_a V_i \frac{dx_i}{dt}$$

(3.19)

Term (1) quantifies the impact of surface condensation and surface drying while term (2) bundles the impact of all sorption active surfaces in the zone. Surface condensation and surface drying both obey:

$$G_{vc/d,ik} = \beta_k A_k (p_{sat,s,k} - p_i) \qquad (3.20)$$

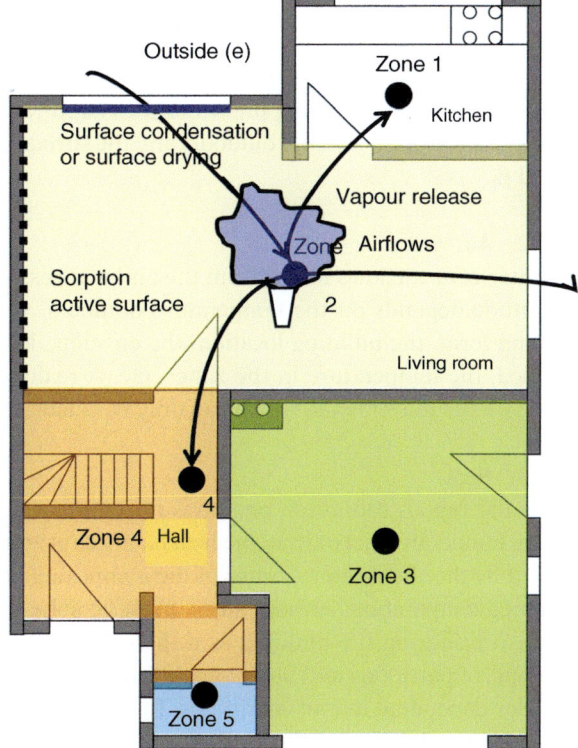

Figure 3.1 Water vapour balance at zone level.

with p_i vapour pressure in the zone and $p_{sat,s,k}$ the vapour saturation pressure on a surface A_k (m²), on which vapour condenses or condensate present evaporates. The equation quantifying the vapour exchange between a sorption active surface (A_l) and the zone air in turn looks:

$$G_{vH,il} = \beta_l A_l (p_{sat,s,l} \varphi_{s,l} - p_i) \tag{3.21}$$

with $\varphi_{s,l}$ the RH and $p_{sat,s,l}$ the vapour saturation pressure on the surface considered.

Both the vapour inflow with the air entering and the outflow with the air leaving write as:

$$xG_a = \frac{R_l p}{R_v(P_a - p)} G_a \approx 6.21 \ 10^{-6} p G_a \tag{3.22}$$

The expression after $\approx$ holds for vapour pressures (p) taking only a few % of the total air pressure (P_a), as is the case in buildings. Inserting the three flow equations in the zone balance gives:

$$-p_i \left[6.21 \ 10^{-6} \sum G_{a,ij} + \sum_{k=1}^{m} (\beta_k A_k) + \sum_{l=1}^{n} (\beta_l A_l) \right] + 6.21 \ 10^{-6} \sum (p_j G_{a,ji})$$

$$+ G_{vP,i} + \sum_{k=1}^{m} (\beta_k A_k p_{sat,s,k}) + \sum_{l=1}^{n} (\beta_l A_l p_{sat,s,l} \varphi_{s,l}) = \frac{V_i}{RT} \frac{dp_i}{dt} \tag{3.23}$$

In it, R is the gas constant of water vapour. Unknown are the vapour pressure p_i in the zone, the vapour pressures p_j in the adjacent zones, the vapour saturation pressure $p_{sat,s,l}$ and RH $\varphi_{s,l}$ on the sorption active surfaces, the vapour saturation pressure $p_{sat,s,k}$ on surfaces suffering from condensation or drying, all airflows and the surfaces A_k and A_l. The suffixes l and k link related quantities to the surfaces concerned. Parameters requiring pre-set or measured values are the vapour release or removal, the vapour pressure outdoors and the surface film coefficients for diffusion β_k and β_l.

3.3.2.2 Air

The inter-zone airflows follow from the air balances per zone, see Chapter 2. Their magnitude depends on the temperature outdoors, wind speed and direction, the building form, the building location, the envelope leakage, the ventilation system installed, the temperature in the zones, etc. A really detailed study also needs all intra-zone air movements, which can only be calculated using CFD.

3.3.2.3 Heat

Fixing the vapour saturation pressures $p_{sat,s,k}$ and $p_{sat,s,l}$ requires knowledge of the surface temperatures. For that, the heat balances at zone and assembly level, the last defined by the thermal properties of the composing layers, their sequence and the reference temperatures at both sides, must be solved. A detailed approach is preferred, considering the building as a three-dimensional volume, consisting of an envelope, of partitions and floors separating spaces and exposed to varying outdoor and user-dependent indoor conditions. To be truly detailed, the air temperature distribution in each zone and the inside surface temperature anywhere on the envelope, the partitions and the floors must be known. For that, the building fabric and all zones should be meshed fine enough to combine CVM at fabric level with CFD at zone level. Such full models are so demanding that a simpler alternative is preferred here. Of course, making things simpler means loss of information.

Buildings so are considered to include several zones, separated from outside by the envelope and inside by partition and floor assemblies. Heat only flows perpendicular to their faces, while the air in all zones is ideally mixed and at air temperature. Once solved and if needed, the zone results are used to evaluate thermal bridging.

In the zone balances, all convective heat flows are injected in the zone nodes, where their sum equals the change in heat stored in the zone air, the furniture and all furnishings:

$$\underbrace{\sum_{m=1}^{m}[h_{ci}A_{m}(\theta_{si,m} - \theta_{j})]}_{(1)} + \underbrace{c_{a}\left[\sum_{j=1}^{j}(G_{a,inf,x}\theta_{x}) - \theta_{j}\sum_{l=1}^{l}G_{a,inf,x}\right]}_{(2)}$$

$$+ \underbrace{\sum(f_{c,i}\Phi_{i,j})}_{(3)} + \underbrace{f_{c,syst}\Phi_{heat/cool,net,j}}_{(4)} = \rho_{a}cV\frac{d\theta_{j}}{dt} \qquad (3.24)$$

Term (1) represents the convective heat exchanged between all assembly faces seeing the zone and the zone air. Term (2) consist of the air-coupled enthalpy flows exchanged with all adjacent zones, with the outdoors and injected or extracted by any ventilation system ($x = $ e for outdoors, $x = $ l for an adjacent zone). Term (3) is the convective fraction ($f_{c,I}$) of the internal gains and term (4) the convective flow delivered by the heating or cooling system with $f_{c,syst} = 1$ for air based and <1 for other systems.

Per m² of opaque envelope part, the heat balance at the outer face is:

$$q_{c,se} + q_{LR,se,} + a_{S,se}q_{S,se} + q_{T,se} = 0$$

In it, $q_{c,se}$ is the convective heat flux, $q_{LR,e}$ the long-wave radiant heat flux with the ambient and the sky, $q_{T,se}$ the conductive flux at the outer face, $a_{S,se}$ the short-wave absorptivity of the outer face and $q_{S,se}$ the solar irradiation. The first three fluxes write as:

Convection $q_{c,se} = h_{c,e}(\theta_{e} - \theta_{se})$

Long wave $q_{se,LR} = 5.67 e_{Le}(F_{se}F_{Tse} + F_{ssk}F_{Tssk})(\theta_{e} - \theta_{se}) - 120 e_{Le}F_{ssk}F_{Tssk}(1 - f_{c})$

Conduction $q_{T,si} = \dfrac{\lambda_{n}}{\Delta x_{n}}(\theta_{n-1} - \theta_{se})$

In the last, λ_{n} is the thermal conductivity of the material layer at the outside, Δx_{n} the mesh width used, θ_{n-1} the mesh temperature and θ_{se} the outside surface temperature.

Per m² of opaque partition and envelope part facing the zone, the heat balance looks:

$$q_{c,si} + q_{R,si} + a_{S,si}q_{S,si} + q_{T,si} = 0$$

with $q_{c,si}$ the convective heat flux, $q_{R,si}$ the radiant heat flux by long wave radiation to all other surfaces in the zone, $a_{S,si}$ the short-wave absorptivity of these surfaces, $q_{S,si}$ the solar radiation transmitted by the transparent envelope parts warming the other surfaces in the zone and $q_{T,si}$ the conductive flux at the assembly's inner face. The four write as:

Convection $\quad q_{c,si} = h_{c,i}(\theta_j - \theta_{si})$

Long wave $\quad q_{R,si} = \dfrac{e_L}{\rho_L}\left[\left(\sum_{k=1}^{m} a_{R,k} H_{b,k}\right) - H_b\right] + \dfrac{\sum[(1-f_{c,I})\Phi_I]}{\sum A}$

$$+ (1 - f_{c,syst})\dfrac{a_L \Phi_{heat/cool,net}}{\sum A}$$

Solar $\quad q_{S,si} = \dfrac{0.9 \cdot 0.95 \sum(\tau_{S,w} f_w r_w A_w q_S)}{\sum\limits_{opaque} A - 0.9 \cdot 0.95 \sum(\tau_{S,w} f_w r_w A_w)}$

Conduction $\quad q_{T,si} = \dfrac{\lambda_1}{\Delta x_1}(\theta_1 - \theta_{si})$

In the long wave equation, m is the number of opaque and transparent assemblies enclosing the zone, ΣA their inside surface, $a_{R,k}$ the fraction of the black body radiation emitted by all other surfaces that is absorbed by the surface considered, Φ_I the internal gains and $\Phi_{heat/cool,net}$ the heating or cooling load. The solar equation assumes that the shortwave radiation entering the zone is reflected so many times in the zone that a diffuse field arises so that each opaque surface absorbs radiation proportionally to its short-wave absorptivity. All transparent envelope assemblies are transmitting part of what is reflected in the zone back to outdoors. Finally, in the conduction equation, λ_1 is the thermal conductivity of the material forming the inner surface, Δx_1 the mesh width, θ_1 the mesh temperature and θ_{si} the inside surface temperature. As envelopes and partitions are composed of flat assemblies, stepping from fluxes to flows is simple: multiply all with their respective surfaces.

Transparent envelope parts react by definition steady state, even at very short time intervals. Under-cooling is accounted for, while solar irradiation is split in transmitted and absorbed by the glass. To simplify things, each pane gets one temperature, which allows describing the combination with solar shading by a system of as many equations as there are panes and shading.

Turning to the meshed flat opaque assemblies, if airtight, the balance per mesh is:

$$\dfrac{\lambda_i(\theta_{i-1} - \theta_i)}{\Delta x_i} + \dfrac{\lambda_{i+1}(\theta_{i+1} - \theta_i)}{\Delta x_{i+1}} = \left[(\rho c)_i \dfrac{\Delta x_i}{2} + (\rho c)_{i+1} \dfrac{\Delta x_{i+1}}{2}\right]\dfrac{\Delta \theta_i}{\Delta t}$$

The system of zone and mesh balances has to be solved per time step using a Cranck-Nicholson scheme.

3.3.2.4 Closing the Loop

Once the air and heat balances per zone are known, the temperatures at all inside faces can be transposed into saturation pressures, which, with the surfaces, become parameters in the zonal vapour balance. Unknowns left are the vapour pressure p_i in the zone, the vapour pressures p_j in the adjacent zones and the RH $\phi_{s,l}$ on all sorption active surfaces. One inaccuracy anyhow remains: coupled to sorption, surface condensation and drying is the release or pick up of latent heat, which changes the inside surface temperatures. Assuming that the impact on these surface temperatures is negligible, allows solving the vapour balances without feedback to the heat balances. If not, all sorption active inside surfaces and those seeing condensation or drying will experience an extra heat flux, so requiring extra loops between the heat, air and vapour balances. For a building counting n zones, the vapour balances generate a system of n equations:

$$|H_v||p| + |1|\left|\sum_{l=1}^{n}(\beta_l A_l p_{\text{sat},s,l}\varphi_{s,l})\right| = -\left|\sum_{k=1}^{n}(\beta_k A_k p_{\text{sat},s,k}) + G_{vP}\right| \quad (3.25)$$

The diagonal terms in the array $|H_v|$ look:

$$H_{v,ii} = -\left[\left(6.21 \cdot 10^{-6}\sum G_{a,ij} + \frac{V_i}{RT}\mathbf{D}\right) + \sum_{k=1}^{n}(\beta_k A_k) + \sum_{l=1}^{n}(\beta_l A_l)\right] \quad (3.26)$$

with $\mathbf{D}$ the differential operator. These outside the diagonal are:

$$H_{v,ij} = H_{v,ji} = \left(6.21 \cdot 10^{-6}\sum G_{a,ij}\right) \quad (3.27)$$

That system of as many equations as zones can be solved if the relations between the vapour pressure in each and the RH on all sorption-active surfaces are known and if all air and heat-related quantities and values are given.

3.3.3 Sorption Active Surfaces and Hygric Inertia

3.3.3.1 In General

A zone i counts l sorption active surface. Known are the heat and air balance, the vapour pressures in all adjacent zones and the vapour pressure outdoors. Unknown in the zone's vapour balance so are the vapour pressure in the air and the RHs on the l surfaces. Besides the zone balance, solving additionally requires as many vapour balances as sorption active surfaces are present. These state that the sum of the vapour fluxes coming from the zone and from inside each assembly to its surface side zone must be zero:

$$\beta_l(p_i - p_{\text{sat},s,i}\varphi_s) + \partial_l \; \mathbf{grad}(p)_s = 0 \quad (3.28)$$

This equation is of no use if $\mathbf{grad}(p)$ is unknown. This demands a quantification of the vapour transfer near an assembly's inside face. In the absence of air transport across, the vapour balance there looks:

$$\mathbf{div}[\delta \; \mathbf{grad}(p)]_s = \frac{\partial w_{H,s}}{\partial t} \quad (3.29)$$

with δ the vapour permeability and w_H the sorption moisture content of the last layer at the zone side. Solving the system formed by the zone and these surface vapour balances gives the l unknown vapour pressures and, with the surface temperature known, the l unknown surface RHs. For a building counting n zones, the whole system so consists of $n + \sum_{n=1}^{n}(nl_j)$ equations.

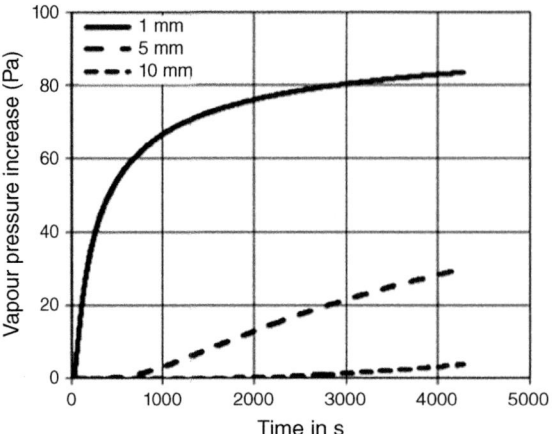

Figure 3.2 Left, gypsum board: transient hygric response when facing a sudden 100 Pa vapour pressure increase on its inner face.

3.3.3.2 Sorption-Active Thickness

For a rather fast changing vapour pressure in the zone, an acceptable solution is found by attributing a sorption active thickness (d_H) to each hygroscopic surface facing the zone. Figure 3.2 underpins this by showing how in a gypsum board finish the vapour pressure evolves after a sudden 100 Pa increase on its inner face.

After one hour, 1 cm deep in the board hardly any change in vapour pressure is seen. Activating hygric inertia so is a very slow process. Defined as active thickness therefore is the distance between the inner face, at which the vapour pressure fluctuates with period T and amplitude 1 Pa and the plane in the material, where the amplitude is 0.368 Pa. With the specific moisture content (ξ_ϕ) and the vapour resistance factor (μ) considered constant and the temperature in that plane and on the inner face the same, this distance, called the sorption active thickness, equals:

$$d_H = \sqrt{\frac{\delta \cdot p_{sat,si} T}{\pi \xi_\phi}} \qquad (3.30)$$

with p_{sat} the saturation pressure on the inner face and T the period considered. For a RH between 30% and 80%, the specific moisture content assumed constant (ξ_ϕ) is a fair simplification, see Figure 3.3.

For a one-hour and a one-day period, the active thickness so becomes:

Hour: $d_H = 33.85\sqrt{\delta \cdot p_{sat,si}/\xi_\phi}$ Day: $d_H = 165.8\sqrt{\delta \cdot p_{sat,si}/\xi_\phi}$

Related hygric capacitances are:

Hour: $C_{\phi,H} = 33.85\sqrt{\delta \cdot \xi_\phi/p_{sat,si}}$ Day: $C_{\phi,H} = 165.8\sqrt{\delta \cdot \xi_\phi/p_{sat,si}}$

with their point of action 0.422 d_H m away from the inside face.

When the sorption-active thickness stretches over several layers, each with its own specific moisture content and vapour resistance factor, the hygric capacitance per layer then becomes:

$$C_{\phi,H,i} = \xi_{\phi,i} d_i / p_{sat,si}$$

Figure 3.3 Measured sorption curve, linear for a relative humidity between 0.2 and 0.8.

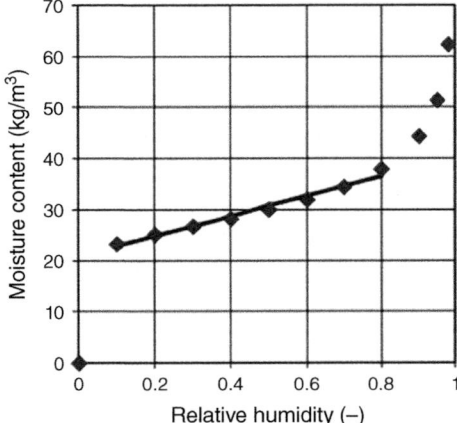

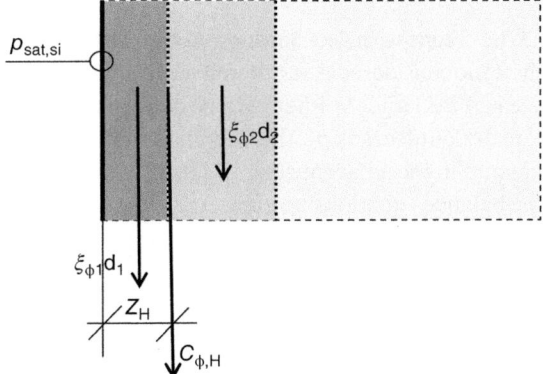

Figure 3.4 Sorption-active thickness including different layers.

with d_i the layer thickness and $p_{sat,si}$ the vapour saturation pressure on the inner face (Figure 3.4).

The summed capacity with as point of action that of application of the resultant of the hygric capacitances, assumed vectorial, of the different layers so becomes:

$$C_{\varphi,H} = \frac{1}{p_{sat,s,i}} \sum (\xi_\varphi d)$$

Past this point of action, seen the thinness of the active thickness, an acceptable approximation is the vapour flux entering the remnant of the assembly equal to the monthly mean steady state one through the assembly.

Still unknown of course is the vapour pressure p_H in the point of action. A mass balance included the summed capacity gives:

$$\frac{p_i - p_H}{Z_H} + \frac{\overline{p}_x - \overline{p}_i}{Z_T} = \left(\frac{C_{\varphi,H}}{p_{sat,si}}\right)\frac{dp_H}{dt} \qquad (3.31)$$

with:

$$\text{One layer: } Z_H = \frac{1}{\beta_1} + 0.422\mu N d_H \quad \text{Multi-layer: } Z_H = \frac{1}{\beta_1} + \sum_{si}^{H}(\mu N d)$$

In the two, Z_H is the diffusion resistance between the indoors and the point of action, p_H the vapour pressure in the point of action, p_i the vapour pressure inside, $\bar{p}_i$ the monthly mean vapour pressure inside, $\bar{p}_x$ the monthly mean vapour pressure in the adjacent zone, outdoors or in the interface where condensate deposits, and Z_T the diffusion resistance of the assembly or the diffusion resistance from indoors to the condensation interface. The second term left in the mass balance has its importance because, indeed, for a given vapour release indoors and a given ventilation flow, a more vapour-permeable enclosure will lower the vapour pressure in the zone. Once the vapour pressure in the point of action is known, the vapour pressure and RH on the inner face follow from:

$$p_{s,i} = p_i - \frac{(p_i - p_w)}{\beta_i Z_H} \qquad \varphi_{s,l} = \frac{p_{s,i}}{p_{sat,s,i}} \tag{3.32}$$

Each hygroscopic surface so adds an equation. Combining with the zonal vapour balance allows calculating the $l + 1$ still unknown vapour pressures.

3.3.3.3 Zone with One Sorption-Active Surface

The zone considered is ventilated with outside air (G_a), while the sorption-active surface is A m² large. Neither surface condensation nor drying takes place. The vapour pressure outdoors is p_e, the vapour release indoors G_{vP}, the monthly mean vapour pressure in the adjacent zones $\bar{p}_x$ and the monthly mean vapour pressure indoors $\bar{p}_i$. The balance equations become:

$$\text{Zone} \quad -p_i\left[6.21\cdot 10^{-6}G_a + \frac{A}{Z_H}\right] + \frac{p_H A}{Z_H} = \frac{V}{RT}\frac{dp_i}{dt} - G_{vP} - 6.21\cdot 10^{-6}G_a p_e$$

$$\text{Surface} \quad \frac{p_i - p_H}{Z_H} = C_H \frac{dp_H}{dt} + \left(\frac{\bar{p}_i - \bar{p}_x}{Z_T}\right)\frac{A_e}{A} \tag{3.33}$$

Unknown are the vapour pressures in the zone (p_i) and in the point of action of the sorption-active surface (p_H). Reshuffling and replacing the value $6.21\cdot 10^{-6}$ by "α," gives:

$$\begin{vmatrix} \left[D + \dfrac{G_a}{\rho_a V} + \dfrac{A}{\alpha\rho_a V Z_H}\right] & -\left(\dfrac{A}{\alpha\rho_a V Z_H}\right) \\ -\left(\dfrac{1}{Z_H C_H}\right) & \left(D + \dfrac{1}{Z_H C_H}\right) \end{vmatrix} \begin{vmatrix} p_i \\ p_H \end{vmatrix} = \begin{vmatrix} \dfrac{G_a}{\rho_a V}p_e + \dfrac{G_{vP}}{\alpha\rho_a V} \\ \left(\dfrac{\bar{p}_x - \bar{p}_i}{Z_T C_H}\right)\dfrac{A_e}{A} \end{vmatrix}$$

or:

$$\begin{vmatrix} D + A_1 + A_2 & -A_2 \\ -B_1 & D + B_1 \end{vmatrix} \begin{vmatrix} p_i \\ p_H \end{vmatrix} = \begin{vmatrix} A_1 p_e + A_3 \\ B_2(\bar{p}_x - \bar{p}_i) \end{vmatrix} \tag{3.34}$$

with D the differential operator, ρ_a the density of air and A_1, A_2, A_3, B_1 and B_2 ratio's given by:

$$A_1 = \frac{G_a}{\rho_a V} \quad A_2 = \frac{A}{\alpha\rho_a V Z_H} \quad A_3 = \frac{G_{vP}}{\alpha\rho_a V} \quad B_1 = \frac{1}{C_H Z_H} \quad B_2 = \left(\frac{1}{C_H Z_T}\right)\frac{A_e}{A}$$

For a sudden increase in vapour release indoors or a sudden increase in vapour pressure outdoors, the solution is:

$$p_i = p_{i,\infty} + C_1\exp(r_1 t) + C_2\exp(r_2 t) \quad p_H = p_{H,\infty} + C_3\exp(r_1 t) + C_4\exp(r_2 t)$$

with $p_{i,\infty}$ and $p_{H,\infty}$ the asymptotes of the vapour pressure indoors and the vapour pressure in the sorption surface's point of action. r_1 and r_2 are the roots of the characteristic equation:

$$r_1 = \frac{1}{2}\left[-(A_1 + A_2 + B_1) + \sqrt{(A_1 + A_2 + B_1)^2 - 4A_1B_1}\right]$$

$$r_2 = \frac{1}{2}\left[-(A_1 + A_2 + B_1) - \sqrt{(A_1 + A_2 + B_1)^2 - 4A_1B_1}\right]$$

The integration constants C_1, C_2, C_3 and C_4 follow from the initial conditions – the vapour pressures $p_{i,o}$ and $p_{H,o}$ at the moment of the sudden increase – and from the relation between both as fixed by the system of differential equations at time 0:

(1) $C_1 + C_2 = p_{i,o} - p_{i,\infty}$
(2) $C_3 + C_4 = p_{H,o} - p_{H,\infty}$
(3) $C_1 r_1 + C_2 r_2 = A_1 p_{i,\infty} - (A_1 + A_2) p_{i,o} + A_2 p_{H,o}$
(4) $C_3 r_1 + C_4 r_2 = B_2(p_x - \bar{p}_i) + B_1 p_{i,o} - B_1 p_{H,o}$

In case the vapour pressure indoors is the same as in the point of action, the initial conditions (3) and (4), simplify to:

(3) $C_1(r_1 + A_1) + C_2(r_2 + A_1) = 0$ (4) $C_3 r_1 + C_4 r_2 = B_2(p_x - \bar{p}_i)$

with as solution:

$$C_1 = \frac{p_{i,o} - p_{i,\infty}}{1 - \frac{r_1 + A_1}{r_2 + A_1}} \quad C_2 = \frac{p_{i,o} - p_{i,\infty}}{1 - \frac{r_2 + A_1}{r_1 + A_1}}$$

Also, a sudden change in ventilation rate at time 0 figures as a possible step change.

3.3.3.4 Zone with Several Sorption-Active Surfaces

Each sorption-active surface adds a mass balance, linked to its point of action. Two surfaces so give a system of three, three of four, four of five, etc., differential equations of first order. Each adds an additional exponential and time constant to the solution at zone level. Calculating the roots of the characteristic equation is only possible if the vapour pressures in all points of action are assumed as having the same value (p_H). Only then, the presence of several sorption-active surfaces requires solving a system of two differential equations of first order with as coefficients A_1, A_2, A_3, B_1 and B_2:

$$A_1 = \frac{G_a}{\rho_a V} \quad A_2 = \frac{1}{a\rho_a V}\sum_{l=1}^{n}\left(\frac{A_1}{Z_{H,l}}\right) \quad A_3 = \frac{G_{vP}}{a\rho_a V} \quad B_1 = \frac{\sum_{l=1}^{n}\left(\frac{A_1}{Z_{H,l}}\right)}{\sum_{l=1}^{n} A_1 C_{H,l}}$$

$$B_2 \underset{\text{if } |\bar{p}_i - p_x| > 0}{=} \frac{\sum_{k=1}^{m}\left(\frac{A_k}{Z_{T,k}}\right)}{\left(\sum_{k=1}^{m} A_1 C_{H,l}\right)}$$

3.3.3.5 Harmonic Analysis

Yearly, the vapour pressure outdoors changes more or less harmonically. With the average ventilation rate and indoor vapour release known, a harmonic analysis is possible. All material properties are thereby assumed known. While the average vapour pressure indoors follows from a steady state balance, the complex vapour flow ($G_{v,s,l}$) to and in an envelope assembly becomes:

$$G_v = A_e \left(\frac{\mathbf{p_e}}{D^n_{g,e}} - Ad^n_{v,e} \mathbf{p_i} \right) \tag{3.35}$$

with $\mathbf{p_e}$ and $\mathbf{p_i}$ the complex vapour pressures out- and indoors, $D^n_{g,e}$ the dynamic diffusion resistance of the assembly and Ad^n_v its hygric admittance. With the vapour pressures in all zones equal, the equation for partitions is:

$$G_v = A_i \left(\frac{\mathbf{p_i}}{D^n_{g,i}} - Ad^n_{v,i} \mathbf{p_i} \right) = A_i \mathbf{p_i} \left(\frac{1}{D^n_{g,i}} - Ad^n_{v,i} \right) \tag{3.36}$$

On average and per harmonic the vapour pressure indoors so becomes:

$$\overline{p}_i = \overline{p}_e + \frac{G_{Pv}}{6.21\ 10^{-6} G_a}$$

$$\hat{p}_i = \hat{p}_e \left(\frac{6.21\ 10^{-6} G_a + \sum_e \left(\frac{A_e}{D^n_{g,e}} \right)}{6.21\ 10^{-6} G_a + \sum_e \left(A_e Ad^n_{v,e} \right) + \sum_i A_i \left(Ad^n_{v,i} - \frac{1}{D^n_{g,i}} \right) + i \left(6.21\ 10^{-6} \rho_a \frac{2\pi}{T} V \right)} \right)$$

3.3.4 Consequences

In steady state, the hygric inertia and related moisture buffering have no impact but transient, it has. To show the consequences, a bedroom is considered having a net floor area of 16 m², an air volume of 40 m³, a 2.25 m² large window and a 2 m² large door bay. The outer cavity wall, the partitions and the ceiling all are finished at the inside with 1.5 cm unpainted gypsum plaster, having as material properties $\mu = 5.8$, $\xi_H = 56$ kg/m³. The floor cover is vapour-tight. Or, without bed and furniture, the sorption-active surface is 51.8 m² large. Of these, 7.75 m² belong to the envelope, 28.05 m² to the partitions and 16 m² to the ceiling. The daily vapour release totals 800 g for a ventilation flow of either 22 or 54 m³/h. Indoors, the temperature is 20 °C year-round, while outside, the weather is temperate and related annual change in vapour pressure is known.

Figure 3.5 illustrates the impact of moisture buffering on the monthly mean in- to outdoor vapour pressure difference.

As the slope of the ovals underline, even on an annual basis the indoor/outdoor vapour pressure difference is dampened, although more ventilation moderates the impact, see the lowest oval. That an oval is formed, follows from the time shift between the vapour pressures in- and outdoors. In fact, despite a constant vapour release and a constant ventilation rate, buffering gives a higher difference between both vapour pressures in the fall than in springtime. That the oval flattens underlines that more ventilation shortens the time shift.

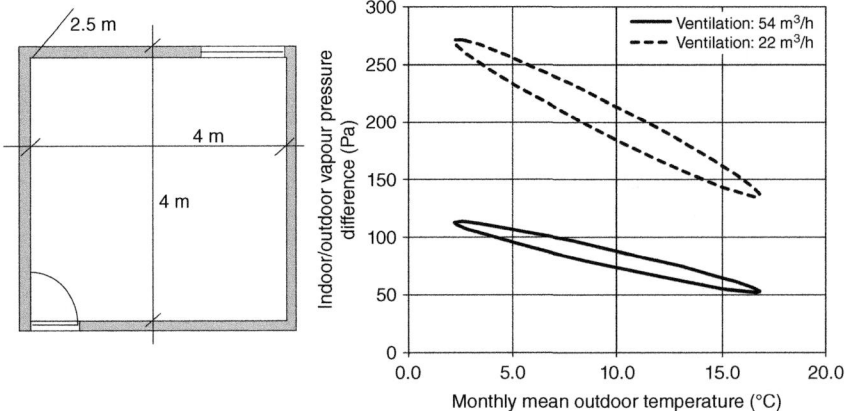

Figure 3.5 Sleeping room, moisture buffering impacting the annual indoor/outdoor vapour pressure difference.

In the bedroom, sleeping eight hours per night gives some 800 g vapour release. Considered is a winter day with outside as mean temperature 2.7 °C and as mean vapour pressure 663 Pa. If the 800 g should be equally spread over the day, a ventilation flow of 22 m^3/h would indoors give a steady state vapour pressure of 867 Pa, while 54 m^3/h would give 746 Pa. In reality, these 800 g convene with 100 g/h during 8 hours. Taking the 9 mm thick sorption-active thickness of the plaster also applicable for that changing vapour release between day and night, the transient for 54 m^3/h looks as traced in Figure 3.6.

When the release starts, air buffering first gives a fast increase in indoor vapour pressure, followed by a strongly retarded rise due to the gypsum plaster adsorbing

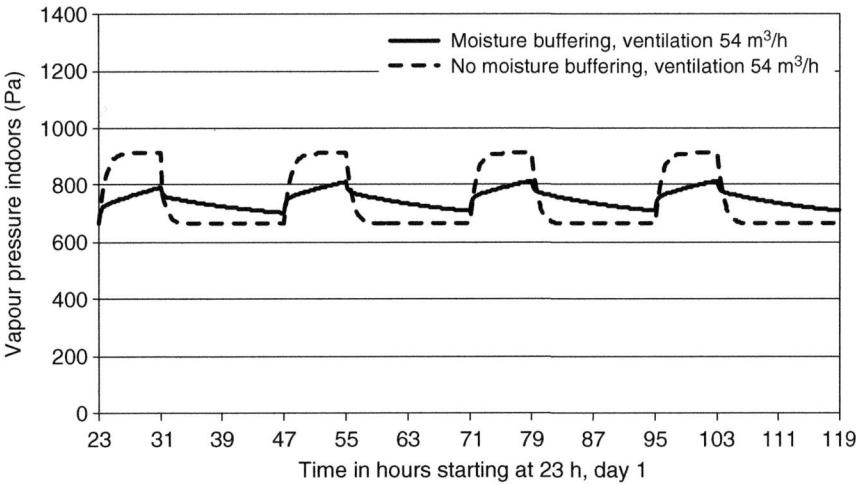

Figure 3.6 Sleeping room, vapour pressure for a ventilation flow of 54 m^3/h and a vapour release of 100 g/h during eight hours at night.

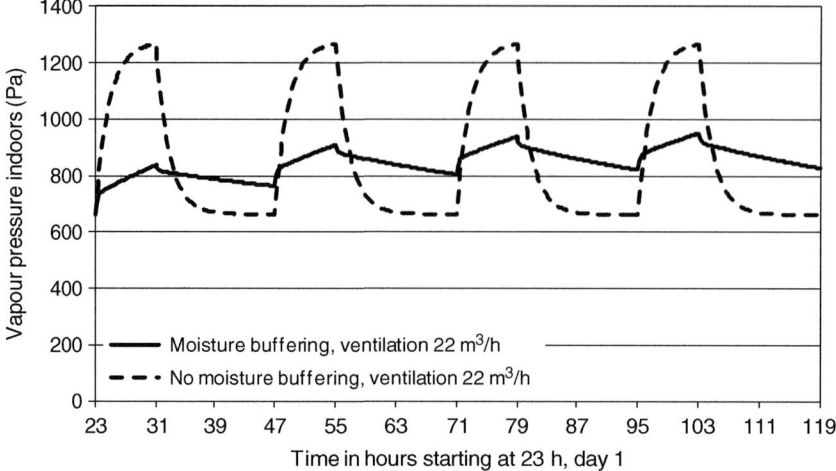

Figure 3.7 Same sleeping room, vapour pressure for a ventilation flow of 22 m³.

and giving a value, which after eight hours is still far away from the 912 Pa the 100 g/h would give without buffering. The same picture is seen in the morning, when the vapour release stops. The average indoor vapour pressure over a series of days anyhow remains 746 Pa. Figure 3.7 shows the results for a ventilation flow of 22 m³/h. The same conclusions remain, though the average over a series of days now increases to 867 Pa.

Figure 3.8 on the other hand illustrates the effect of the window set ajar one hour each morning, giving a 10 ach ventilation peak, and closing it the rest of the day,

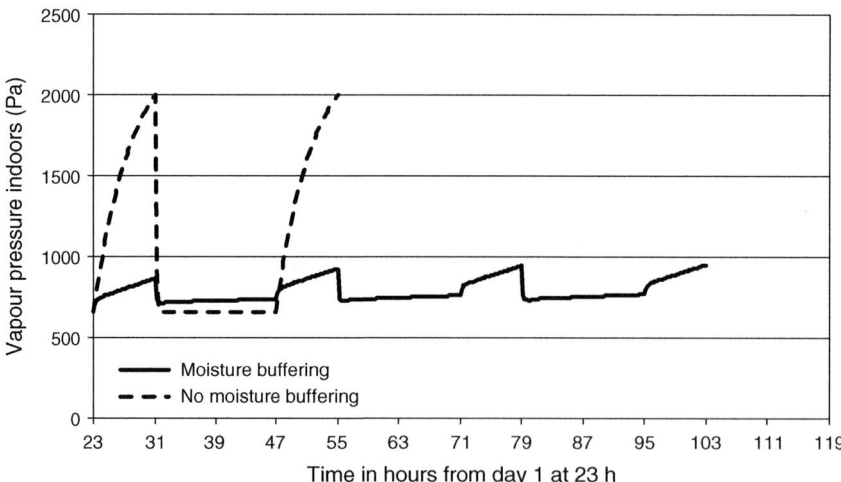

Figure 3.8 Same sleeping room, between 23 am and 7 pm 100 g/h vapour released. Vapour pressure course for the window set ajar between 7 and 8 pm and closed the rest of the day.

so reducing the ventilation rate to a 0.2 ach infiltration. A steep decrease in vapour pressure is seen, followed by a very slow climb to an equilibrium with the hygric memory of the room.

In reality, the increase after closing the window will go faster, mainly because the sorption active thickness underestimates the early, quick vapour release by diffusion. Or, a short peak ventilation after hours of usage has less effect than expected.

Problems and Solutions

Problem 31 A social estate consists of 24 two-story two-family houses with pitched roofs:

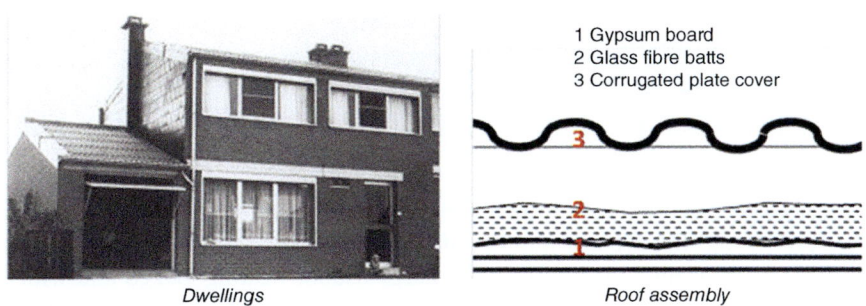

Dwellings   Roof assembly

The 2.5 m high ground floor and equally high first-floor measure $7.2 \times 7.2 \, m^2$. The long pitch of the asymmetric roof has a 17°, the short pitch a 10° slope. An open staircase connects both floors. No house has a purpose-designed ventilation system. Instead, air leakage and window opening have to guarantee an acceptable indoor air quality. The only difference between the 24 is the orientation: 4 have a front looking northwest, 2 a front looking north-west, 8 a front looking northeast, 3 a front looking east, 3 a front looking southeast and 4 a front looking southwest.

A few years after occupation, 41 houses had moisture spots spread over the ceiling of the sleeping room, while tenants complained about moisture dripping in their bed after cold nights. Inspection showed that the gypsum board forming the ceilings were mounted with open joints, that the insulating glass fibre bats with bituminous paper backing were laid out with open joints between the rafters with the flanges not overlapping against their underside. The backside of the corrugated fibre cement plates cover, the rafters and the top of the gypsum boards showed abundant traces of water run-off, while the board's inside face suffered from discolouration along the open joints, see the pictures below.

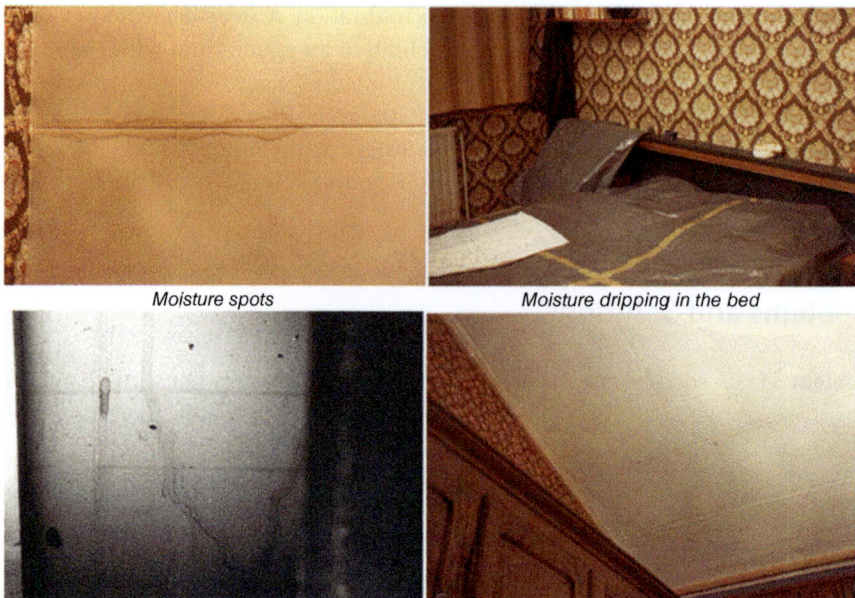

| Moisture spots | Moisture dripping in the bed |
| Run-off at the topside of the gypsum boards | PE air and vapour retarder mounted |

To clarify the randomness of the complaints, a correlation was sought between their severity and the average number of tenants. Looked was whether a cooking hood was present, how the annual end energy use for heating looked and which orientation the front façade had. Only the heating bill correlated with the severeness of the complaints. On average, dwellings with severe ones consumed 128 GJ/a, and those with moderate ones 164 GJ/a. Tenants there either heated more rooms or ventilated more while heating.

During the winter 1981–1982, the inside temperature and RH were monitored in three dwellings, two with severe and one with moderate complaints (1 = moderate; 2 and 3 = severe). Before, in dwelling 2, a polyethene (PE) air and vapour retarder with all joints and overlaps carefully sealed was fixed against the underside of the gypsum board ceiling, see picture above. The table below summarizes the results.

	Parents' bedroom		Children bedroom		Bathroom	
	Temp., °C	Δp_{ie}, Pa	Temp., °C	Δp_{ie}, Pa	Temp., °C	Δp_{ie}, Pa
1	$13.6 + 0.42\theta_e$	$196 - 1.2\theta_e$	$14.1 + 0.42\theta_e$	$159 - 0.9\theta_e$		
2	$13.1 + 0.32\theta_e$	$373 - 14.7\theta_e$	$13.9 + 0.26\theta_e$	$237 - 2.5\theta_e$	$14.3 + 0.21\theta_e$	$457 - 17.7\theta_e$
3	$11.7 + 0.48\theta_e$	$324 - 10.8\theta_e$	$15.6 + 0.06\theta_e$	$411 - 34\theta_e$	$17.7 + 0.25\theta_e$	$395 - 19.4\theta_e$

The two with severe complaints showed the highest in- to outdoor vapour pressure difference; Happily, the PE-foil in dwelling 2 stopped the dripping completely.

Additional data

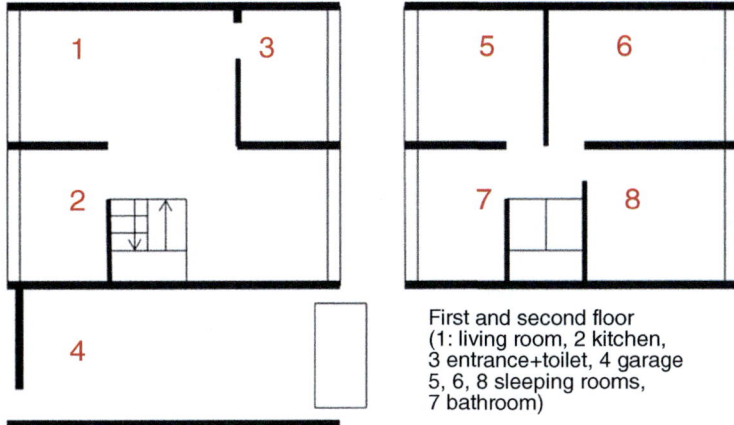

First and second floor
(1: living room, 2 kitchen,
3 entrance+toilet, 4 garage
5, 6, 8 sleeping rooms,
7 bathroom)

The volume out to out is 344.9 m³ of which 149 m³ for the ground floor. The enclosed air volume is 248.3 m³ large, while the surface areas out to out of all envelope assemblies equal:

Opaque envelope	assemblies	Area, m²	Windows and outer door		Area, m²
Floor on grade		51.8	Ground floor	Front door	2.0
Cavity wall	Sidewall 1	45.8		Toilet	0.2
	Sidewall 2	45.8		Living room, front	5.4
	Front	26.7		Living room, back	5.4
	Rear	24.3		Kitchen	5.3
Roof	Large pitch	29.7	First floor	Sleeping room 1	4.5
	Small pitch	25.0		Sleeping room 2	4.5
				Sleeping room 3	4.5
				Bathroom	1.8

The thermal transmittances and air permeances as built are:

Part	U-value, W/(m².K)	K_a, kg/(m².Pa.s)
Façade: non-insulated cavity wall, plastered inside	1.66	0
Roof	0.49	$3.3 \, 10^{-4} \Delta P_a^{-0.33}$
Floor on grade	0.70	0
Window between both roof pitches	3.34	0
Glazing Double at the ground floor	2.70	
Single at the first floor	5.70	
Aluminium frames (20% of the window area)	5.90	

Overall air leakage equals $\dot{V}_a = a\Delta P_a^{0.67}$ (m³/s), while the roof has a section

Layer (all air permeable)	d, m	λ, W/(m.K)	R, m².K/W	μd, –
Corrugated plates	0.006	0.95		0.34
Air space	0.18		0.17	0.00
Thermal insulation	0.06	0.04		0.07
Vapour retarder	–	–		2.30
Air space	0.04		0.17	0.00
Gypsum board	0.0095	0.21		0.12

The inside surface film coefficient is 7.7 W/(m².K) for the outer walls, 6 W/(m².K) for the floor on grade and 10 W/(m².K) for the roof. Outside, the values are 25 W/(m².K) for the outer walls and floor on grade and 17 W/(m².K) for the roof. The inside surface film coefficient for diffusion equals $2.6 \cdot 10^{-8}$ s/m.

The temperature indoors is 18 °C and the air change rate at 50 Pa (n_{50}) 10.4 h⁻¹ with the front and rear facade of the ground and first floor equally air permeable. Vapour release indoors touche ≈13.5 kg/day, while the outdoor climate during the cold week considered gives:

Temperature °C	Temperature roof, °C	Relative humidity for −2.5 °C, %	Mean wind speed, m/s	Wind direction
−2.5	−3.9	98	3.8	NE

The mean wind speed is the one measured in the nearest weather station at a height of 10 m. As the estate forms a closed landscape, the effective terrain roughness of 1 m and a friction velocity of 0.47 m/s changes this value into:

$$v = 1.12 \ln(h + 1) \text{ m/s}$$

with h height above grade. Wind pressure follows from $0.6Cv^2$ with C equal to:

	Wind angle with the normal to the front facade							
	0	45	90	135	180	225	270	315
Front	0.2	0.05	−0.25	−0.3	−0.25	−0.3	−0.25	0.05
Back	−0.25	−0.3	−0.25	0.05	0.2	0.05	−0.25	−0.3
Side left	−0.25	0.05	0.2	0.05	−0.25	−0.3	−0.25	−0.3
Side right	−0.25	−0.3	−0.25	−0.3	−0.25	0.05	0.2	0.05

| | | Wind angle with the normal to the front facade | | | | | | |
		0	45	90	135	180	225	270	315
Roof, <10°	Front	−0.5	−0.5	−0.4	−0.5	−0.5	−0.5	−0.4	−0.5
	Rear	−0.5	−0.5	−0.4	−0.5	−0.5	−0.5	−0.4	−0.5
	Mean	−0.5	−0.5	−0.4	−0.5	−0.5	−0.5	−0.4	−0.5
Roof, 11–30°	Front	−0.3	−0.4	−0.5	−0.4	−0.3	−0.4	−0.5	−0.4
	Rear	−0.3	−0.4	−0.5	−0.4	−0.3	−0.4	−0.5	−0.4
	Mean	−0.3	−0.4	−0.5	−0.4	−0.3	−0.4	−0.5	−0.4

The question is: what causes the complaints?

Solution 31 First, the dwelling's steady state heat, air and moisture response during the cold week is analysed. For the air, the dwelling is simplified to a three-node system, one node 1 m above grade representing the first floor, a second 3.75 m above grade representing the second floor and a third 6 m above grade representing the zone below the roof. The heat and vapour balances instead use a single-node approach. Doors between rooms are open daylong, while the floor on grade, the outer walls, and the deck between both floors are considered vapour tight. All calculations use dimensions out to out.

What concerns the airflows between nodes and from outdoors, per node their sum must be zero, or:

$$\sum G_a = 0.$$

Thermal stack in node 1 is 0. In node 2, 2.75 m above node 1, for $\theta_i = 18\,°C$ and $\theta_e = -2.5\,°C$ it's:

$$p_{T,2} = 2.75 \frac{gP_a}{R_a}\left(\frac{1}{T_i} - \frac{1}{T_e}\right) = -2.48 \text{ Pa}$$

In node 3, 5 m above node 1, for $\theta_i = 18\,°C$ and $\theta_{roof} = -3.9\,°C$ due to undercooling, it is

$$p_{T,3} = 5 \frac{gP_a}{R_a}\left(\frac{1}{T_i} - \frac{1}{T_e^*}\right) = -4.84 \text{ Pa}$$

In these formulas, g is the acceleration by gravity (9.81 m/s²), P_a the atmospheric pressure (some 100 000 Pa) and R_a the gas constant for air (287 Pa.m³/(kg.K)). The ratio $gP_a/R_a = 3462$ Pa.K/m.

Stack and wind now give as vertical and horizontal airflows:

$$G_a = a(P_{a,x} + p_{T,x} - P_{a,y} - p_{T,y})^b \quad (1) \qquad G_a = a(P_{a,x} - P_{a,y})^b \quad (2)$$

The nodal balances form a system of three equations with the air pressures $P_{a,x}$ unknown:

Node 1

$$a_{e1,1}A_{e1}[P_{a,e1} - P_{a,x1}]^{0.67} + a_{e2,1}A_{e2}[P_{a,e2} - P_{a,x1}]^{0.67}$$
$$+ a_{2,1}A_{2,1}[P_{a,x2} + p_{T,2} - P_{a,x1} - p_{T,1}]^{0.5} = 0$$

Node 2

$$a_{e3,2}A_{e3}[P_{a,e3} - P_{a,x2}]^{0.67} + a_{e4,2}A_{e4}[P_{a,e4} - P_{a,x2}]^{0.67}$$
$$+ a_{1,2}A_{1,2}[P_{a,x1} + p_{T,1} - P_{a,x2} - p_{T,2}]^{0.5}$$
$$+ a_{3,2}A_{3,2}[P_{a,x3} + p_{T,3} - P_{a,x2} - p_{T,2}]^{0.5} = 0$$

Node 3

$$a_{e5,3}A_{e5}[P_{a,e5} - P_{a,x3}]^{0.67} + a_{e6,3}A_{e6}[P_{a,e6} - P_{a,x3}]^{0.67}$$
$$+ a_{2,3}A_{2,3}[P_{a,x2} + p_{T,2} - P_{a,x3} - p_{T,3}]^{0.5} = 0$$

Solving this system of three equations requires linearization first, followed by a split between known and unknown terms. For that, it is written as:

$$\begin{vmatrix} C_{11} & C_{12} & C_{13} \\ C_{21} & C_{22} & C_{23} \\ C_{31} & C_{32} & C_{33} \end{vmatrix} \cdot \begin{vmatrix} P_{a,x1} \\ P_{a,x3} \\ P_{a,x3} \end{vmatrix} = \begin{vmatrix} F_1 \\ F_2 \\ F_3 \end{vmatrix}$$

with the coefficients C_{ij}, and the known terms F_i, equal to

Node 1 $C_{11} = -\left(\dfrac{a_{e1,1}A_{e1}}{\text{abs}(P_{a,e1} - P_{x1})^{0.33}} + \dfrac{a_{e2,1}A_{e2}}{\text{abs}(P_{a,e2} - P_{x1})^{0.33}} \right.$
$$\left. + \dfrac{a_{2,1}A_{2,1}}{\text{abs}(P_{a,x2} + p_{T,2} - P_{a,x1} - p_{T,1})^{0.5}} \right)$$

$$C_{12} = \dfrac{a_{2,1}A_{2,1}}{\text{abs}(P_{a,x2} + p_{T,2} - P_{a,x1} - p_{T,1})^{0.5}}$$

$$C_{13} = 0$$

$$F_1 = -\left(\dfrac{a_{e1,1}A_{e1}P_{a,e1}}{\text{abs}(P_{a,e1} - P_{x1})^{0.33}} + \dfrac{a_{e2,1}A_{e2}P_{a,e2}}{\text{abs}(P_{a,e2} - P_{x1})^{0.33}} \right.$$
$$\left. + \dfrac{a_{2,1}A_{2,1}(p_{T,2} - p_{T,1})}{\text{abs}(P_{a,x2} + p_{T,2} - P_{a,x1} - p_{T,1})^{0.5}} \right)$$

Node 2 $C_{21} = \dfrac{a_{1,2}A_{1,2}}{\text{abs}(P_{a,x1} + p_{T,1} - P_{a,x2} - p_{T,2})^{0.5}}$

$$C_{22} = -\left(\dfrac{a_{e3,2}A_{e3}}{\text{abs}(P_{a,e3} - P_{x2})^{0.33}} + \dfrac{a_{e4,2}A_{e4}}{\text{abs}(P_{a,e4} - P_{x2})^{0.33}} \right.$$
$$+ \dfrac{a_{1,2}A_{1,2}}{\text{abs}(P_{a,x1} + p_{T,1} - P_{a,x2} - p_{T,2})^{0.5}}$$
$$\left. + \dfrac{a_{3,2}A_{3,2}}{\text{abs}(P_{a,x3} + p_{T,3} - P_{a,x2} - p_{T,2})^{0.5}} \right)$$

$$C_{22} = \dfrac{a_{3,2}A_{3,2}}{\text{abs}(P_{a,x3} + p_{T,3} - P_{a,x2} - p_{T,2})^{0.5}}$$

$$F_2 = -\begin{pmatrix} \dfrac{a_{e3,2}A_{e2}P_{a,e3}}{abs(P_{a,e3}-P_{x2})^{0.33}} + \dfrac{a_{e2,1}A_{e2}P_{a,e4}}{abs(P_{a,e4}-P_{x2})^{0.33}} \\ + \dfrac{a_{1,2}A_{1,2}(p_{T,1}-p_{T,2})}{abs(P_{a,x1}+p_{T,1}-P_{a,x2}-p_{T,2})^{0.5}} \\ + \dfrac{a_{3,2}A_{3,2}(p_{T,3}-p_{T,2})}{abs(P_{a,x3}+p_{T,2}-P_{a,x2}-p_{T,3})^{0.5}} \end{pmatrix}$$

Node 3 $C_{31} = 0$

$$C_{32} = -\left(\dfrac{a_{e5,3}A_{e5}}{abs(P_{a,e5}-P_{x3})^{0.33}} + \dfrac{a_{e6,3}A_{e6}}{abs(P_{a,e6}-P_{x3})^{0.33}} \right.$$
$$\left. + \dfrac{a_{2,3}A_{2,3}}{abs(P_{a,x2}+p_{T,2}-P_{a,x3}-p_{T,3})^{0.5}} \right)$$

$$C_{33} = \left(\dfrac{a_{2,3}A_{2,3}}{abs(P_{a,x2}+p_{T,2}-P_{a,x3}-p_{T,3})^{0.5}} \right)$$

$$F_3 = -\left(\dfrac{a_{e5,3}A_{e5}P_{a,e5}}{abs(P_{a,e5}-P_{x3})^{0.33}} + \dfrac{a_{e6,3}A_{e6}P_{a,e6}}{abs(P_{a,e6}-P_{x3})^{0.33}} \right.$$
$$\left. + \dfrac{a_{2,3}A_{2,3}(p_{T,2}-p_{T,3})}{abs(P_{a,x2}+p_{T,2}-P_{a,x3}-p_{T,3})^{0.5}} \right)$$

To solve, the known pressures and temperatures are inserted together with guessed values for the three unknown pressures $P_{a,x1}$, $P_{a,x2}$ and $P_{a,x3}$. This allows calculating the coefficients C_{ij} and the known terms F_i. After solving, the C_{ij}'s and F_i's are recalculated with the three-node pressures found and the system is resolved. Iterating so goes on till the root of the summed quadratic deviations between new and previous drops below a pre-set value: $a_{preset} < \sqrt{\Sigma \varepsilon^2}$

With $n_{50} = 10$ ach, quantifying the air permeance of the façade at the front and the rear starts with calculating the roof's mean air permeance coefficient:

$$K_{a,roof} = 3.3\ 10^{-4}/1.2 = 2.75\ 10^{-4}\ m^3/(m^2.s.Pa^b)$$

A ventilation rate 10 ach at 50 Pa for a net air volume of 248.3 m³ means an airflow equal to $(10.4 \cdot 248.3/3600)/50^{0.67} = 0.0502\ m^3/s$ at 1 Pa, of which $(3.3 \cdot 10^{-4}/1.2) \cdot 54.7 = 0.0154\ m^3/s$ passes the roof, leaving 0.037 m³/s for the front and rear. Assuming each half of both façades shows an equal leakage, the result is $0.037/4 = 9.25 \cdot 10^{-3}\ m^3/(s.Pa^b)$.

The flow equations for the open staircase and the link between node 2 and 3 follows from conservation of energy, stating that the difference in pressure and stack should equal the change in kinetic energy of the airflow. This gives as permeance coefficients:

Staircase: $4.0\ m^3/(s.Pa^{0.5})$ First floor to roof zone: $56.2\ m^3/(s.Pa^{0.5})$

The stack pressures are already calculated. When the wind blows NE and the front façade looks NE, those the wind induces become:

Height	P_w, Pa	
	Front facade	Rear facade
1 m	0.072	−0.09
3.75 m	0.37	−0.46
6 m (roof)	−0.85	−0.85

These pressures are much smaller than the stack's, meaning the airflows are mainly buoyancy related. The node system so becomes:

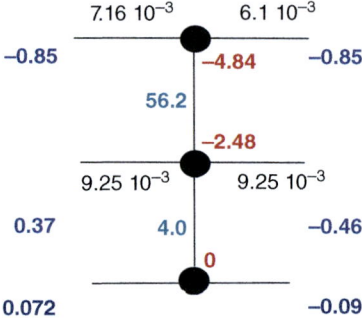

Solving the equations for $n_{50} = 10$ ach now gives as air pressures and airflows (+ is in, − outflow):

	Ground floor		First floor		Roof	
	Rear	Front	Rear	Front	Rear	Front
Air pressure (Pa)	−2.47		0.0086		2.37	
Flow, m³/h	59.5	62.2	−20.0	16.8	−54.2	−64.3

Turning to vapour now, in the single node dwelling, outside air enters and inside air leaves, while of the vapour released, some may condense. Without surface condensation, the vapour pressure indoors should be:

$$p_i = p_e + RT_i \frac{G_{v,P}}{\dot{V}_a}$$

However, surface condensation could deposit either on the window frames (1), on these and the single glass (2) or on the frames, the single and the double glass (3). The three gives as vapour pressure indoors (p_i):

$$(1)\ p_i = \frac{p_e + \dfrac{RT_i}{\dot{V}_a}(G_{v,P} + \beta A_{fr} p_{sat,fr})}{1 + \dfrac{RT_i}{\dot{V}_a}\beta A_{fr}}$$

$$(2) \quad p_i = \frac{p_e + \frac{RT_i}{V_a}(G_{v,P} + \beta A_{fr} p_{sat,fr} + \beta A_{sg} p_{sat,sg})}{1 + \frac{RT_i}{V_a}\beta(A_{fr} + A_{sg})}$$

$$(3) \quad p_i = \frac{p_e + \frac{RT_i}{V_a}(G_{v,P} + \beta A_{fr} p_{sat,fr} + \beta A_{sg} p_{sat,sg} + \beta A_{dg} p_{sat,dg})}{1 + \frac{RT_i}{V_a}\beta(A_{fr} + A_{sg} + A_{dg})}$$

The vapour saturation pressure on the inside of the frame, the single and double glass is 721, 749 and 1298 Pa. The average vapour pressure indoors (p_i) during the cold week with well or no surface condensation on the aluminium frames and the glass and, if so, the weekly amounts deposited and the percentage of the vapour released condensing are listed in the table below.

p_i (Pa)	Surface condensation				Amount condensing kg/week	% of the vapour released
	Aluminium frame	Single glass	Double glass			
852	Yes	Yes	No	Aluminium	14.6	15.4
				Single glass	22.3	23.6

Apparently, on both floors, large amounts of condensate are depositing on the aluminium window frames, in the bedrooms also on the single glass.

Up to the roof now. Combined diffusion and convection may cause condensation against the backside of the corrugated fibre-cement plates. First, the temperature there is calculated:

$$\theta_{cover} = 18 + (\theta_e^* - 18)\frac{1 - \exp(-1008 g_a R_i^{cover})}{1 - \exp(-1008 g_a R_i^e)} = -2.6°C$$

In this formula, g_a is the air flux in kg/(m².s) exfiltrating through the roof, equal to $\rho_a \dot{V}_a/(3600 A_{roof})$. The vapour saturation pressure (p_{sat} cover) there so is:

θ cover	p_{sat} cover
−2.6 °C	493 Pa

The value found lays quite below the vapour pressure inside, equal to 852 Pa. Interstitial condensation against the corrugated fibre-cement plates so is a fact. The amounts depositing equal:

$$g_c = \frac{-6.21 \cdot 10^{-6} g_a}{1 - \exp(-6.21 \cdot 10^{-6} g_a Z_i^x)}(p_{sat,x} - p_i \exp(-6.21 \cdot 10^{-6} g_a Z_i^x))$$
$$- \frac{-6.21 \cdot 10^{-6} g_a}{1 - \exp(-6.21 \cdot 10^{-6} g_a Z_x^{se})}(p_{sat,se} - p_{sat,x} \exp(-6.21 \cdot 10^{-6} g_a Z_x^{se}))$$

3 Heat, Air and Moisture Combined

giving:

Deposit (g_c) kg/(m².week)	Total for the roof kg/week	% of the vapour released
0.98	53.4	56.6

Things are even worse as also the exterior surface of the plates suffers from undercooling condensation. What with the deposit? No dripping demands less than 10 kg a week for the whole roof. Dripping in the beds so will start from the second day on, while the abundant surface condensation on the windows in the sleeping rooms will also annoy, but less.

What causes the complaints? Clearly a too air leaky roof and the easy air coupling by the open staircase between first and second floor. How the impact on the net energy demand for heating during the cold week looks, is calculated according to the EN standards, using the dimensions out to out, though with abstraction of the solar gains. The result is:

Net energy demand without exfiltration across the roof, MJ/week	Net energy demand with exfiltration across the roof, MJ/week	% less
4430	4224	4.6

Due to the high conductive losses and despite the high air leakage at 50 Pa, infiltration only counts for some 15% of the total. Outflow through the roof at the same time lowers the conduction losses and related total demand, as if its thermal transmittance dropped from 0.49 W/(m².K) to 0.21 W/(m².K). The impact anyhow is marginal.

Problem 32 Keep the dwelling and data of problem (31) and redo the analysis in case the n_{50}-value is 6 ach, all other data and the methodology remaining identical.

Solution 32 n_{50} equals 6 ach. Thermal stack still represents:

Height above grade	P_T (Pa)
1 m	0
3.75 m	−2,48
6 m (roof)	−4,84

Air pressures indoors and airflows now are:

	Ground floor		First floor		Roof	
	Rear	Front	Rear	Front	Rear	Front
Air pressure (Pa)	1.29		−1.07		−3.55	
Flow, m³/h	31.1	32.1	9.7	17.3	−41.2	−49.0

The vapour pressure indoors and the amounts of surface condensate on the aluminium window frames on both floors and on the single glass in the sleeping rooms yet equal:

p_i (Pa)		Total kg/week	% of vapour released
897	Aluminium	19.7	20.8
	Single glass	32.2	34.0

In the roof, the temperature (θ_x) and saturation pressure at the backside of the corrugated fibre cement plates ($p_{sat,x}$) become:

θ_x (°C)	$p_{sat,x}$ (Pa)	$<p_i$?
−2.8	486	Yes

The amounts condensing there equal:

g_c kg/(m².week)	Total roof kg/week	% of vapour released
0.837	45.8	48.4

Net energy demand for heating in that one cold week touches:

Without exfiltration MJ/week	With exfiltration MJ/week	Difference %
4226	4059	4.0

The air change rate by infiltration yet does not pass 0.36 ach.

3 Heat, Air and Moisture Combined

Problem 33 Keep the dwelling and data of problem (31) and redo the analysis in case the n_{50}-value is 14 ach, all other data and the methodology remaining identical.

Solution 33 n_{50} equals 14 ach.
Thermal stack represents:

Height above grade	P_T (Pa)
1 m	0
3.75 m	−2,48
6 m (roof)	−4,84

Air pressures indoors and airflows now are:

	Ground floor		First floor		Roof	
	Rear	Front	Rear	Front	Rear	Front
Air pressure (Pa)	−2.14		0.34		2.70	
Flow, m³/h	80.2	84.4	−42.8	4.7	−57.8	−68.7

The vapour pressure indoors and the amounts of surface condensate on the aluminium window frames on both floors and on the single glass in the sleeping rooms yet equal:

p_i (Pa)		Total kg/week	% of vapour released
862	Aluminium	15.7	16.6
	Single glass	24.5	25.9

In the roof, the temperature (θ_x) and saturation pressure at the backside of the corrugated fibre cement plates ($p_{sat,x}$) become:

θ_x (°C)	$p_{sat,x}$ (Pa)	$<p_i$?
−2.5	495	Yes

The amounts condensing there equal:

g_c kg/(m².week)	Total roof kg/week	% of vapour released
1.07	58.7	62.1

Net energy demand for heating in that one cold week touches:

Without exfiltration MJ/week	With exfiltration MJ/week	Difference %
4560	4345	4.7

The air change rate by infiltration yet reaches 0.68 ach.

Problem 34 Keep the dwelling and the data of problem (31), but without heating the first floor, while the average inside temperature of the first floor is 18 °C. Vapour release touches 7.88 kg a day on the first and 5.62 kg a day on the second floor. Besides the 3 air nodes, the dwelling now counts 2 heat and vapour nodes, node 1 coinciding with air node 1 and node 2 uniting the air nodes 2 and 3. The average internal gains on the first floor (node 2) touch 260 W, solar gains are overlooked and the finished floor between first and second floor has a U-value 2.16 W/(m².K) and 48 m² as surface. Additional data are:

First floor		Area, m²	U, W/(m².K)
Cavity wall	Front	1265	166
	Rear	1335	166
Sidewall		229	166
Roof	Large pitch	297	049
	Small pitch	25	049
Floor first floor		48	216

Solution 34 n_{50} equals 10.4 ach and the first floor remains unheated now. The temperature there drops to 6.8 °C and thermal stack left represents:

Height above grade	P_T (Pa)
1 m	0
3.75 m	−1.82
6 m (roof)	−3.11

Air pressures indoors and airflows now are:

	Ground floor		First floor		Roof	
	Rear	Front	Rear	Front	Rear	Front
Air pressure (Pa)	−1.77		0.051		1.34	
Flow, m³/h	47.2	50.2	−21.2	15.5	−41.9	−49.7

The vapour pressure indoors and the amounts of surface condensate on the aluminium window frames on both floors and on the single glass in the sleeping rooms yet equal:

Floor	p_i, Pa	Surface condensation	
		Total kg/week	% of vapour released
Ground floor (condensation on aluminium)	879	7.7	8.1
First floor (aluminium + glass)	777	46.5	49.2

In the roof, temperature (θ_x) and saturation pressure at the backside of the corrugated fibre cement plates ($p_{sat,x}$) become:

θ_x (°C)	$p_{sat,x}$ (Pa)	$<p_i$?
−3.3	463	Yes

The amounts condensing there equal:

Deposit g_c kg/(m².week)	Total roof kg/week	% of vapour released
0.602	32.9	34.8

Net energy demand for heating in that one cold week touches:

Without exfiltration MJ/week	With exfiltration MJ/week	Difference %
3822	3515	8.0

Partially heating looks like an effective way to economize on net energy demand: −24%! Also, the air change rate by infiltration drops: 0.45 instead of 0.56 ach.

Further Reading

Abuku, M. (2009). Moisture stress of wind-driven rain on building enclosures. Doctoral thesis. KU-Leuven.

Arfvidsson, J. (1998). Moisture transport in porous media, modelling based on kirchhoffs potentials. Doctoral dissertation. Lund University.

Blocken B. (2004). Wind-driven Rain on Buildings, Doctoraal proefschrift, K.U. Leuven.

Blocken, B., Hens, H., and Carmeliet, J. (2002). Methods for the quantification of driving rain on buildings. *ASHRAE Transactions* 108, part 2, : 338–350.

Brocken, H. (1998). Moisture transport in brick masonry: the grey area between bricks. Doctoral thesis. TUE, Eindhoven.

Carmeliet, J., Hens, H., and Vermeir, G. (ed.) (2003). *Research in Building Physics*, 1020. Lisse, Abingdon, Exton (PA), Tokyo: Balkema Publishers.

Chaddock, J.B. and Todorovic, B. (1991). *Heat and Mass Transfer in Building Materials and Structures*. New York: Hemisphere Publishing Corporation.

Desta, T. and Roels, S. (2010). *Experimental and Numerical Analysis of Heat, Air and Moisture Transport in a Lightweight Building Wall*. Clearwater Beach (CD-Rom): Proceedings Buildings XI.

Desta, T. and Roels, S. (2010). The influence of air on the heat and moisture transport through a lightweight building wall. *Proceedings CESBP 2010*, (ed. D. Gawin, and T. Kisilewicz), Poland: Technical University of Lodz.

Duforestel T. (1992). Bases métrologiques et modèles pour la simulation du comportement hygrothermique des composants et ouvrages du bâtiment, Thèse de Doctorat, Ecole Nationale des Ponts et des Chaussées, Paris (in French).

Garrecht, H. (1992). *Porenstrukturmodelle für den Feuchtehaushalt von Baustoffen mit und ohne Salzbefrachtung und rechnerische Anwendung auf Mauerwerk*. Doktors Abhandlung: Universität Karlsruhe (in German).

Häupl, P. and Roloff, J. (2002). Proceedings of the 11. Bauklimatisches Symposium, Band 1 & 2, TU Dresden, Institut für Bauklimatik.

Hens, H. (1978). *1981, Bouwfysica, Warmte en Vocht, Theoretische Grondslagen, 1^e en 2^e Uitgave*. Leuven: ACCO (in Dutch).

Hens, H. (1996). Modelling, Final Report Task 1, Annex 24, Vol. 1, ACCO, Leuven, 90.

Holm, A. (2001). Ermittlung der Genauigkeit von instationären hygrothermischen Bauteilberechnungen mittels eines stochastischen Konzeptes, Doktors Abhandlung, Universität Stuttgart (in German).

Janssen, H. (2002). The influence of soil moisture transfer on building heat loss via the ground. Doctoral thesis. KU-Leuven, Leuven.

Janssens, A. (1998). Reliable control of interstitial condensation in lightweight roof systems. Doctoral thesis. KU-Leuven, Leuven.

Koci V., Madera J., Keppert M., and Cerny R. (2010). Mathematical models and computer codes for modelling heat and moisture transport in building materials: a comparison. *Proceedings CESBP 2010*, (ed. D. Gawin, and T. Kisilewicz), Poland: Technical University of Lodz.

Kohonen, R. and Ojanen, S. (1985). Coupled convection and conduction in two dimensional building structures, *4th Conference on Numerical Methods in Thermal Problems*, Swansea.

Kohonen, R. and Ojanen, T. (1987). Coupled diffusion and convection heat and mass transfer in building structures, *Building Physics Symposium*, Lund.

Krus, M. (1995). Feuchtetransport- und Speicherkoeffizienten poröser mineralischer Baustoffe. Theoretische Grundlagen und Neue Meβtechniken, Doktors Abhandlung, Universität Stuttgart (in German).

Künzel, H.M. (1994). Verfahren zur ein- und zweidimensionalen Berechnung des gekoppelten Wärme- und Feuchtetransports in Bauteilen mit einfachen Kennwerten, Doktors Abhandlung, Universität Stuttgart (in German).

Pedersen, C.R. (1990). Combined heat and moisture transfer in building constructions. Ph.D. thesis. Technical University of Denmark.

Roels, S. (2000). Modelling unsaturated moisture transport in heterogeneous limestone. Doctoral thesis. KU-Leuven, Leuven.

Sergio Vera Araya, (2009). Inter-zonal air and moisture transport through large horizontal openings: an integrated experimental and numerical study. Doctoral thesis. Concordia University. Montreal.

Taveirne, W. (1990). Eenhedenstelsels en groothedenvergelijkingen: overgang naar het SI, Pudoc, Wageningen, p. 719, (in Dutch).

Time, B. (1998). Hygroscopic moisture transport in wood. Doctors thesis. NUST Trondheim.

Trechsel, H.R. (ed.) (1994). *Moisture Control in Buildings*, ASTM Manual Series: MNL 18, PCN 28-018094-10, 19103. Philadelphia, PA: ASTM.

Van Mook, J.R. (2003). Driving rain on building envelopes. *TU/e Bouwstenen* 69: 198.

Welty, J., Wicks, C., and Wilson, R. (1969). *Fundamentals of Momentum, Heat and Mass Transfer*, 697. New York: Wiley.

Woloszyn, M. and Rode, C. (2008). Modelling principles and common exercises, Final Report IEA-ECBCS Annex 41 'Whole Building Heat, Air and Moisture Response, ACCO, Leuven.

Xiaochuan, Q. (2003). Moisture transport across interfaces between building materials. Doctoral thesis. Concordia University, Montreal.

Zillig, W. (2009). Moisture transport in wood using a multiscale approach. Doctoral thesis. KU-Leuven.

4

Heat, Air and Moisture Material Property Values

4.1 In General

Solving heat, air and moisture problems is not possible without knowing which material property values to use. That is why this last chapter tabulates a collection of property values as listed in standards and as measured. To get a clear view of all, Table 4.1 shows the whole array of material properties, whose values should be known.

Table 4.1 Array of heat, air and moisture material properties.

Density	The mass per unit volume of dry material, ρ in kg/m³
Open porosity	The volume taken by the pores, accessible for water, Ψ_o in m³/m³

	Heat	**Air**	**Moisture**
Storage	Specific heat capacity c	Specific moisture ratio ξ	Specific air content
	Volumetric specific heat capacity ρc	Volumetric specific moisture ratio $\rho \xi$	
Transport	Thermal conductivity λ	Water vapour permeability δ	Air permeability k_a
	Thermal resistance R	Vapour resistance factor μ	Air permeance K_a
	Radiation:	Diffusion thickness μd	
	Absorptivity α	Moisture permeability k_m	
	Emissivity e	Thermal moisture diffusion coefficient K_θ	
	Reflectivity ρ		
Combined	Thermal diffusivity a	Moisture diffusivity D_w	
	Contact coefficient b	Water sorption coefficient A	
Consequences	Thermal expansion coefficient α	Hygric expansion ε	

Building Physics – Heat, Air and Moisture: Fundamentals, Engineering Methods, Material Properties and Exercises,
Fourth Edition. Hugo Hens.
© 2024 Ernst & Sohn GmbH. Published 2024 by Ernst & Sohn GmbH.

4.2 Dry Air and Water

Water

Density (ρ)	Pressure (bar)	Temperature (°C)	ρ (kg/m³)
	1	0	999.9
	1	10	999.7
	1	20	998.2
	1	40	992.2
	1	60	983.2
	1	80	971.8
	1	100	958.4
Viscosity (η)	Pressure (bar)	Temperature (°C)	η (Pa.s)
	1	0	$1.787 \cdot 10^{-3}$
	1	10	$1.307 \cdot 10^{-3}$
	1	20	$1.002 \cdot 10^{-3}$
	1	40	$0.653 \cdot 10^{-3}$
	1	60	$0.467 \cdot 10^{-3}$
	1	80	$0.354 \cdot 10^{-3}$
	1	100	$0.282 \cdot 10^{-3}$
Surface tension (σ)	$(75.9 - 0.17\theta) \cdot 10^{-3}$ N/m with θ temperature in °C		
Specific heat capacity (c)	4187 J/(kg·K)		
Heat of evaporation (l_b)	2 500 000 J/kg at 0 °C		
Heat of solidification (l_s)	33 400 J/kg at 0°C		
Thermal conductivity	Pressure (bar)	Temperature (°C)	λ (W/(m·K))
	1	0	0.54
	1	60	0.67
Long-wave emissivity	0.95		

Dry air

Gas constant	287.055 J/(kg·K)		
Atmospheric pressure	Normally 101325 Pa, so, close to 1 bar. Higher in cyclonic, lower in anti-cyclonic weather		
Viscosity (η)	Pressure (bar)	Temperature (°C)	Viscosity (Pa.s)
	1	0	0.0000174
	1	50	0.0000199
	1	100	0.0000222
Specific heat capacity (c_p)	1007 J/(kg·K)		
Thermal conductivity (λ)	Pressure (bar)	Temperature (°C)	λ (W/(m·K))
	1	0	0.024
	1	100	0.031

4.3 Thermal Properties

4.3.1 Definitions

The values given for the thermal conductivity (λ) are either:

Declared (λ_D)	Start is the thermal conductivity of the dry material, measured in a certified laboratory at an average temperature of 10°C on a collection of samples large enough to allow a statistical analysis of the results (≥ 20). The declared value coincides with the 90% percentile, ensuring that 90% of what's produced does better
Certified versus non-certified	A material gets the term certified when the declared thermal conductivity is known. It is considered as non-certified when the thermal conductivity was measured by a non-certified laboratory.
Design (λ_U)	Are linked to the declared values, the equations being: $$\lambda_U = \lambda_D \exp[f_u(X_2 - X_1)] \quad \lambda_U = \lambda_D \exp[f_\psi(\Psi_2 - \Psi_1)] \quad (4.1)$$ with X_2 and Ψ_2 moisture ratio in kg/kg or in m³/m³ when used and X_1 and Ψ_1 moisture ratio in kg/kg or in m³/m³ during testing (normally 0). f_u and f_ψ are standardised conversion factors

4.3.2 Standard Values

4.3.2.1 Regardless of Being on the In- or on the Outside of the Thermal Insulation

Group	Material	Density (ρ) (kg/m³)	Specific heat capacity (c) (J/(kg·K))	Thermal conductivity (λ) (W/(m·K))
Metals	Aluminium alloys	2800	880	220
	Duralumin	2800	880	160
	Brass	8400	380	120
	Bronze	8700	380	65
	Copper	8900	380	380
	Iron	7900	450	75
	Iron, cast	7500	450	50
	Steel	7800	450	50
	Stainless steel	7900	460	17
	Lead	11 300	130	35
	Zinc	7100	380	110

Wood and wood-based materials	Softwood	500	1600	0.13
	Hardwood	700	1600	0.18
	Plywood (from low to high density)	300	1600	0.09
		500	1600	0.13
		700	1600	0.17
		1000	1600	0.24
Wood and wood-based materials	Particle board			
	Soft	300	1700	0.10
	Semi-hard	600	1700	0.14
	Hard	900	1700	0.18
	Particle board, cement bounded	1200	1500	0.23
	OSB	680	1700	0.13
Wood and wood-based materials	Fibreboard			
	Soft	400	1700	0.10
	Semi-hard	600	1700	0.14
	Hard	800	1700	0.18
Gypsum	Gypsum blocks	600	1000	0.18
		900	1000	0.30
		1200	1000	0.43
		1500	1000	0.56
	Gypsum board	900	1050	0.25
Mortars	Cement mortar, mixed on site	1800	1100	0.9
		1900	1100	1.0
Plasters	Gypsum plaster			
	Light	600	1000	0.18
	Normal	1000	1000	0.40
	Heavy	1300	1000	0.57
	Lime or gypsum plus sand	1600	1000	0.70
	Cement plus sand	1700	1000	1.0
Concrete	Medium to high density	1800	1000	1.15
		2000	1000	1.35
		2200	1000	1.65
		2400	1000	2.00
	Reinforced 1% steel	2300	1000	2.30
	2% steel	2400	1000	2.50
Stone	Crystalline rock	2800	1000	3.50
	Sedimentary rock	2600	1000	2.30
	Sedimentary, light	1500	1000	0.85

	Lava	1600	1000	0.55
	Basalt	2700–3000	1000	3.50
	Gneiss	2400–2700	1000	3.50
	Granite	2500–3000	1000	2.80
	Marble	2800	1000	3.50
	Slate	2000–2800	1000	2.20
	Limestone			
	Extra soft	1600	1000	0.85
	Soft	1800	1000	1.10
	Semi-hard	2000	1000	1.40
	Hard	2200	1000	1.70
	Extra hard	2600	1000	2.30
	Sandstone	2600	1000	2.30
	Natural pumice	400	1000	0.12
	Artificial stone	1750	1000	1.30
Soils	Clay or silt	1200–1800	1670–2500	1.5
	Sand and gravel	1700–2200	910–1180	2.0
Water, ice, snow	Ice			
	$-10\,°C$	920	2000	2.30
	$0\,°C$	900	2000	2.20
	Water			
	$10\,°C$	1000	4187	0.60
	$40\,°C$	990	4187	0.63
	Snow			
	Fresh	100	2000	0.05
	Soft	200	2000	0.12
	Compacted	300	2000	0.23
		500	2000	0.60
Synthetics, solid	Acrylic	1050	1500	0.20
	Polycarbonates	1200	1200	0.20
	PFTE	2200	1000	0.25
	PVC	1390	900	0.17
	PMMA	1180	1500	0.18
	Polyacetate	1410	1400	0.30
	Polyamide (nylon)	1150	1600	0.25
	Nylon, 25% glass fibre	1450	1600	0.30
	PE, High density	980	1800	0.50
	PE, low density	920	2100	0.33

	Polystyrene	1050	1300	0.16
	Polypropylene (PP)	910	1800	0.22
	PP, 25% glass fibre	1200	1800	0.25
	Polyurethane	1200	1800	0.25
	Epoxy resin	1200	1200	0.20
	Phenol resin	1300	1700	0.30
	Polyester resin	1400	1200	0.19
Rubbers	Natural	910	1100	0.13
	Neoprene	1240	2140	0.23
	Butyl	1200	1400	0.24
	Foam rubber	60–80	1500	0.06
	Hard rubber	1200	1400	0.17
	EPDM	1150	1000	0.25
	Polyisobutylene	920	1130	0.13
	Polysulphide	1700	1000	0.43
	Butadiene	980	1000	0.25
Glass	Quartz glass	2200	750	1.4
	Normal glass	2500	750	1.0
	Glass mosaic	2000	750	1.2
Gases	Air	1.23	1008	0.025
	Argon	1.70	519	0.017
	Carbon dioxide	1.95	820	0.014
	Sulphur hexafluoride	6.36	614	0.013
	Krypton	3.56	245	0.009
	Xenon	5.68	160	0.0054
Thermal breaks, sealants, weather stripping	Silica gel (desiccant)	720	1000	0.13
	Silicone foam	750	1000	0.12
	Silicone pure	1200	1000	0.35
	Silicone, filled	1450	1000	0.50
	Urethane/PUR (thermal break)	1300	1800	0.21
	PVC, flexible with 40% softener	1200	1000	0.14
	Elastomeric foam, flexible	60–80	1500	0.05
	Polyurethane foam	70	1500	0.05
	Polyethene foam	70	2300	0.05
Roofing	Asphalt	2100–2300	1000	0.7
	Bitumen, pure	1050	1000	0.17

	Bitumen felt	1100	1000	0.23	
	Clay tiles	2000	800	1.00	
	Concrete tiles	2100	1000	1.50	
Floor covering	Rubber	1200	1400	0.17	
	Plastic	1700	1400	0.25	
	Linoleum	1200	1400	0.17	
	Carpet/textile	200	1300	0.06	
	Underlay, rubber	270	1400	0.10	
	Underlay, felt	120	1300	0.05	
	Underlay, wool	200	1300	0.06	
	Underlay, cork	200	1500	0.05	
	Ceramic tiles	2300	840	1.30	
	Plastic tiles	1000	1000	0.20	
	Cork tiles, light	200	1500	0.050	
	Cork tiles, heavy	500	1500	0.065	

4.3.2.2 Depending on Being on the In- or on the Outside of the Thermal Insulation

Inside prevails for all layers at the back of the thermal insulation with as design value λ_{Ui} the one at 23 °C and 50% RH. Outside applies to all layers in front of the thermal insulation, with the design value λ_{Ue} fixed at 10 °C and a moisture ratio of 0.75 times the critical value for capillary materials (see Table 4.2) and at 10 °C and 80% RH for non-capillary materials.

Table 4.2 Moisture ratio in masonry at indoor and outdoor conditions.

Masonry	Density, ρ (kg/m³)	X_i (kg/kg)	Moisture ratio 23°C, 50% RH Ψ_i (m³/m³)	X_e (kg/kg)	Moisture ratio 0.75(X_{cr} or Ψ_{cr}) Ψ_e (m³/m³)	Conversion factors f_u (kg/kg)	f_Ψ (m³/m³)
Bricks, perforated	700–2100		0.007		0.075		10
Lime-sand stone	900–2200		0.012		0.090		10
Concrete blocks (massive, perforated)							
Heavy	1600–2400		0.025		0.090		4
Expanded clay	400–1700		0.020		0.090		4
Lightweight	500–1800		0.030		0.090		4
Cellular concrete	300–1000	0.026		0.150		4	

Values

Material	ρ (kg/m³)	c (J/(kg·K))	λ_{Ui} (W/(m·K))	λ_{Ue} (W/(m·K))
Metals				
Lead	11340	130	35	35
Copper	$8300 \leq \rho \leq 8900$	390	384	384
Steel	7800	480–530	45	45
Aluminium, 99%	2700	880	203	203
Iron, cast	7500	530	56	56
Zinc	7000	390	113	113
Stone				
Heavy (granite, gneiss, basalt, porphyry)	$2750 \leq \rho \leq 3000$	1000	3.49	3.49
Limestone, hard	2700	1000	2.91	3.49
Marble	2750	1000	2.91	3.49
Sandstone				
Hard	2550	1000	2.21	2.68
Semi	2350	1000	1.74	2.09
Soft	2200	1000	1.40	1.69

Glued masonry, joint width ≤3 mm

Bricks and perforated large format bricks

		Certified		Not certified	
ρ (kg/m³)	c (J/(kg·K))	λ_{Ui} (W/(m·K))	λ_{Ue} (W/(m·K))	λ_{Ui} (W/(m·K))	λ_{Ue} (W/(m·K))
≤700	1000	0.20	0.39	0.22	0.43
$700 < \rho \leq 800$	1000	0.23	0.45	0.25	0.49
$800 < \rho \leq 900$	1000	0.26	0.51	0.28	0.56
$900 < \rho \leq 1000$	1000	0.29	0.57	0.32	0.63
$1000 < \rho \leq 1100$	1000	0.32	0.64	0.35	0.70
$1100 < \rho \leq 1200$	1000	0.35	0.70	0.39	0.77
$1200 < \rho \leq 1300$	1000	0.39	0.76	0.42	0.84
$1300 < \rho \leq 1400$	1000	0.43	0.85	0.47	0.93
$1400 < \rho \leq 1500$	1000	0.46	0.91	0.51	1.00
$1500 < \rho \leq 1600$	1000	0.50	0.99	0.55	1.09

$1600 < \rho \leq 1700$	1000	0.55	1.08	0.60	1.19
$1700 < \rho \leq 1800$	1000	0.59	1.16	0.65	1.28
$1900 < \rho \leq 1900$	1000	0.64	1.27	0.71	1.40
$1900 < \rho \leq 2000$	1000	0.69	1.35	0.76	1.49
$2000 < \rho \leq 2100$	1000	0.74	1.46	0.81	1.61

Lime-sandstone

≤ 900	1000	0.33	0.71	0.36	0.78
$900 < \rho \leq 1000$	1000	0.34	0.74	0.37	0.81
$1000 < \rho \leq 1100$	1000	0.36	0.79	0.40	0.87
$1100 < \rho \leq 1200$	1000	0.41	0.89	0.45	0.97
$1200 < \rho \leq 1300$	1000	0.46	1.01	0.51	1.11
$1300 < \rho \leq 1400$	1000	0.52	1.13	0.57	1.24
$1400 < \rho \leq 1500$	1000	0.60	1.30	0.66	1.43
$1500 < \rho \leq 1600$	1000	0.69	1.50	0.76	1.65
$1600 < \rho \leq 1700$	1000	0.79	1.72	0.87	1.89
$1700 < \rho \leq 1800$	1000	0.91	1.99	1.00	2.19
$1800 < \rho \leq 1900$	1000	1.04	2.26	1.14	2.49
$1900 < \rho \leq 2000$	1000	1.18	2.58	1.30	2.84
$2000 < \rho \leq 2100$	1000	1.35	2.95	1.49	3.25
$2100 < \rho \leq 2200$	1000	1.54	3.37	1.70	3.71

Normal concrete blocks

≤ 1600	1000	0.97	1.26	1.07	1.39
$1600 < \rho \leq 1700$	1000	1.03	1.33	1.13	1.47
$1700 < \rho \leq 1800$	1000	1.12	1.45	1.23	1.59
$1800 < \rho \leq 1900$	1000	1.20	1.56	1.33	1.72
$1900 < \rho \leq 2000$	1000	1.32	1.71	1.45	1.88
$2000 < \rho \leq 2100$	1000	1.44	1.86	1.58	2.05
$2100 < \rho \leq 2200$	1000	1.57	2.04	1.73	2.24
$2200 < \rho \leq 2300$	1000	1.72	2.24	1.90	2.46

Expanded clay concrete blocks

$400 < \rho \leq 500$	1000	0.16		0.18	
$500 < \rho \leq 600$	1000	0.19	0.26	0.21	0.28
$600 < \rho \leq 700$	1000	0.23	0.30	0.25	0.33
$700 < \rho \leq 800$	1000	0.27	0.36	0.30	0.39
$800 < \rho \leq 900$	1000	0.30	0.40	0.33	0.44
$900 < \rho \leq 1000$	1000	0.35	0.46	0.38	0.50
$1000 < \rho \leq 1100$	1000	0.39	0.52	0.43	0.57
$1100 < \rho \leq 1200$	1000	0.44	0.59	0.49	0.65
$1200 < \rho \leq 1300$	1000	0.50	0.66	0.55	0.73
$1300 < \rho \leq 1400$	1000	0.55	0.73	0.61	0.80

ρ (kg/m³)	c	λ_{Ui}		λ_{Ue}	
$1400 < \rho \leq 1500$	1000	0.61	0.80	0.67	0.88
$1500 < \rho \leq 1600$	1000	0.68	0.90	0.75	0.99
$1600 < \rho \leq 1700$	1000	0.76	1.00	0.83	1.10

Concrete blocks with other lightweight aggregates

ρ (kg/m³)	c	λ_{Ui}		λ_{Ue}	
≤ 500	1000	0.27		0.30	
$500 < \rho \leq 600$	1000	0.30	0.39	0.33	0.43
$600 < \rho \leq 700$	1000	0.34	0.43	0.37	0.47
$700 < \rho \leq 800$	1000	0.37	0.47	0.41	0.52
$800 < \rho \leq 900$	1000	0.42	0.53	0.46	0.58
$900 < \rho \leq 1000$	1000	0.46	0.59	0.51	0.65
$1000 < \rho \leq 1100$	1000	0.52	0.66	0.57	0.73
$1100 < \rho \leq 1200$	1000	0.59	0.75	0.64	0.82
$1200 < \rho \leq 1300$	1000	0.65	0.83	0.72	0.91
$1300 < \rho \leq 1400$	1000	0.74	0.95	0.82	1.04
$1400 < \rho \leq 1500$	1000	0.83	1.06	0.92	1.17
$1500 < \rho \leq 1600$	1000	0.94	1.19	1.03	1.31
$1600 < \rho \leq 1800$	1000	1.22	1.55	1.34	1.70

Autoclaved aerated concrete blocks, air-dry (contains production moisture when fresh)

ρ (kg/m³)	c	λ_{Ui}		λ_{Ue}	
≤ 300	1000	0.09		0.10	
$300 < \rho \leq 400$	1000	0.12		0.13	
$400 < \rho \leq 500$	1000	0.14		0.16	
$500 < \rho \leq 600$	1000	0.18	0.29	0.20	0.32
$600 < \rho \leq 700$	1000	0.20	0.33	0.22	0.36
$700 < \rho \leq 800$	1000	0.23	0.38	0.26	0.42
$800 < \rho \leq 900$	1000	0.27	0.44	0.29	0.48

Masonry with mortar joints

The design value λ_U is calculated as:

$$\lambda_U = \lambda_{U,block} f_{block} + \lambda_{U,mortar}(1 - f_{block}) \tag{4.2}$$

with $\lambda_{U,block}$ and $\lambda_{U,mortar}$ the thermal conductivity of blocks and mortar and f_{block} the surface ratio taken by the blocks. If only block density is known, use the values for glued masonry.

Normal concrete (The values are low compared to what's measured and given in ISO 10456)

Material	ρ (kg/m³)	c (J/(kg·K))	λ_{Ui} (W/(m·K))	λ_{Ue} (W/(m·K))
Reinforced	2400	1000	1.7	2.2
Not reinforced	2200	1000	0.3	1.7

Lightweight concrete for slabs or screeds

Material	ρ (kg/m³)	c (J/(kg·K))	λ_{Ui} (W/(m·K))	λ_{Ue} (W/(m·K))
Concrete with expanded clay, furnace slag, vermiculite, cork, perlite or polystyrene pearls as aggregates	350	1000	0.12	
	$300 < \rho \leq 400$	1000	0.14	
	$400 < \rho \leq 450$	1000	0.15	
	$450 < \rho \leq 500$	1000	0.16	
	$500 < \rho \leq 550$	1000	0.17	
	$550 < \rho \leq 600$	1000	0.18	
	$600 < \rho \leq 650$	1000	0.20	0.31
Cellular concrete	$650 < \rho \leq 700$	1000	0.21	0.34
	$700 < \rho \leq 750$	1000	0.22	0.36
	$750 < \rho \leq 800$	1000	0.23	0.38
	$800 < \rho \leq 850$	1000	0.24	0.40
	$850 < \rho \leq 900$	1000	0.25	0.43
	$900 < \rho \leq 950$	1000	0.27	0.45
	$950 < \rho \leq 1000$	1000	0.29	0.47
	$1000 < \rho \leq 1100$	1000	0.32	0.52
	$1100 < \rho \leq 1200$	1000	0.37	0.58

Gypsum with and without lightweight aggregates

ρ (kg/m³)	c (J/(kg·K))	λ_{Ui} (W/(m·K))	λ_{Ue} (W/(m·K))
800	1000	0.22	
$800 < \rho \leq 1100$	1000	0.35	
>1100	1000	0.52	

Plaster

Material	ρ (kg/m³)	c (J/(kg·K))	λ_{Ui} (W/(m·K))	λ_{Ue} (W/(m·K))
Cement plaster	1900	1000	0.93	1.5
Lime plaster	1600	1000	0.70	1.2
Gypsum plaster[a]	1300	1000	0.52	

a) A heavy-weight gypsum plaster. Most ready-to-mix gypsum plasters have a density <1000 kg/m³ (see gypsum).

Wood and wood-based materials

Material	ρ (kg/m³)	c (J/(kg·K))	λ_{Ui} (W/(m·K))	λ_{Ue} (W/(m·K))
Timber	≤ 600	1880	0.13	0.15
	>600	1880	0.18	0.20
Plywood	≤ 400	1880	0.09	0.11
	$400 < \rho \leq 600$	1880	0.13	0.15
	$600 < \rho \leq 850$	1880	0.17	0.20
	>850	1880	0.24	0.28
Particle board	≤ 450	1880	0.10	
	$450 < \rho \leq 750$	1880	0.14	
	>750	1880	0.18	
Fibre-cement board	1200	1470	0.23	
OSB	650	1880	0.13	
Fibreboard	≤ 375	1880	0.07	
	$375 < \rho \leq 500$	1880	0.10	
	$500 < \rho \leq 700$	1880	0.14	
	>700	1880	0.18	

Insulation materials (may under no circumstances become humid, only λ_{Ui} is given)

Material	ρ (kg/m³)	c (J/(kg·K))	λ_{Ui} (W/(m·K))
			Not certified
Cork	90–160	1560	0.05
Glass fibre and mineral fibre	10–200	1030	0.045
EPS	10–50	1450	0.045
Polyethylene (PE)	20–65	1450	0.045
Phenol foam	20–50	1400	0.045
PUR, lined	28–55	1400	0.035
XPS	20–65	1450	0.040
Cellular glass	100–140	1000	0.055
Perlite board	200–300	900	0.060
Vermiculite	50–170	1080	0.065
Vermiculite board		900	0.090
			Certified
Glass and mineral fibre	10–200	1030	0.040
EPS	10–50	1450	0.040
Phenol foam	20–50	1400	0.025

PUR, lined	28–55	1400	0.028
XPS	20–65	1450	0.034
Cellular glass	100–140	1000	0.048
Perlite board	200–300	900	0.055
VIPs			0.004–0.008

Miscellaneous

Material	ρ (kg/m³)	c (J/(kg·K))	λ_{Ui} (W/(m·K))	λ_{Ue} (W/(m·K))
Glass	2500	750	1.00	1.00
Clay tiles	1700	1000	0.81	1.00
Grès tiles	2000	1000	1.20	1.30
Rubber	1500	1400	0.17	0.17
Linoleum, PVC	1200	1400	0.19	
Fibre cement	$1400 < \rho < 1900$	1000	0.35	0.50
Asphalt	2100	1000	0.70	0.70
Bitumen	1100	1000	0.23	0.23

Perforated blocks, floor elements, gypsum board

Material	Thickness d (cm)	c (J/(kg·K))	R_{Ui} (m²·K/W)
Masonry, perforated concrete blocks			
Normal concrete ($\rho > 1200$ kg/m³)	14	1000	0.11
	19	1000	0.14
	29	1000	0.20
Lightweight concrete ($\rho < 1200$ kg/m³)	14	1000	0.30
	19	1000	0.35
	29	1000	0.45
Clay floor elements			
One cavity along span	8	1000	0.08
	12	1000	−0.11
Two cavities along span	12	1000	−0.13
	16	1000	0.16
	20	1000	0.19
Concrete floor elements	12	1000	0.11
	16	1000	0.13
	20	1000	0.15

| Gypsum board | <1.4 | 1000 | 0.05 |
| | ≥1.4 | 1000 | 0.08 |

4.3.3 Surfaces, Radiant Properties

Long wave emissivity (e_L) and shortwave absorptivity (a_S) are as follows:

Material, surface	e_L	α_S
Snow		0.15
White paint	0.85	0.25
Black paint	0.97	
Oil paint	0.94	
Whitewashed surface		0.30
Light colours, polished aluminium		0.3–0.5
Yellow brick	0.93	0.55
Red brick	0.93	0.75
Concrete, light-coloured floor cover		0.6–0.7
Grass and leaves		0.75
Dark-coloured floor cover		0.8–0.9
Carpet		0.8–0.9
Moist bottom		0.90
Dark grey slates		0.90
Bituminous felt	0.92	0.93
Gold, silver and copper polished	0.02	
Copper, oxidized	0.78	
Aluminium, polished	0.05	
Aluminium, oxidized	0.30	
Steel, hot casted	0.77	
Steel, oxidized	0.61	
Steel with silver finish	0.26	
Steel, polished	0.27	
Lead, oxidized	0.28	
Glass	0.92	
Porcelain	0.92	
Plaster	0.93	
Timber, unpainted	0.90	
Marble, polished	0.55	
Paper	0.93	
Water	0.95	
Ice	0.97	

4.3.4 Measured Values

4.3.4.1 Thermal Conductivity, Test Methods

Commonly used are the Poensgen and the heat flow meter method. The set up for the Poensgen method consists of a $30 \times 30\,\text{cm}^2$ large heating plate composed of a 10×10^2 cm large central core, which acts as measuring surface, surrounded by an equally thick 10 cm wide perimeter plate, both heated electrically After mounting a $30 \times 30\,\text{cm}^2$ large material sample at each side against that heating plate, the two get a $30 \times 30\,\text{cm}^2$ water cooled plate as cover, after which the whole is packed in a thermally insulated box, see Figure 4.1. Both the perimeter and central heating plate are kept at a same temperature, for example 20 °C, while both cooling plates are kept at a temperature of for example 0 °C. Once the heat delivered to the $30 \times 30\,\text{cm}^2$ heating plate turns steady state, the amount of heat (Q) and time span (Δt) logged allow to calculate the thermal conductivity from:

$$\lambda = Q/(0.2\Delta t)\ \text{W}/(\text{m·K})$$

The heat flow meter method uses a same approach, though the $30 \times 30\,\text{cm}^2$ heating plate has no central core now but is at both sides covered by a rubber foil with centrally a calibrated heat flow meter with diameter 10 cm and same thickness as the foil embedded. Again, a $30 \times 30\,\text{cm}^2$ large material sample is mounted against both sides of that heating plate and gets a 30×30 cm cooling plate as cover. The whole then is packed in a thermally insulated box, after which the heating plate is electrically warmed to for example 20 °C, while both cooling plates are liquid cooled to for example 0 °C. Once the heat flow meters at both sides show steady state is reached, the flux delivered (q) is noted and the thermal conductivity is calculated:

$$\lambda = q/20\ \text{W}/(\text{m·K})$$

Figure 4.1 The Poensgen method.

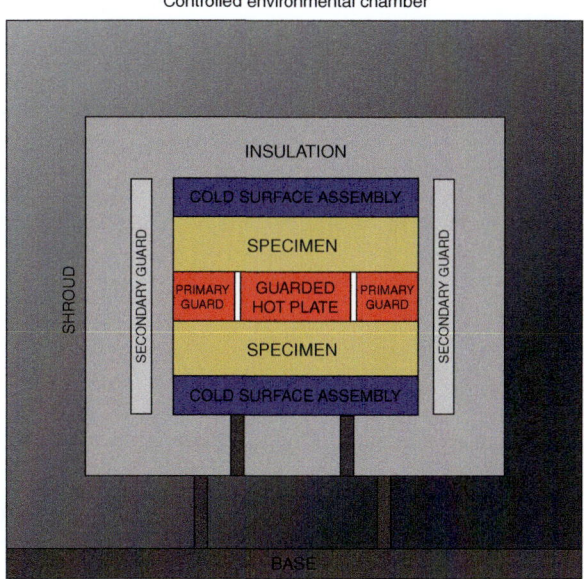

4.3.4.2 Test Results
4.3.4.2.1 Building Materials
Concrete

Density	kg/m³	2176, $\sigma = 40.5$		
		Mean for 39 samples		
Specific heat capacity (dry)	J/(kg·K)	840		
Thermal conductivity	W/(m·K)	$2.74 + 0.0032w$, w: moisture content in kg/m³		
Absorptivity, reflectivity	–	T (K)	300	6000
		a	0.88	0.6
		ρ	0.12	0.4
Thermal expansion coefficient	°C^{-1}	$12 \cdot 10^{-6}$		

Lightweight concrete

Density	kg/m³	$644 \leq \rho \leq 1187$	
Specific heat capacity (dry)	J/(kg·K)	840	
Thermal conductivity	W/(m·K)	$644 \leq \rho \leq 1187$ kg/m³ $\theta = 20°C$, $w = 0$ kg/m³	$0.0414 \exp(0.00205\rho)$
		$1158 \leq \rho \leq 1187$ kg/m³ $\theta = 20°C$, $w \leq 74$ kg/m³	$0.511 + 0.0026w$
		$1130 \leq \rho \leq 1138$ kg/m³ $\theta = 20°C$, $w \leq 144$ kg/m³	$0.371 + 0.001w$
		$644 \leq \rho \leq 674$ kg/m³ $\theta = 20°C$, $w \leq 39$ kg/m³	$0.161 + 0.0015w$

Autoclaved aerated concrete

Density	kg/m³	$455 \leq \rho \leq 800$	
Specific heat capacity (dry)	J/(kg·K)	840	
Thermal conductivity	W/(m·K)	$598 \leq \rho \leq 626$ kg/m³ $\theta = 10°C$, $w \leq 425$ kg/m³	$0.176 + 0.0008w$
		$598 \leq \rho \leq 626$ kg/m³ $\theta = 20°C$, $w \leq 425$ kg/m³	$0.177 + 0.001w$
		$455 \leq \rho \leq 492$ kg/m³ $\theta = 20°C$, $w \leq 298$ kg/m³	$0.138 + 0.0009w$
As a function of density, temperature and moisture content		$\lambda = 0.172 - 1.67 \cdot 10^{-3}w - 9.34 \cdot 10^{-3}\theta$ $-2.97 \cdot 10^{-6}\rho + 3.77 \cdot 10^{-6}\rho w + 1.16 \cdot 10^{-4}w\theta$ $+1.6 \cdot 10^{-5}\rho\theta - 1.62 \cdot 10^{-7}\rho w\theta$	

Concrete with expanded polystyrene pearls as aggregate

Property	Unit				
Density	kg/m³	$259 \leq \rho \leq 792$			
Specific heat capacity (dry)	J/(kg·K)	Depends on the concentration of EPS-pearls:			
		$\rho = 259$ kg/m³ $c = 1370$ J/(kg·K)			
		$\rho = 792$ kg/m³ $c = 1018$ J/(kg·K)			
Thermal conductivity	W/(m·K)	$259 \leq \rho \leq 792$ kg/m³	$0.041 \exp(0.00232\rho)$		
		$\theta = 10\,°C, w = 0$ kg/m³			
		$\theta = 20\,°C,$ $w \leq 425$ kg/m³	$A_1 + A_2 w$		
			ρ (kg/m³)	A_1	$A_2 \cdot 10^{-4}$
			259–335	0.074	4.8
			357–382	0.111	4.6
			407–456	0.126	6.1
			641	0.151	7.9
			792	0.213	1.0
		$\rho = 422$ kg/m³ $0 \leq \theta \leq 30\,°C$	$B_1 + B_2 \theta$		
			w (kg/m³)	B_1	$B_2 \cdot 10^{-4}$
			0	0.112	1.3
			94	0.171	12
			262	0.231	15

Lightweight and normal cement mortars

Property	Unit				
Density	kg/m³	$1055 \leq \rho \leq 1822$			
Specific heat capacity (dry)	J/(kg·K)	840			
Thermal conductivity	W/(m·K)	$1055 \leq \rho \leq 1822$ kg/m³	$0.088 \exp(0.00125\rho)$		
		$\theta = 20\,°C, w = 0$ kg/m³	$0.177 + 0.001w$		
			$0.138 + 0.0009w$		
		$\theta = 20\,°C,$ $w \leq 330$ kg/m³	$A_1 + A_2 w$		
			ρ (kg/m³)	A_1	$A_2 \cdot 10^{-3}$
			1072	0.346	1.2
			1512	0.526	3.1
			1800	0.854	4.5

Lime-sandstone masonry

Property	Unit					
Density	kg/m³	$1170 \leq \rho \leq 1230$				
Specific heat capacity (dry)	J/(kg·K)	840				
Thermal resistance	m²·K/W	$\theta = 20\,°C$	$1/(A_1 + A_2 u)$			
		$u \leq u_c$, u in %kg/kg	d (m)	ρ (kg/m³)	A_1	A_2
			0.14	1140	4.07	0.24

Large format perforated brick masonry

Density	kg/m³	$860 \leq \rho \leq 1760$	
Specific heat capacity (dry)	J/(kg·K)	840	
Thermal resistance	m²·K/W	$d = 14$ cm	$1/[0.98 \exp(0.001\rho)]$
		$860 \leq \rho \leq 1430$ kg/m³	
		$\theta = 20°C, w = 0$ kg/m³	
		$d = 19$ cm	$1/[0.59 \exp(0.0012\rho)]$
		$830 \leq \rho \leq 1630$ kg/m³	
		$\theta = 20°C, w = 0$ kg/m³	
		$\theta = 20°C$	$1/(A_1 + A_2 u)$
		$u \leq u_c$, u in %kg/kg	

d (m)	ρ (kg/m³)	A_1	A_2
0.09	1470	7.94	0.40
0.14	863	2.09	0.09
	1100	3.35	0.18
	1120	2.72	0.28
	1180	3.37	0.31
	1200	3.13	0.32
	1240	4.17	0.38
	1360	2.96	0.34
	1430	4.17	0.42
0.19	800	1.53	0.10
	830	1.51	0.08
	880	1.83	0.13
	1100	2.41	0.09
	1140	2.26	0.13
	1650	4.00	0.34

Concrete block masonry

Density	kg/m³	$860 \leq \rho \leq 1650$	
Specific heat capacity (dry)	J/(kg·K)	840	
Thermal resistance	m²·K/W	$d = 14$ cm	$1/[0.73 \exp(0.0014\rho)]$
		$860 \leq \rho \leq 1650$ kg/m³	
		$\theta = 20°C, w = 0$ kg/m³	
		$\theta = 20°C$	$1/(A_1 + A_2 u)$

$u \leq u_c$, u in %kg/kg	d (m)	ρ (kg/m³)	A_1	A_2
	0.12	980	2.92	0.12
	0.14	1080	3.12	0.24
	0.14	1115	3.37	0.16
	0.19	860	2.35	0.08

Autoclaved cellular concrete masonry

Density	kg/m³	$518 \leq \rho \leq 660$			
Specific heat capacity (dry)	J/(kg·K)	840			
Thermal resistance	m².K/W	$\theta = 20°C$	$1/(A_1 + A_2 u)$		
		$w \leq w_c$, w in kg/m³	d (m)	ρ (kg/m³)	A_1	A_2
			Mortar			
			0.15	524	1.23	0.007
				660	1.44	0.007
			Glue			
			0.15	518	1.13	0.006
				634	1.20	0.007
			Mortar			
			0.18	550	1.25	0.009

Gypsum plaster

Density	kg/m³	975	
Specific heat capacity (dry)	J/(kg·K)	840	
Thermal conductivity	W/(m·K)	$569 \leq \rho \leq 981$ kg/m³ $\theta = 20°C$, w in kg/m³	$0.263 + 0.001w$

Outside rendering

Density	kg/m³	$878 \leq \rho \leq 1736$	
Specific heat capacity (dry)	J/(kg·K)	840	
Thermal expansion coefficient	K⁻¹	$\rho = 1736$ kg/m³	$10.9 \cdot 10^{-6}$

Wood

Density	kg/m^3	400 (pine) $\leq \rho \leq$ 690 (beech)	
Specific heat capacity (dry)	J/(kg·K)	1880	
Thermal conductivity	W/(m·K)	Pine $\theta = 20\,°C, w = 0\,kg/m^3$	0.11

Particle board

Density	kg/m^3	$570 \leq \rho \leq 800$	
Specific heat capacity (dry)	J/(kg·K)	1880	
Thermal conductivity	W/(m·K)	$500 \leq \rho \leq 702\,kg/m^3$ $\theta = 20\,°C, w = 0\,kg/m^3$	$0.098 + 0.0001(\rho - 590)$
		$587 \leq \rho \leq 702\,kg/m^3$ $\theta = 20\,°C, w$ in kg/m^3	$0.106 + 1.3 \cdot 10^{-4} w$ $+ 3.310^{-7} w^2$

Plywood

Density	kg/m^3	$445 \leq \rho \leq 799$	
Specific heat capacity (dry)	J/(kg·K)	1880	
Thermal conductivity	W/(m·K)	$445 \leq \rho \leq 692\,kg/m^3$ $\theta = 20\,°C, w = 0\,kg/m^3$	$0.020 + 1.7 \cdot 10^{-4} \rho$
		$445 \leq \rho \leq 799\,kg/m^3$ $\theta = 20\,°C, w$ in kg/m^3	$0.113 + 3.1 \cdot 10^{-4} w$

Fibre cement

Density	kg/m^3	$823 \leq \rho \leq 2052$	
Specific heat capacity (dry)	J/(kg·K)	840	
Thermal conductivity	W/(m·K)	$823 \leq \rho \leq 866\,kg/m^3$	$0.14 + 5.8 \cdot 10^{-4} w$
		$\rho = 1495\,kg/m^3$	
		Both $\theta = 20\,°C$	$0.42 + 1.2 \cdot 10^{-3} w$

Gypsum board

Weight	kg/m^2	$6.5 \leq \rho \leq 13$	
Specific heat capacity (dry)	J/(kg·K)	840	
Equivalent thermal conductivity	W/(m·K)	$\theta = 20\,°C, d = 9.5\,mm\ w$ in kg/m^3	$0.07 + 2.1 \cdot 10^{-4} w$

4.3.4.2.2 Insulation Materials

Cork

Density	kg/m³	111	
Specific heat capacity (dry)	J/(kg·K)	1880	
Thermal conductivity	W/(m·K)	$\theta = 20\,°C$, $w = 0\,kg/m^3$	0.042

Cellular glass

Density	kg/m³	$113 \leq \rho \leq 140$	
Specific heat capacity (dry)	J/(kg·K)	840	
Thermal conductivity	W/(m·K)	$113 \leq \rho \leq 139\,kg/m^3$ $\theta = 20\,°C$, $\psi = 0\%\,m^3/m^3$	$0.037 + 8.8 \cdot 10^{-5}\rho$
		$126 \leq \rho \leq 134\,kg/m^3$ $\theta = 20\,°C$, ψ in $\%m^3/m^3$	$0.047 + 8.2 \cdot 10^{-4}\psi$
		$\rho = 129\,kg/m^3$ $0 \leq \theta \leq 35$	$0.046 + 2.4 \cdot 10^{-4}\theta$

Glass fibre

Density	kg/m³	$11.6 \leq \rho \leq 136$	
Specific heat capacity (dry)	J/(kg·K)	840	
Thermal conductivity	W/(m·K)	$\theta = 20°C$, $\psi = 0\%\,m^3:m^3$	$0.0268 + 4.9 \cdot 10^{-5}\rho + 0.178/\rho$

Mineral fibre

Density	kg/m³	$32 \leq \rho \leq 191$	
Specific heat capacity (dry)	J/(kg·K)	840	
Thermal conductivity	W/(m·K)	$\theta = 20\,°C$, $\psi = 0\%\,m^3/m^3$	$0.0317 + 2.6 \cdot 10^{-5}\rho + 0.206/\rho$
		$\theta = 10\,°C$, $\psi = 0\%\,m^3/m^3$	$0.026 + 5.5 \cdot 10^{-5}\rho + 0.331/\rho$

EPS (expanded polystyrene)

Density	kg/m³	$13 \leq \rho \leq 40$				
Specific heat capacity (dry)	J/(kg·K)	1470				
Thermal conductivity	W/(m·K)	$\psi = 0\%$ m³/m³	θ (°C)	A_1	$A_2 \cdot 10^{-4}$	A_3
		$A_1 + A_2\rho + A_3/\rho$	10	0.017	1.9	0.258
			20	0.021	1.2	0.235
		$\theta = 20°C$, ψ in % m³/m³	ρ (kg/m³)	B_1	$B_2 \cdot 10^{-3}$	
		$B_1 + B_2\psi$	15	0.0390	2.0	
			20	0.0348	1.9	
			25	0.0326	2.7	
			30	0.0331	1.2	
		$\rho = 15$ kg/m³, d in m	$0.029 + 0.017 d^{0.25}$			
		θ in °C	$0.0354 + 1.6 \cdot 10^{-4} \theta$			

XPS (extruded polystyrene)

Density	kg/m³	$25 \leq \rho \leq 55$	
Specific heat capacity (dry)	J/(kg·K)	1470	
Thermal conductivity	W/(m·K)	$\theta = 10°C$, $\psi = 0\%$ m³/m³	$0.0174 + 1.6 \cdot 10^{-5}\rho$ $+ 0.263/\rho$
		$\theta = 20°C$, $\psi = 0\%$ m³/m³	$0.0404 + 3.9 \cdot 10^{-4}\rho$ $+ 0.029/\rho$
		$\theta = 10°C$, $\rho = 35$ kg/m³, ψ in 0% m³/m³	$0.0240 + 1.6 \cdot 10^{-4}\psi$ $+ 5.8 \cdot 10^{-5}\psi^2$
		$\theta = 20°C$, $\rho = 35$ kg/m³, ψ in 0% m³/m³	$0.0251 + 5.2 \cdot 10^{-5}\psi$ $+ 7.0 \cdot 10^{-5}\psi^2$

PUR/PIR (polyurethane foam, polyisocyanurate foam)

Density	kg/m³	$20 \leq \rho \leq 40$	
Specific heat capacity (dry)	J/(kg·K)	1470	
Thermal conductivity	W/(m·K)	$\theta = 10°C$, $\psi = 0\%$ m³/m³	$-0.112 + 1.9 \cdot 10^{-3}\rho$ $+ 2.36/\rho$
		$\theta = 20°C$, $\psi = 0\%$ m³/m³	$-0.008 + 5.1 \cdot 10^{-4}\rho$ $+ 0.436/\rho$

Perlite board

Density	kg/m³	$135 \leq \rho \leq 215$				
Specific heat capacity (dry)	J/(kg·K)	1000				
Thermal conductivity	W/(m·K)	$\theta = 20\,°C$, $\psi = 0\%\ m^3/m^3$	$0.046 + 0.00014(\rho - 100)$			
		$\theta = 20\,°C$, ψ in % m³/m³	$B_1 + B_2 \psi$			
			ρ (kg/m³)	$\theta(°C)$	B_1	$B_2 \cdot 10^{-3}$
			142	10	0.052	1.5
			142	20	0.053	1.8
			171	10/20	0.059	2.9
			212	20	0.058	5.7
		$\rho = 140$ kg/m³	$D_1 + D_2 \theta$			
			ψ (%m³/m³)	D_1	$D_2 \cdot 10^{-4}$	
			0	0.047	1.2	
			1.6	0.056	5.7	

4.4 Air Properties

4.4.1 Standard Values

No standards advance lists with values. The 2021 ASHRAE Handbook Fundamentals, Chapter 26, contains a Table 4 with air permeability values for different materials, but all constants without any air pressure dependency ($b = 1$).

4.4.2 Measured Values

4.4.2.1 Air Permeance, Test Method

Measuring the air permeance is done by building $0.9 \times 0.9\ m^2$ large masonry, roof cover, underlay foil, insulation or internal lining parts in a wooden frame with air-tightened outside face, see Figure 4.2.

The filled frames are then mounted airtight against an equally large airtight box, at its back linked to an air duct with fan, whose blowing at increasing rpm allows measuring the airflow and the pressure difference between box and environment. This way, known the surface, the data needed to calculate the air permeance by using an exponential least square approach to link the air fluxes to the air pressure differences are collected. The exponential turns linear if it's air permeability that characterises the d m thick part (see Figure 4.3):

$$M_a/A = G_a = K_a \Delta P_a \xrightarrow{\text{Yields}} K_a = a\Delta P_a^{b-1} \text{ or, if } b = 1, \text{ then } a = k_a/d = G_a d/(\Delta P_a A)$$

Are large samples to be tested, a hot box with measuring area $2 \times 2\ m^2$ is used the way just described (Figure 4.4).

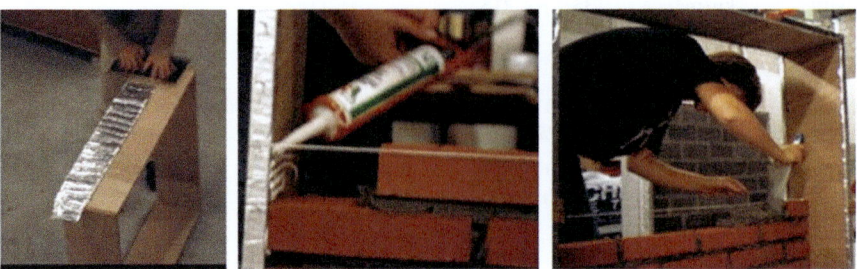

Figure 4.2 Preparing wall samples for measuring the air permeance.

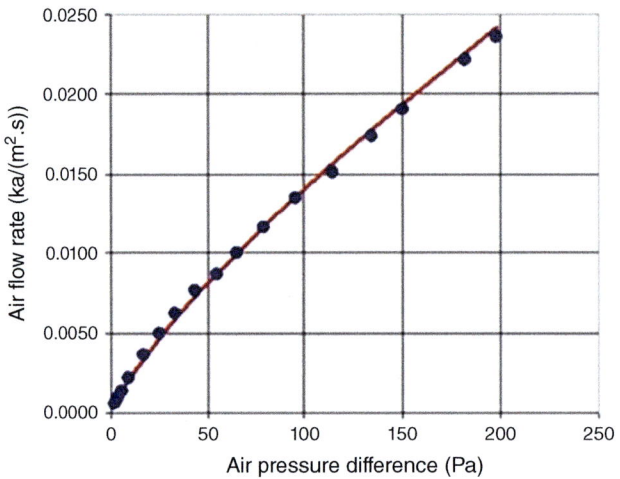

Figure 4.3 Left up, air permeance measurement, least square fit.

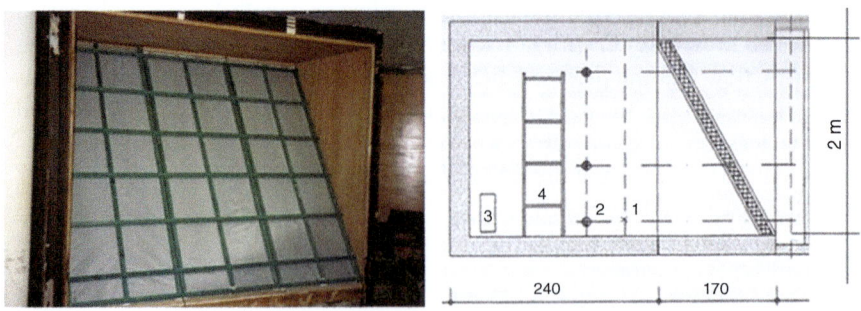

Figure 4.4 Hot box/cold box.

4.4.2.2 Test Results
4.4.2.2.1 Building Materials and Building Parts

Masonry, air permeance per m² $\left(K_a = a\Delta P_a^{b-1}\right)$

Component	a (kg/(s.m².Pa^b))	b
9 cm thick facing brick veneers		
Bricks $19 \times 9 \times 4.5$ cm³, joints not pointed, badly filled	$3.24 \cdot 10^{-4}$	0.69
Bricks $19 \times 9 \times 4.5$ cm³, joints not pointed, well filled	$1.08 \cdot 10^{-4}$	0.75
Bricks $19 \times 9 \times 4.5$ cm³, joints pointed, 1	$3.48 \cdot 10^{-5}$	0.80
Bricks $19 \times 9 \times 6.5$ cm³, joints pointed, 2	$3.60 \cdot 10^{-5}$	0.78
Bricks $19 \times 9 \times 6.5$ cm³, joints not pointed, badly filled	$7.20 \cdot 10^{-4}$	0.68
Bricks $19 \times 9 \times 6.5$ cm³, joints not pointed, well filled	$1.56 \cdot 10^{-4}$	0.71
Bricks $19 \times 9 \times 6.5$ cm³, joints pointed, 1	$3.60 \cdot 10^{-5}$	0.81
Bricks $19 \times 9 \times 6.5$ cm³, joints pointed, 2	$3.48 \cdot 10^{-5}$	0.82

Masonry, air permeance per m² $\left(K_a = a\Delta P_a^{b-1}\right)$

Component	a (kg/(s.m².Pa^b))	b
9 cm thick concrete block veneers		
Concrete blocks $19 \times 9 \times 9$ cm³, 1955 kg/m³, joints pointed	$1.44 \cdot 10^{-4}$	0.88
Concrete blocks $19 \times 9 \times 9$ cm³, 1927 kg/m³, joints pointed	$1.92 \cdot 10^{-4}$	0.86
Concrete blocks $19 \times 9 \times 9$ cm³, 1881 kg/m³, joints pointed	$2.76 \cdot 10^{-4}$	0.82
Concrete blocks $19 \times 9 \times 9$ cm³, 2109 kg/m³, joints pointed	$2.42 \cdot 10^{-5}$	0.85
Concrete blocks $19 \times 9 \times 9$ cm³, 2153 kg/m³, joints pointed	$2.29 \cdot 10^{-5}$	0.83
Concrete blocks $19 \times 9 \times 9$ cm³, 2260 kg/m³, joints pointed	$4.27 \cdot 10^{-5}$	0.77
Concrete blocks $19 \times 9 \times 9$ cm³, 2091 kg/m³, joints pointed	$6.73 \cdot 10^{-5}$	0.86
14 cm thick brick inside leafs		
Lightweight brickwork $19 \times 14 \times 14$ cm³, joints not pointed 1	$2.64 \cdot 10^{-3}$	0.59
Lightweight brickwork $19 \times 14 \times 14$ cm³, joints not pointed 2	$4.20 \cdot 10^{-3}$	0.57
Lightweight brickwork $19 \times 14 \times 14$ cm³, joints pointed 1	$2.88 \cdot 10^{-5}$	0.72
Lightweight brickwork $19 \times 14 \times 14$ cm³, joints pointed 2	$2.04 \cdot 10^{-5}$	0.81
Lightweight brickwork $19 \times 14 \times 14$ cm³, joints pointed 3	$1.68 \cdot 10^{-5}$	0.82
Lightweight brickwork $19 \times 14 \times 14$ cm³, plastered at the inside 1	$3.72 \cdot 10^{-7}$	0.96
Lightweight brickwork $19 \times 14 \times 14$ cm³, plastered at the inside 2	$2.76 \cdot 10^{-7}$	0.97

Masonry, air permeance per m² $\left(K_a = a\Delta P_a^{b-1}\right)$

14 cm thick concrete block inside leafs

Hollow concrete blocks $39 \times 19 \times 14\,\text{cm}^3$, $987\,\text{kg/m}^3$, joints not pointed	$3.96 \cdot 10^{-3}$	0.58
Hollow concrete blocks, $39 \times 19 \times 14\,\text{cm}^3$, $954\,\text{kg/m}^3$, joints not pointed	$5.04 \cdot 10^{-3}$	0.56
Hollow concrete blocks $39 \times 19 \times 14\,\text{cm}^3$, $910\,\text{kg/m}^3$, joints not pointed	$6.00 \cdot 10^{-3}$	0.55
Hollow concrete blocks $39 \times 19 \times 14$, $987\,\text{kg/m}^3$, joints pointed	$1.92 \cdot 10^{-4}$	0.91
Hollow concrete blocks $39 \times 19 \times 14$, $954\,\text{kg/m}^3$, joints pointed	$3.72 \cdot 10^{-4}$	0.79
Hollow concrete blocks $39 \times 19 \times 14$, $910\,\text{kg/m}^3$, joints pointed	$6.00 \cdot 10^{-4}$	0.70
Hollow concrete blocks $39 \times 19 \times 14$, $987\,\text{kg/m}^3$, plastered at the inside	$3.48 \cdot 10^{-7}$	0.96
Hollow concrete blocks $39 \times 19 \times 14$, $954\,\text{kg/m}^3$, plastered at the inside	$2.64 \cdot 10^{-7}$	0.95
Hollow concrete blocks $39 \times 19 \times 14$, $910\,\text{kg/m}^3$, plastered at the inside	$3.12 \cdot 10^{-7}$	0.97

14 cm thick aerated concrete leafs

Aerated concrete $60 \times 24 \times 14\,\text{cm}^3$, $510\,\text{kg/m}^3$, joints glued, not pointed, 1	$2.28 \cdot 10^{-3}$	0.61
Aerated concrete $60 \times 24 \times 14\,\text{cm}^3$, $510\,\text{kg/m}^3$, joints glued, not pointed, 2	$1.44 \cdot 10^{-3}$	0.62
Aerated concrete $60 \times 24 \times 14\,\text{cm}^3$, $510\,\text{kg/m}^3$, joints glued and pointed, 1	$9.60 \cdot 10^{-5}$	0.64
Aerated concrete $60 \times 24 \times 14\,\text{cm}^3$, $510\,\text{kg/m}^3$, joints glued and pointed, 2	$1.02 \cdot 10^{-4}$	0.63
Aerated concrete $60 \times 24 \times 14\,\text{cm}^3$, $510\,\text{kg/m}^3$, joints glued, plastered	$4.08 \cdot 10^{-7}$	0.97

Masonry, air permeance per m² $\left(K_a = a\Delta P_a^{b-1}\right)$

14 cm thick Lime-sandstone leafs

Lime-sandstone, $29 \times 14 \times 14\,\text{cm}^3$, $1140\,\text{kg/m}^3$, joints not pointed	$3.24 \cdot 10^{-3}$	0.61
Lime-sandstone, $29 \times 14 \times 14\,\text{cm}^3$, $1140\,\text{kg/m}^3$, joints not pointed	$4.20 \cdot 10^{-3}$	0.57
Lime-sandstone, $29 \times 14 \times 14\,\text{cm}^3$, $1140\,\text{kg/m}^3$, joints pointed	$2.28 \cdot 10^{-5}$	0.75
Lime-sandstone, $29 \times 14 \times 14\,\text{cm}^3$, $1140\,\text{kg/m}^3$, joints pointed	$1.80 \cdot 10^{-5}$	0.80
Lime-sandstone, $29 \times 14 \times 14\,\text{cm}^3$, $1140\,\text{kg/m}^3$, plastered at the inside	$3.00 \cdot 10^{-7}$	0.95

Metal construction, air permeance per m² $\left(K_a = a\Delta P_a^{b-1}\right)$

	a (kg/(s.m².Pa^b))	b
Wall made of horizontal sheet metal elements: see Figure 4.4		
No special care for airtightness	$7.9 \cdot 10^{-5}$	0.97
Screw holes at the columns sealed	$6.7 \cdot 10^{-5}$	0.90
Screw holes sealed and joints between metal boxes taped	$1.6 \cdot 10^{-5}$	0.92

Roof covers, air permeance per m² $\left(K_a = a\Delta P_a^{b-1}\right)$

Layer	a (kg/(s.m².Paᵇ))	b
Ceramic tiles, single lock, 9.2 m run of joints per m² (Roman tiles)	0.014	0.55
Ceramic tiles, double lock, 9.2 m run of joints per m² (Roman tiles)	0.019	0.50
Ceramic tiles, double lock, 9.2 m run of joints per m² (Roman tiles), 1 ventilation tile per m²	0.016	0.50
Ceramic tiles, double lock, 8.3 m run of joints per m² (Pan tiles)	0.015	0.50
Concrete tiles, double lock and overlap, 6.2 m run of joints and overlaps per m²	0.011	0.68
Quarry slates, 3 slates thick	0.014	0.62
Fibre cement slates, 3 slates thick	0.0081	0.54
Metal tiles, single lock, 3.5 m run of locked joints per me	0.011	0.54
Fibre cement corrugated boards	0.0042	0.79
Metal roof with standing seam, outside the fixing zone	0.0054	0.64
Metal roof with standing seam, in the fixing zone (extends 15 cm at both sides of the fixation)	0.0014	0.70
Roof sandwich element, composed of (top to bottom): – corrugated 0.8 mm thick aluminium sheet as cover – 50 mm EPS with a top surface that follows the corrugations of the aluminium sheet – 0.17 mm thick aluminium vapour barrier Total surface: 4.93 m². Contains two joints, parallel to the slope, and one overlap, orthogonal to the slope. The parallel joints are formed by one corrugation overlapping at the cover side and a profiled insert inside	0.0017	0.79

Underlays, air permeance per m² $\left(K_a = a\Delta P_a^{b-1}\right)$

	a (kg/(s.Paᵇ))	b
Fibre/cellulose cement plate, $d = 3.2$ mm, perfectly closed overlap	0.00042	0.66
Fibre/cellulose cement plate, $d = 3.2$ mm, overlap 1.4 mm open	0.00042	0.69
Fibre/cellulose cement plate, $d = 3.2$ mm, overlap 3.6 mm open	0.0032	0.60
Fibre/cellulose cement plate, $d = 3.2$ mm, overlap 13 mm open	0.01	0.55
Spun-bonded PE $d = 0.42$ mm, 140 g/m²		
Sample 1	$1.5 \cdot 10^{-6}$	0.97
Sample 2	$2.5 \cdot 10^{-6}$	1.0
Sample 3	$2.8 \cdot 10^{-6}$	1.0
Spun-bonded PE, 11 perforations $\phi 1.2$ mm/m²	$4.6 \cdot 10^{-6}$	1.0
Spun-bonded PE, 11 perforations $\phi 1.7$ mm/m²	$7.6 \cdot 10^{-6}$	0.96
Spun-bonded PE, 11 perforations $\phi 2.7$ mm/m²	$1.7 \cdot 10^{-5}$	0.85
Spun-bonded PE, 11 perforations $\phi 3.0$ mm/m²	$2.2 \cdot 10^{-5}$	0.82
Spun-bonded PE, 11 perforations $\phi 5.0$ mm/m²	$9.2 \cdot 10^{-5}$	0.66

Spun-bonded PE, 11 perforations $\phi 8.5$ mm/m²	$3.2 \cdot 10^{-4}$	0.63
Spun-bonded PE, with overlap	0.000013	0.67
Spun-bonded PE, 11 staples per m²	$2.8 \cdot 10^{-6}$	1.0
Spun-bonded PE, 33 staples per m²	$3.2 \cdot 10^{-6}$	1.0
Spun-bonded PE, 55 staples per m²	$3.3 \cdot 10^{-6}$	1.0
Spun-bonded PE, 111 staples per m²	$3.3 \cdot 10^{-6}$	1.0
Spun-bonded PE, overlapping, overlaps continuous sealed		
Sample 1	$0.8 \cdot 10^{-6}$	1.0
Sample 2	$9.5 \cdot 10^{-6}$	0.94
Sample 3	$3.1 \cdot 10^{-6}$	1.0
Bitumen impregnated polypropylene foil, $d = 0.6$ mm, 496 g/m²	$7.9 \cdot 10^{-7}$	0.43
Bitumen impregnated polypropylene foil, $d = 0.6$ mm, 496 g/m², overlap	0.000065	0.37
Bituminous underlay foil		
Sample 1	$5.3 \cdot 10^{-7}$	0.99
Sample 2	$6.1 \cdot 10^{-7}$	0.97
Bituminous underlay foil, 11 perforations $\phi 1.2$ mm/m²	$6.9 \cdot 10^{-6}$	0.74
Bituminous underlay foil, 11 perforations $\phi 1.7$ mm/m²	$1.7 \cdot 10^{-5}$	0.67
Bituminous underlay foil, 11 perforations $\phi 2.7$ mm/m²	$5.1 \cdot 10^{-5}$	0.60
Bituminous underlay foil, 11 perforations $\phi 3.0$ mm/m²	$7.3 \cdot 10^{-5}$	0.59
Bituminous underlay foil, 11 perforations $\phi 5.0$ mm/m²	$2.1 \cdot 10^{-4}$	0.58
Bituminous underlay foil, 11 perforations $\phi 8.5$ mm/m²	$7.0 \cdot 10^{-4}$	0.54
Micro-perforated, glass-fibre reinforced plastic foil	0.00009	0.77
with overlap of 20 cm, per m run	0.005	0.2
with overlap of 30 cm, per m run	0.0025	0.43
with overlap of 50 cm, per m run	0.0012	0.43

4.4.2.2.2 Insulation Materials and Insulating Layers

Air permeability

Material	Air permeability (k_a), kg/(m.s.Pa)
Fiberglas (ρ: density in kg/m³)	$4.2 \, 10^{-3} \rho^{-1.24}$
Mineral fibre (ρ: density in kg/m³)	$1.8 \, 10^{-2} \rho^{-1.44}$
Fiberglas 18 kg/m³, along the thickness	$7.2 \cdot 10^{-5}$
Fiberglas, 42 kg/m³, along the thickness	$5.2 \cdot 10^{-5}$
Fiberglas, 42 kg/m³, orthogonal to the thickness	$8.1 \cdot 10^{-5}$
Fiberglas, 78 kg/m³, along the thickness	$3.9 \cdot 10^{-5}$
Fiberglas, 78 kg/m³, orthogonal to the thickness	$5.4 \cdot 10^{-5}$
Fiberglas, 147 kg/m³, along the thickness	$2.2 \cdot 10^{-5}$
Fiberglas, 147 kg/m³, orthogonal to the thickness	$2.9 \cdot 10^{-5}$

Air permeance per m² $\left(K_a = a\Delta P_a^{b-1}\right)$

Material	a (kg/(s.m².Pa^b))	b
EPS, $d = 64.4$ mm, $\rho = 10.2$ kg/m³	0.0047	0.50
EPS, $d = 63.6$ mm, $\rho = 10.1$ kg/m³	0.017	0.823
EPS, $d = 39.1$ mm, $\rho = 11.6$ kg/m³	$7.3 \cdot 10^{-6}$	0.94
EPS, $d = 39.1$ mm, $\rho = 11.5$ kg/m³	$8.4 \cdot 10^{-6}$	0.93
EPS, $d = 50.1$ mm, $\rho = 14.1$ kg/m³	0.000045	0.94
EPS, $d = 50.2$ mm, $\rho = 14.6$ kg/m³	0.000053	0.90
EPS, $d = 40.9$ mm, $\rho = 45.1$ kg/m³	0.00003	0.95
EPS, $d = 39.3$ mm, $\rho = 64.2$ kg/m³	0.000058	0.90
EPS, $d = 49.0$ mm, $\rho = 64.5$ kg/m³	0.000028	0.92
EPS, $d = 49.6$ mm, $\rho = 63.7$ kg/m³	$5.5 \cdot 10^{-6}$	0.93
EPS, $d = 80$ mm, $\rho = 19.8$ kg/m³	0.000018	0.93
EPS, $d = 80$ mm, $\rho = 20.2$ kg/m³	0.000019	0.86
EPS, $d = 100$ mm, $\rho = 20.6$ kg/m³	0.000016	0.90
EPS, $d = 100$ mm, $\rho = 21.2$ kg/m³	0.000032	0.88
EPS, $d = 120$ mm, $\rho = 20.5$ kg/m³	0.000023	0.97
EPS, $d = 120$ mm, $\rho = 19.7$ kg/m³	0.000015	0.93
EPS, $d = 140$ mm, $\rho = 19.9$ kg/m³	$6.8 \cdot 10^{-6}$	0.91
EPS, $d = 140$ mm, $\rho = 20.3$ kg/m³	$6.7 \cdot 10^{-6}$	0.91
EPS, $d = 80$ mm, $\rho = 28.3$ kg/m³	$9.4 \cdot 10^{-6}$	0.89
EPS, $d = 80$ mm, $\rho = 28.5$ kg/m³	0.000013	0.83
EPS, $d = 100$ mm, $\rho = 30.5$ kg/m³	0.000015	0.91
EPS, $d = 100$ mm, $\rho = 30.9$ kg/m³	0.000013	0.87
EPS, $d = 120$ mm, $\rho = 29.7$ kg/m³	0.000017	0.89
EPS, $d = 120$ mm, $\rho = 27.7$ kg/m³	$8.9 \cdot 10^{-6}$	0.91
EPS, $d = 140$ mm, $\rho = 29.0$ kg/m³	$4.5 \cdot 10^{-6}$	0.81
EPS with aluminium foil backing, $d = 80$ mm, $\rho = 28.7$ kg/m³	$9.0 \cdot 10^{-6}$	0.74
EPS with aluminium foil backing, $d = 100$ mm, $\rho = 29.7$ kg/m³	$8.2 \cdot 10^{-7}$	0.97
EPS with aluminium foil backing, $d = 120$ mm, $\rho = 30.0$ kg/m³	$7.2 \cdot 10^{-6}$	0.81
EPS with aluminium foil backing, $d = 140$ mm, $\rho = 29.9$ kg/m³	$6.6 \cdot 10^{-6}$	0.80
Glass-fibre bat, $d = 6$ cm, bituminous paper backing, overlaps taped	0.000065	0.71
	0.000092	0.70

Air permeance per m² $((K_a = a\Delta P_a^{b-1}))$

Material	a (kg/(s.m².P$_a^b$))	b
Glass-fibre bat, $d = 6$ cm, bituminous paper backing, overlaps taped, staples alongside the tape	0.000075	0.74
Glass-fibre bat, $d = 6$ cm, bituminous paper backing, overlaps taped, 1 nail $\phi 3$ mm/m² perforating the bat	0.000071	0.75
	0.000097	0.64
Glass-fibre bat, $d = 6$ cm, bituminous paper backing, overlaps taped, bat ripped ($L = 40$ mm, B-40 mm, 1/m²)	0.0002	0.61
Glass-fibre bat, $d = 6$ cm, bituminous paper backing, overlaps held together with nailed lath	0.001	0.56
	0.00033	0.76
Glass-fibre bat, $d = 6$ cm, bituminous paper backing, overlaps not overlapping	0.0032	1.00
	0.00089	0.85
Glass-fibre bat, $d = 6$ cm, aluminium paper backing, overlaps taped	0.000047	1.00
Glass-fibre bat, $d = 6$ cm, aluminium paper backing, overlaps taped, staples alongside the tape	0.000048	1.00
Glass-fibre bat, $d = 6$ cm, aluminium paper backing, overlaps taped, 1 nail $\phi 3$ mm/m² perforating the bat	0.00005	1.00
Glass-fibre bat, $d = 6$ cm, aluminium paper backing, overlaps taped, bat ripped ($L = 40$ mm, B-40 mm, 1/m²)	0.000085	1.00
Mineral fibre, dense boards, 170 kg/m³, $d = 10$ cm	0.00023	0.89
XPS, $d = 5$ cm, mounted tongue and groove	0.00040	0.59
XPS, $d = 5$ cm, mounted with well-closed joints	0.00047	0.54
XPS, $d = 5$ cm, mounted with open joints of 2 mm	0.002	0.55
EPS, $d = 12$ m, 9.3 kg/m³, plain board	0.000077	0.91
Composite element particle board 4 mm/EPS 88 mm/particle board 4 mm with EPS/EPS joints		
0.5 mm wide	0.00071	0.56
7.5 mm wide	0.0023	0.60
15 mm wide	0.010	0.53
Composite element particle board 4 mm/PUR 61 mm/particle board 4 mm with timber/timber joints		
0.3 mm wide	0.00065	0.56
3.5 mm wide	0.00074	0.63
8.0 mm wide	0.0011	0.61
16 mm wide	0.0015	0.55
Composite element: particle board 4 mm/PUR 88 mm/particle board 4 mm with PUR/PUR joints		
0.2 mm wide	0.00015	0.66
4.3 mm wide	0.00014	0.65
16 mm wide	0.015	0.53

XPS board, $d = 60$ mm, $\rho = 39$ kg/m³

Sample 1	$5.7 \cdot 10^{-7}$	1.00
Sample 2	$4.4 \cdot 10^{-7}$	1.00
Sample 3	$6.7 \cdot 10^{-7}$	1.00

XPS-board, $d = 60$ mm, $\rho = 39$ kg/m³ with gypsum board, glued to the XPS, as inside lining

Sample 1	$5.6 \cdot 10^{-7}$	1.00
Sample 2	$6.6 \cdot 10^{-7}$	1.00
Sample 3	$4.0 \cdot 10^{-7}$	1.00
Sample 4	$9.0 \cdot 10^{-7}$	1.00

Air permeance per running metre $\left(K_a = a\Delta P_a^{b-1}\right)$

Layer	a (kg/(s.m.Pa^b))	b
Composite element: plywood/30 mm air cavity/120 mm mineral fibre insulation/aluminium vapour retarder/plywood. The elements have timber/timber joints, with a wooden insert close to the outside surface and a special joint profile which should guarantee air tightness		
Element loaded with 65 kg/m²		
Joint perfectly closed	0.000018	0.87
Joint perfectly closed, joint profile removed	0.00024	0.64
Joint 1 mm open	0.000086	0.69
Joint 2 mm open	0.00048	0.68
Joint 3 mm open	0.0013	0.57
Element unloaded		
Joint 1 mm open	0.00024	0.66
Joint 2 mm open	0.00032	0.75
Joint 5 mm open	0.0017	0.57
XPS, $d = 50$ mm (manufacturer 1), mounted tongue and groove		
(α)		
Joint 0 mm open	0.00017	0.70
Joint 5 mm open	0.00060	0.60
Joint 10 mm open	0.0014	0.50

Layer	a	b
XPS, $d = 50$ mm (man. 2), mounted tongue and groove (α)		
Joint width 0 mm	0.000025	1.0
Joint width 5 mm	0.00019	0.70
Joint width 10 mm	0.00036	0.60
XPS, $d = 50$ mm (manufacturer 2, other batch), mounted tongue and groove (α)		
Joint width 0 mm	0.000042	1.0
Joint width 5 mm	0.00014	0.70
Joint width 10 mm	0.00035	0.60
XPS, $d = 50$ mm (manufacturer 3), mounted tongue and groove (α)		
Joint width 0 mm	$8.4 \cdot 10^{-6}$	1.0
Joint width 5 mm	0.00012	0.70
Joint width 10 mm	0.00096	0.60
EPs, $d = 50$ mm (manufacturer 1), mounted tongue and groove (α)		
Joint width 0 mm	0.000012	1.0
Joint width 5 mm	0.000096	1.0
Joint width 10 mm	0.00096	0.60
EPs, $d = 40$ mm (manufacturer 2), mounted tongue and groove (α)		
Joint width 0 mm	$4.8 \cdot 10^{-6}$	1.0
Joint width 5 mm	0.000026	1.0
Joint width 10 mm	0.00024	0.60
Two rebated XPS boards on a rafter, joints between rafter and boards and between boards open		
Sample 1	$6.7 \cdot 10^{-5}$	1.0
Sample 2	$1.8 \cdot 10^{-3}$	0.75

Air permeance per running metre ($(K_a = a \Delta P_a^{b-1})$)

Layer	a (kg/(s.m.Pa^b))	b
Two rebatted XPS boards on a rafter, joints between rafter and boards sealed with a bituminous tape		
Sample 1	$2.2 \cdot 10^{-5}$	0.98
Sample 2	$1.1 \cdot 10^{-4}$	0.95
Sample 3	$2.7 \cdot 10^{-5}$	0.80
Sample 4	$4.1 \cdot 10^{-5}$	1.0
Two rebated XPS boards on a rafter, joints between rafter and boards and at the top, between the two boards, sealed with a bituminous tape		
Sample 1	$6.6 \cdot 10^{-6}$	0.96
Sample 2	$6.8 \cdot 10^{-6}$	0.95
Sample 3	$11.9 \cdot 10^{-6}$	0.95
Sample 4	$4.1 \cdot 10^{-6}$	0.99

Two rebated XPS boards on a rafter, joints between rafter and boards and at the top, between the two boards, sealed with silicone joint filler

Sample 1	$1.3 \cdot 10^{-6}$	0.97
Sample 2	$8 \cdot 10^{-7}$	1.00
Sample 3	$2.1 \cdot 10^{-6}$	0.91
Sample 4	$2.5 \cdot 10^{-6}$	0.92

4.4.2.2.3 Air Barriers

Air permeance per m² ($K_a = a\Delta P_a^{b-1}$)

Layers	a (kg/(s.m².Pa^b))	b
PE-foil, $d = 0.2$ mm		
Sample 1	$5.5 \cdot 10^{-11}$	1.0
Sample 2	$3.9 \cdot 10^{-7}$	1.0
Sample 3	$4.8 \cdot 10^{-7}$	1.0
Sample 4	$4.3 \cdot 10^{-7}$	1.0
PE-foil, $d = 0.2$ mm, 11 perforations $\phi 1.2$ mm/m²	$1.1 \cdot 10^{-6}$	0.91
PE-foil, $d = 0.2$ mm, 11 perforations $\phi 1.7$ mm/m²	$2.5 \cdot 10^{-6}$	0.82
PE-foil, $d = 0.2$ mm, 11 perforations $\phi 2.7$ mm/m²	$9.2 \cdot 10^{-6}$	0.71
PE-foil, $d = 0.2$ mm, 11 perforations $\phi 3.0$ mm/m²	$2.4 \cdot 10^{-5}$	0.65
PE-foil, $d = 0.2$ mm, 11 perforations $\phi 5.0$ mm/m²	$1.1 \cdot 10^{-4}$	0.57
PE-foil, $d = 0.2$ mm, 11 perforations $\phi 8.5$ mm/m²	$4.7 \cdot 10^{-4}$	0.53
PE-foil, $d = 0.2$ mm, with overlap, overlap sealed		
Sample 1 (badly done)	$6.0 \cdot 10^{-6}$	0.77
Sample 2	$4.9 \cdot 10^{-7}$	1.0
Sample 3	$4.4 \cdot 10^{-7}$	1.0
Plastic foil SD2		
Sample 1	$4.6 \cdot 10^{-7}$	1.0
Sample 2	$5.1 \cdot 10^{-7}$	1.0
Sample 3	$4.7 \cdot 10^{-7}$	1.0
Plastic foil SD2, 11 perforations $\phi 1.2$ mm/m²	$1.9 \cdot 10^{-6}$	0.86
Plastic foil SD2, 11 perforations $\phi 1.7$ mm/m²	$6.6 \cdot 10^{-6}$	0.73
Plastic foil SD2, 11 perforations $\phi 2.7$ mm/m²	$2.4 \cdot 10^{-5}$	0.63
Plastic foil SD2, 11 perforations $\phi 3.0$ mm/m²	$4.6 \cdot 10^{-5}$	0.60
Plastic foil SD2, 11 perforations $\phi 5.0$ mm/m²	$1.2 \cdot 10^{-4}$	0.59
Plastic foil SD2, 11 perforations $\phi 8.5$ mm/m²	$3.5 \cdot 10^{-4}$	0.60
Plastic foil SD2, 22 staples per m²	$6.5 \cdot 10^{-7}$	1.0
Plastic foil SD2, 44 staples per m²	$7.6 \cdot 10^{-7}$	0.96

Plastic foil SD2, 111 staples per m²		$1.1 \cdot 10^{-6}$	0.94
Plastic foil SD2			
Sample 1		$7.3 \cdot 10^{-7}$	0.97
Sample 2		$7.1 \cdot 10^{-7}$	1.0
Sample 3 (not well done)		$5.3 \cdot 10^{-6}$	0.80

4.4.2.2.4 Internal Linings

Air permeance per m² $\left(K_a = a\Delta P_a^{b-1}\right)$

Layers	a (kg/(s.m².Pa^b))	b
Lathed ceiling, $d = 1$ cm, tongue and groove		
Sample 1	0.00027	0.63
Sample 2	0.00041	0.68
Lathed ceiling, $d = 1$ cm, tongue and groove, leak ϕ20 mm/m²	0.00076	0.63
Gypsum board, no joint	$1.5 \cdot 10^{-7}$	0.89
Gypsum board, no joint, painted	$1.2 \cdot 10^{-7}$	0.82
Gypsum board, joints reinforced and plastered	0.000031	0.81
Gypsum board, joints reinforced and plastered, leak ϕ20 mm/m²	0.00038	0.61
Gypsum board, joint not plastered	0.00033	0.73
Gypsum board, joint not plastered, leak ϕ20 mm/m²	0.00063	0.73
Gypsum board with alu backing, joints reinforced + plastered	0.000013	1
Gypsum board with alu backing, joints as above, leak ϕ20 mm/m²	0.00047	0.53
Gypsum board with alu backing, joints not plastered, leak ϕ20 mm/m²	0.00056	0.59

4.5 Moisture Properties

4.5.1 Standard Values

4.5.1.1 Building and Finishing Materials

Group	Material	Density (ρ) (kg/m³)	Vapour resistance factor μ (–)	
			Dry cup	Wet cup
Metals	Aluminium alloys	2800	∞	∞
	Duralumin	2800		
	Brass	8400		
	Bronze	8700		
	Copper	8900		

4.5 Moisture Properties

Group	Material	Density (ρ) (kg/m³)	Vapour resistance factor μ (–)	
	Iron	7900		
	Iron, cast	7500		
	Steel	7800		
	Stainless steel	7900		
	Lead	11300		
	Zinc	7100		
Wood and wood-based materials	Softwood	500	50	20
	Hardwood	700	200	50
	Plywood (from low to high density)	300	150	50
		500	200	70
		700	220	90
		1000	250	110
	Particle board			
	Soft	300	50	10
	Semi-hard	600	50	15
	Hard	900	50	20
	OSB	680	50	30
	Fibreboard			
	Soft	400	10	5
	Semi-hard	600	20	12
	Hard	800	30	20
Gypsum	Gypsum blocks	600	10	4
		900	10	4
		1200	10	4
		1500	10	4
	Gypsum board	900	10	4
Mortars	Normal mortar, mixed at site	1800	10	6
Plasters	Gypsum, Light	600	10	6
	Normal	1000	10	6
	Heavy	1300	10	6
	Lime or gypsum, sand	1600	10	6
	Cement, sand	1700	10	6
Concrete	Medium to high density	1800	100	60
		2000	100	60
		2200	120	70
		2400	130	80

4 Heat, Air and Moisture Material Property Values

Group	Material	Density (ρ) (kg/m³)	Vapour resistance factor μ (–)	
	Reinforced 1% steel	2300	130	80
	2% steel	2400	130	80
Stone	Crystalline rock	2800	10000	10000
	Sedimentary rock	2600	250	200
	Sedimentary, light	1500	30	20
	Lava	1600	20	15
	Basalt	2700–3000	10000	10000
	Gneiss	2400–2700	10000	10000
	Granite	2500–3000	10000	10000
	Marble	2800	10000	10000
	Slate	2000–2800	1000	800
	Limestone			
	Extra soft	1600	30	20
	Soft	1800	40	25
	Semi-hard	2000	50	40
	Hard	2200	200	150
	Extra hard	2600	250	200
	Sandstone	2600	40	302.30
	Natural pumice	400	8	6
	Artificial stone	1750	50	40
Soils	Clay or silt	1200–1800	50	50
	Sand and gravel	1700–2200	50	50
Water, ice, snow	Ice, –10 °C	920	No data	No data
	Ice, 0 °C	900		
	Water, 10 °C	1000		
	Water, 40 °C	990		
	Snow			
	Fresh (<30 mm)	100		
	Soft (30–70 mm)	200		
	Slightly compacted (70–100 mm)	300		
	Compacted (>200 mm)	500		
Plastics, solid	Acrylic	1050	10 000	10 000
	Polycarbonates	1200	5000	5000
	PFTE	2200	10 000	10 000
	PVC	1390	50 000	50 000
	PMMA	1180	50 000	50 000
	Polyacetate	1410	100 000	100 000
	Polyamide (Nylon)	1150	50 000	50 000

Group	Material	Density (ρ) (kg/m³)	Vapour resistance factor μ (–)	
V Plastics, solid	Nylon, 25% glass fibre	1450	50 000	50 000
	PE. High density	980	100 000	100 000
	PE, low density	920	100 000	100 000
	Polystyrene	1050	100 000	100 000
	Polypropylene (PP)	910	10 000	10 000
	PP, 25% glass fibre	1200	10 000	10 000
	Polyurethane	1200	6000	6000
	Epoxy resin	1200	10 000	10 000
	Phenol resin	1300	100 000	100 000
	Polyester resin	1400	10 000	10 000
Rubbers	Natural	910	10 000	10 000
	Neoprene	1240	10 000	10 000
	Butyl	1200	200 000	200 000
	Foam rubber	60–80	7000	7000
	Hard rubber	1200	∞	∞
	EPDM	1150	6000	6000
	Polyisobutylene	920	10 000	10 000
	Polysulphide	1700	10 000	10 000
	Butadiene	980	100 000	100 000
Glass	Quartz glass	2200	∞	∞
	Normal glass	2500	∞	∞
	Glass mosaic	2000	∞	∞
Gases	Air	1.23	1	1
	Argon	1.70	1	1
	Carbon dioxide	1.95	1	1
	Sulphur hexafluoride	6.36	1	1
	Krypton	3.56	1	1
	Xenon	5.68	1	1
Thermal breaks, sealants, weather stripping	Silica gel (desiccant)	720	∞	∞
	Silicone foam	750	10 000	10 000
	Silicone pure	1200	5000	5000
	Silicone, filled	1450	5000	5000
	PUR (thermal break)	1300	60	60
	PVC, flexible with 40% softener	1200	100 000	100 000
	Elastomeric foam, flexible	60-80	10 000	10 000
	Polyurethane foam	70	60	60

Group	Material	Density (ρ) (kg/m³)	Vapour resistance factor μ (−)	
Roofing materials	Polyethene foam	70	100	100
	Asphalt	2100–2300	50 000	50 000
	Bitumen, pure	1050	50 000	50 000
	Bitumen felt	1100	50 000	50 000
	Clay tiles	2000	40	30
	Concrete tiles	2100	100	60
Floor coverings	Rubber	1200	10 000	10 000
	Plastic	1700	10 000	10 000
	Linoleum	1200	1000	800
	Carpet/textile	200	5	5
	Underlay, rubber	270	10 000	10 000
	Underlay, felt	120	20	15
	Underlay, wool	200	20	15
	Underlay, cork	200	20	15
	Ceramic tiles	2300	∞	∞
	Plastic tiles	1000	10 000	10 000
	Cork tiles, light	200	20	15
	Cork tiles, heavy	500	40	20

4.5.1.2 Insulation Materials

Material	ρ (kg/m³)	μ (−)
Cork	90–160	10
Glass and mineral fibre	10–200	1
EPS	10–50	30–70
XPS	25–65	100–300
PUR	28–55	50–100
VIPs		∞

4.5.2 Measured Values

4.5.2.1 Diffusion Resistance Factor (μ), Test Method

Used is the cup test, described in the American Society for Testing and Materials (ASTM) and EN-ISO standards. A sample of the material to be tested with known thickness and the surface adapted to the cup used is tightly sealed as cover on a non-corroding, vapour tight cup, partly filled with either a desiccant, a saturated salt or distilled water, see Table 4.3. When ready, the cup is placed in a temperature and RH controlled climate chamber and weighed at registered time intervals. Once enough data gained, the vapour permeability is calculated from the slope of the least square line linking the weight measured to time, see Figure 4.5.

4.5 Moisture Properties

Table 4.3 Equilibrium RH in % above desiccant, saturated salt solutions and distilled water.

Desiccant (silicagel)	RH (%)	0					
Lithium chloride	Temperature (°C)	0.2	9.6	19.2	29.6	39.6	46.8
LCl H$_2$O	RH (%)	14.7	13.4	12.4	11.8	11.8	11.4
Magnesium chloride	Temperature (°C)	0.4	9.8	19.5	30.2	40.0	48.1
MgCl 6 H$_2$O	RH (%)	35.2	34.1	33.4	33.2	32.7	31.4
Sodium dichromate	Temperature (°C)	0.6	10.1	19.8	30.0	37.4	47.3
Na$_2$Cr$_2$O$_2$ 2 H$_2$O	RH (%)	60.4	57.8	55.5	52.4	50.4	48.0
Magnesium nitrate	Temperature (°C)	0.4	9.9	19.6	30.5	40.2	48.1
Mg(NO$_3$)$_2$ 6H$_2$O	RH (%)	60.7	57.5	55.8	51.6	49.7	46.2
Sodium chloride	Temperature (°C)	0.9	10.2	20.3	30.3	39.2	48.3
MaCl	RH (%)	75.0	75.3	75.5	75.6	74.6	74.9
Ammonium sulphate	Temperature (°C)	0.4	10.1	20.0	30.9	40.0	48.0
(NH$_4$)$_2$SO$_4$	RH (%)	83.7	81.8	80.6	80.0	80.1	79.2
Potassium nitrate	Temperature (°C)	0.6	10.2	20.0	30.7	40.4	48.1
KNO$_3$	RH (%)	97.0	95.8	93.1	90.6	88.0	85.6
Potassium sulphate	Temperature (°C)	0.5	10.1	19.8	30.4	39.9	48.1
K$_2$SO$_4$	RH (%)	99.0	98.0	97.1	96.8	96.1	96.0
Distilled water	RH (%)	100					

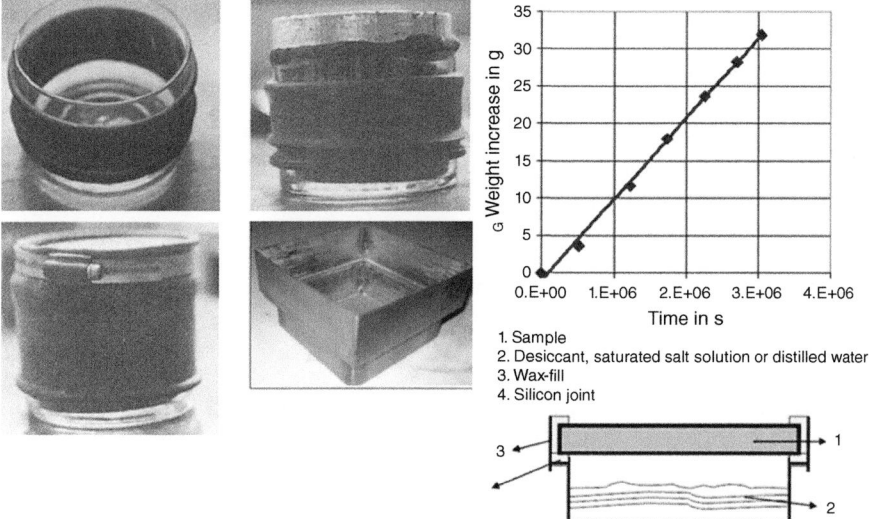

1. Sample
2. Desiccant, saturated salt solution or distilled water
3. Wax-fill
4. Silicon joint

Figure 4.5 Measuring the vapour resistance factor. The round glass cup has a seal of aluminium around the sample, a special mastic joint between sample and cup and a rubber cover around the cup. The stainless-steel quadratic cup has a wax-filled seal between sample and the somewhat broader cup collar. Calculating the vapour permeability.

Figure 4.6 Wall made of horizontal sheet metal elements.

The correctness of the test depends on several conditions that must be met. Absolute vapour tightness of the seal between sample and cup is a first, see Figure 4.6. That the vapour flow through the sample must be one-dimensional is a second and that the results measured must be independent of the flow direction, inward for the vapour pressure in the climate chamber higher than in the cup, outward if the reverse a third.

Isothermal means characterized by a vapour pressure gradient aligned to the RH gradient, which acts as driving force activating surface flow in the water molecule layers adsorbed by the pore walls present in the sample tested. Therefore, cup tests on hygroscopic materials give a vapour permeability, increasing with average RH. Commonly used as formula for this is:

$$\delta = a + b \, \exp(c\phi)$$

with a, b and c constants and ϕ the RH on a scale from 0 to 1. In what follows, the vapour permeability is transposed into an (equivalent) vapour resistance factor (μ):

$$\mu = \delta_a / \delta$$

with δ_a the vapour permeability of stagnant air.

4.5.2.2 Test Results
4.5.2.2.1 Building and Finishing Materials

Stone

Bamberger							
Density	kg/m³	$\rho = 1980$					
Hygroscopic curve	kg/m³	$\phi = 0.1$	0.3	0.5	0.65	0.8	0.9
Sorption			8.5	17.6	27.5	35.6	43.1
Capillary moisture content	kg/m³	210					
Saturation moisture content	kg/m³	230					

Diffusion resistance factor	–	$\phi = 0.265$	0.535	0.715	0.85	
		20	17	14	8.8	
Capillary water absorption coeff.	kg/(m²·s^0.5)	0.044				

Obernkirchner

Density	kg/m³	$\rho = 2150$				
Hygroscopic curve		$\phi = 0.1$	0.3	0.5	0.65	0.8	0.9
Sorption	kg/m³	0.6	1.3		2.6	3.4	4.3
Capillary moisture content	kg/m³	110				
Saturation moisture content	kg/m³	140				
Diffusion resistance factor	–	$\phi = 0.265$	0.535	0.715	0.85	
		32	30	28	18	
Capillary water absorption coeff.	kg/(m²·s^0.5)	0.046				

Rüthener

Density	kg/m³	$\rho = 1950$				
Hygroscopic curve		$\phi = 0.1$	0.3	0.5	0.65	0.8	0.9
Sorption	kg/m³	1.8	4.5		8.0	12.4	16.9
Capillary moisture content	kg/m³	200				
Saturation moisture content	kg/m³	240				
Diffusion resistance factor	–	$\phi = 0.265$	0.535	0.715	0.85	
		17	16	13	9.4	
Capillary water absorption coeff.	kg/(m²·s^0.5)	0.30				

Sander

Density	kg/m³	$\rho = 2120$				
Hygroscopic curve		$\phi = 0.1$	0.3	0.5	0.65	0.8	0.9
Sorption	kg/m³	4.4		10.2	15.2		22.6
Capillary moisture content	kg/m³	130				
Saturation moisture content	kg/m³	170				
Diffusion resistance factor	–	$\phi = 0.265$	0.535	0.715	0.85	
		33	30	22	13	
Capillary water absorption coeff.	kg/(m²·s^0.5)	0.02				

Savonnières

Density	kg/m³	$\rho = 1661$ ($\sigma = 34$, 120 samples)
Capillary moisture content	kg/m³	160, ($\sigma = 21.5$, 120 samples)
Saturation moisture content	kg/m³	382 ($\sigma = 14.2$, 120 samples)
Capillary water absorption coeff.	kg/(m²·s^0.5)	0.085 ($//$, $\sigma = 0.062$), 0.054 ($\perp$, $\sigma = 0.038$)

Concrete

Property	Units	Range	Value
Density	kg/m³		$\rho \leq 2176$
Hygroscopic curve	kg/m³		
Sorption		$0.2 \leq \phi \leq 0.98$	$147.5\left(1 - \dfrac{\ln \phi}{0.0453}\right)^{-\frac{1}{1.67}}$
Desorption		$0.2 \leq \phi \leq 0.98$	$147.5\left(1 - \dfrac{\ln \phi}{0.570}\right)^{-\frac{1}{0.64}}$
Critical moisture content	kg/m³		100–110
Capillary moisture content	kg/m³		110
Saturation moisture content	kg/m³		153
Diffusion resistance factor		$-0.2 \leq \phi \leq 0.98$	$\dfrac{1}{6.8 \cdot 10^{-3} + 8.21 \cdot 10^{-5} \exp(5.66\phi)}$
Capillary water absorption coeff.	kg/(m²·s^{0.5})		0.018
Moisture diffusivity (±)	m²/s, w in kg/m³		$1.8 \cdot 10^{-11} \exp(0.058w)$

Lightweight concrete

Property	Units	Range	Value
Density	kg/m³		$644 \leq \rho \leq 1442$
Hygroscopic curve	kg/m³	$938 \leq \rho \leq 1442$ kg/m³	$110\left(1 - \dfrac{\ln \phi}{0.0277}\right)^{-\frac{1}{2.14}}$
Sorption		$0.2 \leq \phi \leq 0.98$	
Desorption		$0.2 \leq \phi \leq 0.98$	$110\left(1 - \dfrac{\ln \phi}{0.0221}\right)^{-\frac{1}{2.91}}$
Critical moisture content	kg/m³, $\rho = 935$ kg/m³		140
Capillary moisture content	kg/m³, $872 \leq \rho \leq 980$ kg/m³		97–190
Saturation moisture content	kg/m³, $\rho = 973$ kg/m³		584
Diffusion resistance factor		$-0.2 \leq \phi \leq 0.98$, $\rho = 975$ kg/m²	$\dfrac{1}{6.76 \cdot 10^{-2} + 1.21 \cdot 10^{-3} \exp(3.94\phi)}$
Capillary water absorption coeff.	kg/(m²·s^{0.5}) $\rho = 975$ kg/m³ $\rho = 1410$ kg/m³		0.08 0.029
Moisture diffusivity (±)	m²/s, w in kg/m³ $\rho = 975$		$1.3 \cdot 10^{-9} \exp(0.035w)$

4.5 Moisture Properties

Aerated concrete 1

Density	kg/m³	$455 \leq \rho \leq 800$
Hygroscopic curve	kg/m³	
Sorption	$465 \leq \rho \leq 621$ kg/m³	$300\left(1 - \dfrac{\ln\phi}{0.0011}\right)^{-\frac{1}{1.99}}$
	$0.2 \leq \phi \leq 0.98$	
Desorption		$300\left(1 - \dfrac{\ln\phi}{0.0038}\right)^{-\frac{1}{1.32}}$
Critical moisture content	kg/m³	180
Capillary moisture content	kg/m³	$109 + 0.383\rho$
Saturation moisture content	kg/m³	$972 - 0.350\rho$
Diffusion resistance factor (−)	$0.2 \leq \phi \leq 0.98$	$\dfrac{1}{1.16 \cdot 10^{-1} + 6.28 \cdot 10^{-3} \exp(4.19\phi)}$
	$458 \leq \rho \leq 770$ kg/m³	
Capillary water absorption coeff.	kg/(m²·s⁰·⁵)	0.02–0.08
Moisture diffusivity (±)	m²/s, w in kg/m³	$9.2 \cdot 10^{-11} \exp(0.0215w)$
	$\rho = 511$ kg/m³	

Aerated concrete 2

Density	kg/m³	$\rho = 600$					
Hygroscopic curve		$\phi = 0.1$	0.3	0.5	0.65	0.8	0.9
Sorption	kg/m³			7.3	12.5	17	38
Capillary moisture content	kg/m³	290					
Saturation moisture content	kg/m³	720					
Diffusion resistance factor	–	$\phi = 0.265$	0.535		0.715	0.85	
		7.6			6.7		
Capillary water absorption coeff.	kg/(m²·s⁰·⁵)	0.09					

Polystyrene concrete

Density	kg/m³	$259 \leq \rho \leq 792$
Hygroscopic curve	kg/m³	
Sorption	$\rho \leq 422$ kg/m³	$235\left(1 - \dfrac{\ln\phi}{0.0097}\right)^{-\frac{1}{1.55}}$
	$0.2 \leq \phi \leq 0.98$	
Saturation moisture content	kg/m³, $\rho \leq 422$ kg/m³	489
Diffusion resistance factor	$-0.2 \leq \phi \leq 0.98$	$\dfrac{1}{8.18 \cdot 10^{-2} + 3.16 \cdot 10^{-3} \exp(2.88\phi)}$

	$357 \leq \rho \leq 425$ kg/m³	
Capillary water absorption coeff.	kg/(m²·s^{0.5})	0.026
	$360 \leq \rho \leq 457$ kg/m³	
Moisture diffusivity (±)	m²/s, w in kg/m³	$4.6 \cdot 10^{-10} \exp(0.064w)$
	$\rho = 422$ kg/m³	
Hygric strain	m/m	$0.0024 \cdot 10^{-4} \phi^2$
	$\rho = 422$ kg/m³	

Mortar

Density	kg/m³	$1050 \leq \rho \leq 1940$
Hygroscopic curve	kg/m³	
Sorption	$\rho = 1940$	$283\left(1 - \dfrac{\ln\phi}{0.029}\right)^{-\frac{1}{1.39}}$
	$0.2 \leq \phi \leq 0.98$	
Desorption		$300\left(1 - \dfrac{\ln\phi}{0.061}\right)^{-\frac{1}{1.77}}$
Capillary moisture content	kg/m³	283
Diffusion resistance factor	$-0.2 \leq \phi \leq 0.98$	$\dfrac{1}{7.69 \cdot 10^{-2} + 2.43 \cdot 10^{-3} \exp(3.61\phi)}$
Capillary water absorption coeff.	kg/(m²·s^{0.5})	0.042–0.80
Moisture diffusivity (±)	m²/s, w in kg/m³	$C_1 \exp(C_2 w)$

ρ (kg/m³)	C_1	C_2
1072	$2.0 \cdot 10^{-9}$	0.022
1500	$2.7 \cdot 10^{-9}$	0.020
1807	$1.4 \cdot 10^{-9}$	0.027

Brick 1

Density	kg/m³	$\rho = 1700$					
Hygroscopic curve		$\phi = 0.1$	0.3	0.5	0.65	0.8	0.9
Sorption	kg/m³				7.5	8.4	18	34
Capillary moisture content	kg/m³	270					
Saturation moisture content	kg/m³	380					
Diffusion resistance factor	–	$\phi = 0.265$	0.535		0.715	0.85	
		9.5	8.8		8.0	6.9	
Capillary water absorption coeff.	kg/(m²·s^{0.5})	0.25					

Brick 2

Property	Units	Range	Value/Formula
Density	kg/m³		$1505 \leq \rho \leq 2047$
Hygroscopic curve	kg/m³		$200\left(1 - \dfrac{\ln \phi}{1.46 \cdot 10^{-4}}\right)^{-\frac{1}{1.59}}$
Sorption		$0.2 \leq \phi \leq 0.98$	
Critical moisture content	kg/m³		100
Capillary moisture content	kg/m³	$1505 \leq \rho \leq 2047$ kg/m³	$730.3 - 0.287\rho$
Saturation moisture content	kg/m³, ρ idem		$1033 - 0.404\rho$
Diffusion resistance factor		$-0.2 \leq \phi \leq 0.98$ $1505 \leq \rho \leq 1860$ kg/m³	$\dfrac{1}{5.6 \cdot 10^{-2} + 4.67 \cdot 10^{-3} \exp(2.79\phi)}$
Capillary water absorption coeff.	kg/(m²·s$^{0.5}$)		
Clay		$1505 \leq \rho \leq 2000$ kg/m³	$0.653 - 0.00030\rho$
Loam		$1628 \leq \rho \leq 1868$ kg/m³	$1.954 - 0.00087\rho$
Moisture diffusivity (±)	m²/s, w in kg/m³		$C_1 \exp(C_2 w)$

ρ (kg/m³)	C_1	C_2
1529	$2.1 \cdot 10^{-9}$	0.032
1619	$1.9 \cdot 10^{-8}$	0.022
1918	$7.4 \cdot 10^{-9}$	0.032

Lime-sandstone 1

Property	Units	Range	Value/Formula
Density	kg/m³		$1685 \leq \rho \leq 1807$
Hygroscopic curve	kg/m³		
Sorption		$0.2 \leq \phi \leq 0.98$	$210\left(1 - \dfrac{\ln \phi}{3.56 \cdot 10^{-3}}\right)^{-\frac{1}{2.39}}$
Desorption		$1685 \leq \rho \leq 1726$ kg/m³	$330\left(1 - \dfrac{\ln \phi}{6.58 \cdot 10^{-3}}\right)^{-\frac{1}{1.81}}$
Critical moisture content	kg/m³ $\rho \leq 1807$		120
Capillary moisture content	kg/m³ $1711 \leq \rho \leq 1777$ kg/m³		233
Diffusion resistance factor		$-0.2 \leq \phi \leq 0.98$ $1505 \leq \rho \leq 1860$ kg/m³	$\dfrac{1}{4.3 \cdot 10^{-2} + 4.56 \cdot 10^{-5} \exp(9.86\phi)}$
Capillary water absorption coeff.	kg/(m²·s$^{0.5}$) $\rho \leq 1807$ kg/m³		0.042
Moisture diffusivity (±)	m²/s, w in kg/m³ $\rho \leq 1807$ kg/m³		$2.2 \cdot 10^{-10} \exp(0.027w)$

Lime-sandstone 2

Density	kg/m³	$\rho = 1900$					
Hygroscopic curve	kg/m³	$\phi = 0.1$	0.3	0.5	0.65	0.8	0.9
Sorption				17	18	24.9	40.2
Capillary moisture content	kg/m³	250					
Saturation moisture content	kg/m³	290					
Diffusion resistance factor	–	$\phi = 0.265$ 28	0.535 24		0.715 18	0.85 13	
Capillary water absorption coeff.	kg/(m²·s⁰·⁵)	0.045					

Gypsum plaster 1

Density	kg/m³	975
Hygroscopic curve	kg/m³	$310 \left(1 - \dfrac{\ln \phi}{3.21 \cdot 10^{-2}}\right)^{-\frac{1}{1.59}}$
Sorption	$0.2 \leq \phi \leq 0.98$	
Desorption		$-0.004\phi^2 + 0.01\phi - 0.0004$
Capillary moisture content	kg/m³	310
Diffusion resistance factor	$-0.2 \leq \phi \leq 0.98$	$\dfrac{1}{0.133 + 7.45 \cdot 10^{-4} \exp(5.1\phi)}$
Capillary water absorption coeff.	kg/(m²·s⁰·⁵)	0.155
Moisture diffusivity (±)	m²/s, w in kg/m³	$1.7 \cdot 10^{-9} \exp(0.0206w)$

Gypsum plaster 2

Density	kg/m³	$\rho = 850$					
Hygroscopic curve		$\phi = 0.1$	0.3	0.5	0.65	0.8	0.9
Sorption	kg/m³				3.6	5.2	6.3	11
Capillary moisture content	kg/m³	400					
Saturation moisture content	kg/m³	650					
Diffusion resistance factor	–	$\phi = 0.265$ 8.3	0.535		0.715 7.3	0.85	
Capillary water absorption coeff.	kg/(m²·s⁰·⁵)	0.29					

Stucco

Density	kg/m³	$878 \leq \rho \leq 1736$		
Capillary moisture content	kg/m³, $878 \leq \rho \leq 1736$ kg/m³	185		
Diffusion resistance factor	–	$1.8 \exp(0.0016\rho)$		
	$880 \leq \rho \leq 1709$ kg/m³	ρ (kg/m³)	ϕ_m (%)	μ (–)
	$\phi = 86\%$	1680	86	43
			95	15

4.5 Moisture Properties

Capillary water absorption coeff.	kg/(m²·s$^{0.5}$)	$0.0039 \leq A \leq 0.029$ On average 0.0128		
Moisture diffusivity (±)	m²/s$^{0.5}$	$C_1 \exp(C_2 w)$		
		ρ (kg/m³)	C_1	C_2
		878	$4.4 \cdot 10^{-12}$	0.027
		1341	$2.1 \cdot 10^{-10}$	0.019

Timber

Density	kg/m³	$400 \leq \rho \leq 690$		
Hygroscopic curve	kg/m³	$100\left(1 - \dfrac{\ln\phi}{0.642}\right)^{-\frac{1}{0.64}}$		
Sorption	$0.2 \leq \phi \leq 0.98$			
Desorption		$120\left(1 - \dfrac{\ln\phi}{0.248}\right)^{-\frac{1}{1.22}}$		
Diffusion resistance factor	–	ρ (kg/m³)	ϕ_m (%)	μ (–)
		Oak, 640	30	110
			87	21
		Beech, 435	25	180
			75	20
		Spruce, 390	53	57
			86	13
Capillary water absorption coeff.	kg/(m²·s$^{0.5}$) $\rho = 390$ kg/m³ (spruce)	0.004 $\perp$ fibres 0.016 // fibres		

Woodwool-cement boards

Density	kg/m³	$314 \leq \rho \leq 767$
Hygroscopic curve	%kg/kg	$15\left(1 - \dfrac{\ln\phi}{0.172}\right)^{-\frac{1}{0.84}}$
Sorption	$0.2 \leq \phi \leq 0.98$ $\rho = 767$ kg/m³	
Capillary moisture content	kg/m³	180
Saturation moisture content		240
Diffusion resistance factor		≈4
Capillary water absorption coeff.	kg/(m²·s$^{0.5}$)	0.007
Moisture diffusivity (±)	m²/s, w in kg/m³	$6.2 \cdot 10^{-12} \exp(0.027w)$

Particle board

Density	kg/m³	$570 \leq \rho \leq 800$
Hygroscopic curve	%kg/kg	$35\left(1 - \dfrac{\ln \phi}{0.0328}\right)^{-\frac{1}{1.89}}$
Sorption	$0.2 \leq \phi \leq 0.98$	
Desorption		$35\left(1 - \dfrac{\ln \phi}{0.081}\right)^{-\frac{1}{1.63}}$
Critical moisture content	%kg/kg	85
Capillary moisture content	%kg/kg	90
Saturation moisture content	%kg/kg	99
Diffusion resistance factor	–	$A_1 \exp(A_2 \rho)$

ϕ (%)	A_1	A_2	r^2
25	0.543	0.00757	0.70
86	0.508	0.00676	0.75

Capillary water absorption coeff.	kg/(m²·s^0.5) UF, glue used	(UF, UMF) 0.0035 (FF) 0.022
Moisture diffusivity (±)	m²/s w in kg/m³	(UF, UNF) $2.3 \cdot 10^{-13} \exp(0.01w)$ (FF) $4.5 \cdot 10^{-12} \exp(0.01w)$

Plywood

Density	kg/m³	$445 \leq \rho \leq 799$
Hygroscopic curve	%kg/kg	$75\left(1 - \dfrac{\ln \phi}{6.14 \cdot 10^{-3}}\right)^{-\frac{1}{1.91}}$
Sorption	$0.2 \leq \phi \leq 0.98$	
Desorption		$75\left(1 - \dfrac{\ln \phi}{9.63 \cdot 10^{-3}}\right)^{-\frac{1}{1.93}}$
Critical, capillary and saturation moisture content	%kg/kg	75
Diffusion resistance factor	$-437 \leq \rho \leq 591$ kg/m³ $\phi = 86\%$	$15.3 \exp[0.0045(\rho - 437)]$

ρ (kg/m³)	ϕ_m (%)	μ (–)
548–580	54	97
437–591	86	24

Capillary water absorption coeff.	kg/(m²·s^0.5)	0.003
Moisture diffusivity (±)	m²/s^0.5	$3.2 \cdot 10^{-13} \exp(0.015w)$

Fibre cement boards

Density	kg/m³	$823 \leq \rho \leq 2052$
Hygroscopic curve	kg/m³	
Sorption	$0.2 \leq \phi \leq 0.98$	
	$\rho = 990$ kg/m³	$300\left(1 - \dfrac{\ln \phi}{0.0077}\right)^{-\frac{1}{1.93}}$
	$\rho = 1495$ kg/m³	$358\left(1 - \dfrac{\ln \phi}{0.0415}\right)^{-\frac{1}{1.36}}$
Desorption	$\rho = 990$ kg/m³	$350\left(1 - \dfrac{\ln \phi}{0.1076}\right)^{-\frac{1}{1.22}}$
	$\rho = 1495$ kg/m³	$358\left(1 - \dfrac{\ln \phi}{0.197}\right)^{-\frac{1}{0.8}}$
Critical moisture content	kg/m³, $\rho = 840$	350
Capillary moisture content	kg/m³, $\rho = 1495$	358
Saturation moisture content	kg/m³, $\rho = 1495$	430
Diffusion resistance factor	$-$, $0.2 \leq \phi \leq 0.98$	
	$\rho = 840$ kg/m³	$\dfrac{1}{0.0565 + 5.58 \cdot 10^{-5} \exp(7.85\phi)}$
	$\rho = 1495$ kg/m³	$\dfrac{1}{0.00642 + 1.4 \cdot 10^{-4} \exp(4.92\phi)}$
Capillary water absorption coeff.	kg/(m²·s$^{0.5}$)	0.024
Moisture diffusivity ($\pm$)	m²/s$^{0.5}$, w in kg/m³ $840 \leq \rho \leq 1495$ kg/m³	$3.4 \cdot 10^{-11} \exp(0.018w)$

Masonry walls

Bricks

Density	kg/m³	$830 \leq \rho \leq 1760$			
Equivalent diffusion thickness		d (m)	wall	ϕ_m (%)	μd (m)
		0.09	Bricks	54	1.20
			6.5 × 9 × 19 cm	86	0.51
		0.09	Idem, painted with acrylic paint	54	2.20
		0.09	Idem, water repellent treated	86	0.65
		0.09	Glazed bricks	86	4.00
		0.19	Bricks	59	1.60
			6.5 × 9 × 19 cm	81	0.61
		0.14	Hollow blocks	58	1.30

			14 × 19 × 29 cm	84	0.84
		0.19	Bricks	88	0.53
			6.5 × 9 × 19 cm		

Lime-sandstone

Density	kg/m³	$860 \leq \rho \leq 1650$			
Equivalent diffusion thickness	d (m)		Wall	ϕ_m (%)	μd (m)
		0.14	$\rho = 1140 \text{ kg/m}^3$	25	3.70
			14 × 14 × 29 cm		

Concrete blocks

Density	kg/m³	$860 \leq \rho \leq 1650$			
Equivalent diffusion thickness	d (m)		wall	ϕ_m (%)	μd (m)
		0.14	Blocks, $\rho = 960$	60	1.30
				86	0.54
			14 × 14 × 29 cm		
		0.14	Blocks	61	0.61
			$\rho = 1450 \text{ kg/m}^3$	64	0.58
			14 × 14 × 29 cm	83	0.61
				90	0.28

4.5.2.2.2 Insulation Materials

Cork

Density	kg/m³	$\rho \leq 111$
Hygroscopic curve	%kg/kg	$60\left(1 - \dfrac{\ln \phi}{1.02 \cdot 10^{-5}}\right)^{-\frac{1}{3.64}}$
Sorption	$0.2 \leq \phi \leq 0.98$	
Critical moisture content	kg/m³	60
Diffusion resistance factor		≈ 22

Cellular glass

Density	kg/m³	$114 \leq \rho \leq 140$
Diffusion resistance factor	–	5000 to 70 000

Glass fibre

Density	kg/m³	$12 \leq \rho \leq 133$
Diffusion resistance factor	19–102 kg/m³	1.2

Mineral fibre

Density	kg/m^3	$32 \leq \rho \leq 191$
Diffusion resistance factor	148–172 kg/m^3	1.5

Expanded polystyrene (EPS)

Density	kg/m^3	$13 \leq \rho \leq 40$
Diffusion resistance factor	$\phi = 0.86$, 15–40 kg/m^3	$4.9 + 1.97\rho$

Extruded polystyrene (XPS)

Density	kg/m^3	$25 \leq \rho \leq 55$
Diffusion resistance factor	$\phi = 0.86$, 25–53 kg/m^3	$48.6 + 3.35\rho$

PUR/PIR

Density	kg/m^3	$20 \leq \rho \leq 40$
Diffusion resistance factor	$\phi = 0.8$, 20–40 kg/m^3	$1.67 \exp(0.088\rho)$

Expanded perlite

Density	kg/m^3	$135 \leq \rho \leq 215$
Hygroscopic curve	kg/m^3	
Sorption	$0.2 \leq \phi \leq 0.98$	$150\left(1 - \dfrac{\ln \phi}{1.15 \cdot 10^{-4}}\right)^{-\frac{1}{2.63}}$
	187–213 kg/m^3	
Critical moisture content	kg/m^3	150–200
Capillary moisture content	kg/m^3	550
Diffusion resistance factor	$0 \leq \phi \leq 1$	$\dfrac{1}{0.0143 + 6.54 \cdot 10^{-8} \exp(16.5\phi)}$

4.5.2.2.3 Finishes
Groove and tongue timber lathing

Weight per m^2	kg/m^2	4.0 ($d = 10$ mm)		
Hygroscopic curve	%kg/kg			
Sorption	$0.2 \leq \phi \leq 0.98$	$50\left(1 - \dfrac{\ln \phi}{0.0213}\right)^{-\frac{1}{1.96}}$		
Desorption		$50\left(1 - \dfrac{\ln \phi}{0.0397}\right)^{-\frac{1}{1.86}}$		
		$\phi\,(-)$	Sorption X (%kg/kg)	desorption X (%kg/kg)
		0.33	5.8	7.1

		0.52	9.4	12.3
		0.75	13.7	15.4
		0.86	17.6	20.8
		0.97	31.4	37.3
		0.98	34.5	
Diffusion thickness	m, $\phi = 0.55$	0.86 ($\sigma = 0.12$)		

Wallpaper

Weight per m², thickness	kg/m², mm	Type	kg/m²	d (mm)
		1. textile	0.291	0.425
		2. vinyl	0.216	0.325
		3. textile	0.333	0.700
		4. vinyl	0.212	0.450
		5. paper	0.168	0.280
		6. paper	0.151	0.280

Hygroscopic curve	%kg/kg						
	Paper ϕ (−) ↓	1	2	3	4	5	6
Sorption	0.33	3.2	1.4	2.9	1.2	1.6	1.8
	0.52	5.5	2.8	5.5	2.3	2.9	4.0
	0.75	7.9	5.0	6.4	3.3	4.6	5.5
	0.86	11.2	8.2	9.8	4.8	7.2	6.8
	0.97	21.4	40.0	24.9	13.3	15.8	16.9
Desorption	0.33	5.3	4.6	4.0	7.0	2.7	3.9
	0.52	8.5	5.6	6.6	8.3	4.2	6.0
	0.75	11.8	8.7	9.9	9.8	6.9	9.4
	0.86	16.1	10.7	14.0	10.0	9.7	12.3
	0.97	42.8	52.2	38.3	17.9	23.8	36.2

Diffusion thickness	m						
	Paper ϕ (−) ↓	1	2	3	4	5	6
	0.42	0.280	2.14	0.155	0.09	0.035	0.025
	0.75	0.006	0.18	0.019	0.025	0.012	0.008

Gypsum board

Weight per m²	kg/m²	$6.5 \leq \rho \leq 13$
Hygroscopic curve	kg/m³	
Sorption	$0.2 \leq \phi \leq 0.98$	$150 \left(1 - \dfrac{\ln \varphi}{2.99 \cdot 10^{-4}}\right)^{-\frac{1}{4.81}}$
Desorption		$150 \left(1 - \dfrac{\ln \phi}{0.026}\right)^{-\frac{1}{7.86}}$
Diffusion resistance factor	$-d = 9.5, d = 12.5$ mm	$\dfrac{1}{0.0712 + 2.81 \cdot 10^{-3} \exp(4.1\varphi)}$

Paints

Diffusion thickness			$\phi\,(-)$		
			0.425	0.750	0.86
	On gypsum board				
	Primer + two layers latex 1 paint				0.17
	Primer + two layers latex 2 paint		4.5		1.10
	Primer + two layers acrylic paint				0.46
	Primer + two layers synthetic paint		3.20		1.00
	Primer + two layers oil paint				0.76
	On cellular concrete				
	Primer + two layers acrylic paint			0.43	
	Structured paint			1.10	

Carpet

Weight per m²	kg/m²			2.18
Hygroscopic curve	%kg/kg			
Sorption	$0.2 \leq \phi \leq 0.98$	$\phi(-)$		X (%kg/kg)
		0.33		8.0
		0.52		9.9
		0.75		13.4
		0.86		17.2
		0.97		29.7
		0.98		29.8

4.5.2.2.4 Miscellaneous

Newspaper

Weight per m²	kg/m²		0.041
Hygroscopic curve	%kg/kg $0.2 \leq \phi \leq 0.98$	$\phi\,(-)$	Sorption X (%kg/kg)
		0.33	5.3
		0.52	9.0
		0.75	12.9
		0.86	16.9
		0.97	32.1
		0.98	40.9

Journal

Weight per m²	kg/m²		0.047
Hygroscopic curve	%kg/kg $0.2 \leq \phi \leq 0.98$	$\phi\,(-)$	Sorption X (%kg/kg)
		0.33	2.9
		0.52	4.7
		0.75	6.5
		0.86	8.4
		0.97	16.6
		0.98	19.0

4.5.2.2.5 Vapour Retarders

Diffusion resistance factor

PE-foil	$\phi\,(-)=$	0.28	0.52	0.70	0.75	0.86
	$\mu=$	321 000			289 000	271 000

Diffusion thickness μd (m)

	$\phi\,(-)=d$ (mm) $=\downarrow$	0.52	0.70	0.75	0.86
Bituminous paper	0.1				1.80
					2.80
	0.2	0.70			
	1.4			1.70	
				6.90	
	0.4		3.90		
			8.10		
Aluminium paper	0.1				2.00
					2.80
	0.2				0.17
					0.33
	0.24		17.80		
			77.30		
	-				6.80
					17.80
Glass fibre reinforced alupaper	0.4		3.80		
			4.70		
Glass fibre reinforced PVC-foil	0.4				12.0
					29.0
Stapled PE-foil	0.15			7.70	

Further Reading

ASHRAE (2021). *Handbook of Fundamentals*, SI Edition. Atlanta, GA: Tullie Circle.

Bomberg, M. and Kumaran, K. (1986). A test method to determine air flow resistance of exterior membranes and sheathings. *Journal of Thermal Insulation* 9: 224–235.

CEN/TC 89, Comité Européen de Normalisation (European Committee for Standardization) (1998). Building materials and products-Hygrothermal properties-Tabulated design values, working draft.

CEN/TR 14613 (2003). Thermal performance of building materials and components-principles for the determination of thermal properties of moist materials and components, Final draft.

Hagentoft, C.E. (2001). *Introduction to Building Physics*. Lund: Studentlitteratur.

Hens, H. (1975). Theoretische en experimentale studie van het hygrothermisch gedrag van bouwen isolatiematerialen bij inwendige condensatie en droging met toepassing op de platte daken (Theoretical and experimental study of the hygrothermal response of building and insulating materials during interstitial condensation and drying with application on low-sloped roofs), Ph.D., K.U.Leuven, p. 311, (in Dutch).

Hens, H. (1984). Cataloog van hygrothermische eigenschappen van bouw- en isolatiematerialen (Catalogue of hygrothermal properties of building and insulating materials), rapport E/VI/2, R-D Energy, 100 p + appendices, (in Dutch).

Hens, H. (1991). Catalogue of Material Properties, Final Report, IEA-EXCO on Energy Conservation in Buildings and Community Systems, Annex 14 'Condensation and Energy'.

ISO 10456, International Organization for Standardization (2007). Building materials and products-Hygrothermal properties-Tabulated design values and procedures for determining declared and design thermal values, p. 24.

Krus, M. (1995). Feuchtetransport- und Speicherkoeffizienten poröser mineralischer Baustoffe. Theoretische Grundlagen und Neue Meßtechniken, Doctor Abhandlung, Universität Stuttgart (in German).

KULeuven, Department of Civil Engineering, Unit Building Physics (1978–2008). Reports on material property measurements (in Dutch and in English).

Kumaran, K. (1996), Material Properties, Final Report, IEA, EXCO on Energy Conservation in Buildings and Community Systems, Annex 24 'Heat, Air and Moisture Transfer in Insulated Envelope Parts, Task 3.

Kumaran, K. (2002). A Thermal and Moisture Property Database for Common Building and Insulating Materials, Final report ASHRAE Research Project 1018-RP, p. 231.

NBN B62-002/A1 (2001). Berekening van de warmtedoorgangscoëfficiënten van wanden van gebouwen, 2^e uitgave (in Dutch).

Roels, S. (2000). Modelling unsaturated moisture transport in heterogeneous limestone. Doctoral thesis. KULeuven.

Roels, S., Carmeliet, J., and Hens H. (2003). Moisture transfer properties and material characterisation, Final Report EU HAMSTAD project.

Roels S. (2008). Experimental Analysis of moisture buffering, Final Report, IEA, EXCO on Energy Conservation in Buildings and Community Systems, Annex 41 'Whole Building Heat, Air and Moisture Response', Task 2

Sagelsdorff, R. (1989). *Wasserdampfdiffusion, Grundlagen, Berechnungsverfahren, Diffusionsnachweis*. SIA SIA-Dokumentation D 018, p. 90 p (in German).

Stichting Bouwresearch (1974). Eigenschappen van bouw- en isolatiematerialen (Properties of building and insulating materials), 2^e, gewijzigde druk (in Dutch).

Trechsel, H.R. and Bomberg, M. (ed.) (1989). *Water Vapor Transmission Through Building Materials and Systems*, 1039. Philadelphia, PA: ASTM.

Postscript

Any evaluation of the heat, air and moisture performance of building assemblies and whole buildings with its consequences for design, construction and renovation should be based on a sound understanding of the theory. That knowledge, however, encompasses only a small part of what a performance-based approach needs. In fact, from a building physics point of view, buildings and their parts must also guarantee good acoustics, pleasant lighting, energy efficiency and a comfortable and healthy indoors. Furthermore, architects, engineers, builders and developers must ensure that buildings are aesthetical and functional, while structural safety, durability, fire safety, maintenance, cost and sustainability with low carbon as important requirement are as important.

Studying heat, air and moisture tolerance helps understanding some important facts. Heat transfer includes conduction, convection and radiation. Convection combines conduction in gases and liquids at surfaces and the flow patterns playing. Because of its electromagnetic nature, radiation and related laws differ from those governing conduction and convection. The three anyhow define to a large extent the heat exchange between buildings and the outdoors, in spaces indoors, in air and gas layers and through porous materials. Related surface film coefficients introduced and the thermal transmittance as defined are a mathematical expediency that facilitates calculation.

Thermal insulation is very effective in lowering the heat losses and gains, though the effectiveness added drops with each extra cm of thickness. However, an air-permeable insulation layer in an assembly that allows air to mitigate not only loses its thermal efficiency, but interstitial condensation could also turn up in layers and interfaces at the cold side. Not preventing air in- and outflow, air looping and indoor air washing may so have really negative consequences for the thermal performance and moisture tolerance of envelope parts. Therefore, when evaluating vapour-related moisture damages, not only equivalent diffusion but also lack of airtightness with potentially vapour-loaded air ingress needs attention. Air flows displace more vapour than diffusion does over quite a short time. Moisture transport of course is not restricted to vapour. In open-porous materials, at higher moisture content, unsaturated water flow takes over. Without knowledge of the governing laws, building moisture, rain penetration, rising damp, accidental moisture uptake and drying cannot be understood and excluded. Opposed to a 'construction as an

Building Physics – Heat, Air and Moisture: Fundamentals, Engineering Methods, Material Properties and Exercises, Fourth Edition. Hugo Hens.
© 2024 Ernst & Sohn GmbH. Published 2024 by Ernst & Sohn GmbH.

art' approach, application of these knowledges during design and construction is the key to a correct 'performance-based building design'.

The subjects discussed in this book resulted in several software packages that help designers and constructors to take the correct heat, air and moisture-related decisions, of course on condition the embedded limitations are known. A profound understanding of the heat, air and moisture transport based on simulating, testing and practice in fact takes precedence. Many software tools for example do not consider air flow through, air looping around and indoor air washing in front of insulation layers, so, often produce outcomes that lack realism. They for example do not cope with rain run-off-related gravity and pressure flow through cracks, leaks and voids. They miss the real geometry with its roughness, cracks, voids, etc. Moreover, many software packages miss a well-established, calculable linkage between the heat, air and moisture transport phenomena modelled and the many moisture-related durability issues encountered in practice.

Index

a

acoustical comfort   2
adsorption constant   170
air displacement   134
airflow in, open-porous materials
  at building level
    applications   145–148
    conservation law   143–145
    defined   141–142
    large openings   142–143
    thermal stack   142
  combined heat and airflow
    air-permeable layers, joints and leaks   153
    heat balance   148
    non-steady-state, flat assemblies   152–153
    non-steady-state, two and three dimensions   153
    steady-state, flat assemblies   148–152
    steady-state, two and three dimensions   152
    vented cavities   153–155
  conservation law   135–138
  one dimension, flat assemblies   138–139
  through assemblies   140–141
  two and three dimensions   139–140
air flux   242
air permeance   134, 137
air pressure differentials
  fans   133
  stack effect   130–133
  wind   130
air properties, measured values
  air permeance, test method   325–326
  test results
    air barriers   335–336
    building materials and building parts   327–330
    insulation materials and insulating layers   330–335
    internal linings   336
air resistance   138
American Society for Testing and Materials (ASTM)   340
American Society of Heating and Ventilation Engineers (ASHVE)   9
anisotropic materials   140
applied physics
  acoustics   8–9
  heat, air, moisture   5–8
  lighting   9
average sol-air temperature   80

b

balance equations
  air   276
  closing the loop   278–279
  heat   276–278
  vapour   275–276

barometric relation   132
black surface constant   68
Brunauer–Emmett–Teller/BET-equation   170
building materials
  autoclaved cellular concrete masonry   321
  concrete   344
    aerated concrete   345
    autoclaved aerated   318
    block masonry   320–321
    with expanded polystyrene pearls   319
    lightweight   318, 344
    polystyrene   345–346
  fibre cement   322, 351
  gypsum board   322
  gypsum plaster   321, 348
  lightweight and normal cement mortars   319
  lime-sandstone   319, 347–348
  masonry walls   351–352
  mortar and brick   346–347
  outside rendering   321
  particle board   322, 350
  perforated brick masonry   320
  plywood   322
  stone   342–343
  stucco   348–349
  timber   349
  wood   322
  woodwool-cement boards   349
building physics
  constraints
    architecture and materials   3
    comfort   2–3
    economy   3
    health and well-being   3
    sustainability   3–4
  defined   1–2
  history   5, 12–13
    applied physics   5–9
    building design and construction   11–12
    building services   11
    indoor air quality and thermal comfort   9–10
    KULeuven and universities, in low countries   13–14
  importance   4–5
building-related applications, heat transfer
  heat balances   100–101
  local inside surface film coefficients   91–92
  non-steady-state
    any boundary conditions, thermal bridges   105
    periodic boundary conditions, flat assemblies   101
    periodic boundary conditions, spaces   101–105
  steady-state, flat assemblies
    average thermal transmittance, parallel envelope parts   83–84
    electrical analogy   84
    interface temperatures   86–87
    solar transmittance   88–90
    thermal resistance of, non-ventilated cavities   84–86
    thermal transmittance   80–83
    thicker insulation layers, on thermal transmittance   87–88
  surface film coefficients and reference temperatures
    indoors   76–78
    methodology   76
    outdoors   78–80
  two and three dimensions steady-state
    building envelopes   99–100
    floors, on grade   93–94
    pipes   92–93
    thermal bridges   94–98
    windows   98–99
buoyancy   130

c

capillary hygroscopic materials   125
capillary non-hygroscopic materials   125

capillary water absorption coefficient 215
capillary water flux 233
Carrier, W. 11
central control volumes 95
CFD *see* computerized fluid dynamics (CFD)
chlorine corrosion 173
coloured surfaces 74
combined heat and airflow
    heat balance 148
    non-steady-state, flat assemblies 152–153
    non-steady-state, two and three dimensions 153
    steady-state, flat assemblies
        equation 148–149
        multi-layered 151–152
        single-layered 149–151
    steady-state, two and three dimensions 152
    vented cavities 153–155
computerized fluid dynamics (CFD) 54, 266
conduction 17
    laws
        first law 20–21
        second law 22
    non-steady-state 38–39
        boundary conditions, flat assemblies 48–51
        periodic boundary conditions, flat assemblies 39–48
        two and three dimensions, thermal bridges 52
    steady-state 26
        one dimension, flat assemblies 27–33
        two and three dimensions, thermal bridges 35–38
        two dimensions, cylinder symmetric 33–34
conservation of mass 266–267
control volume method (CVM) 35, 153, 274
convection
    convective surface film coefficient, quantifying 55–58
    forced 59
    typology
        driving forces 55
        flow types 55
convective heat flux 76
convective surface film coefficient
    quantifying
        analytical 55
        dimensional 56–58
        numerical 55–56
    values for
        cavities 60–62
        flat surfaces 58–60
        pipes 62–63
Cranck–Nicholson scheme 278
Cremer, L. 9
CVM *see* control volume method (CVM)

d

De Grave, A. 13
diffusion constant 180
diffusion resistance, of unvented cavities 198
dirac impulses 48
drying flux 235
dynamic thermal resistance 43

e

electrical analogy 31
electrical resistors 31
electromagnetic waves 63
emissivity 72
enclosures 99
energy conservation 268
EN-ISO standards 340
enthalpy and vapour saturation pressure 271
envelope
    assemblies 44, 78, 86, 87
    building 99–100
    leaky 151, 152
    transparent 278

equivalent conduction 269
equivalent diffusion 248
equivalent hydraulic circuit 141
expanded polystyrene (EPS) 324, 353
extruded polystyrene (XPS) 324, 353

f

fibrous insulation materials 26
Fick, A. 8
Fick's empirical diffusion law 173
fine porous, hygroscopic material assemblies 274
floor heating system 118
fluid's kinematic viscosity 57
fluid's velocity 57
flux equations
    heat 270
    mass, air 270
    mass, moisture 270–271
forced convection 59
force equilibrium 216
Fourier, J. 5
Fourier's law 148, 270
friction factor 135, 136

g

gas constant 135
Grasshof number 57, 58
grey surfaces 72–74
groove and tongue timber lathing 353–354
gypsum transformation 173

h

heat absorption coefficient 51
heat, air and moisture combined
    material and assembly level
        assumptions 265
        conservation of energy 267–269
        conservation of mass 266–267
        equations of state 271
        flux equations 270–271
        simplified models 272–274
        solution 266
        start, boundary and contact conditions 271–272
    whole building level 274–275
        balance equations 275–279
        consequences 284–287
        sorption active surfaces and hygric inertia 279–280
heat, air and moisture material property values 303
    air properties
        measured values 325–336
        standard values 325
    dry air and water 304
    moisture properties
        measured values 340–356
        standard values 336–340
    thermal properties
        defined 305
        measured values 317–325
        standard values 305–316
        surfaces, radiant properties 316
heat balances 112
heat conduction 267
heat exchange, at surfaces 52–53
    convection 53–54
        convective surface film coefficient, quantifying 55–58
        typology 55
    radiation
        defined 63–64
        radiant surfaces 66–74
        reflection, absorption and transmission 64–66
        thermal radiation 63
heat flux equation 29
heating, ventilating, air conditioning (HVAC) 11, 17
heat of evaporation 170
heat transfer
    building-related applications
        heat balances 100–101
        local inside surface film coefficients 91–92
        non-steady-state 101–105
        steady-state, flat assemblies 80–90

surface film coefficients and
    reference temperatures 76–80
  two and three dimensions
    steady-state 92–100
conduction
  conduction laws 20–22
  conservation of energy 19–20
  non-steady-state 38–52
  steady-state 26–38
  thermal conductivity 22–26
  defined 19
  heat 17
  heat exchange, at surfaces 52–53
    convection 53–63
    radiation 63–76
  latent heat 18
  sensible heat 17–18
  temperature 18
heat transmission 5
HVAC *see* heating, ventilating, air conditioning (HVAC)
hygroscopic behaviour 126
hygroscopic equilibrium 235
hygroscopic materials 175, 187–189

i

IAQ *see* indoor air quality (IAQ)
IEQ *see* indoor environmental quality (IEQ)
indoor air quality (IAQ) 10
indoor environmental quality (IEQ) 3
indoor/outdoor vapour pressure difference 241
in- or outside, of thermal insulation 305–309
  glued masonry, joint width 310–312
  gypsum, with and without lightweight aggregates 313
  insulation materials 314–315
  lightweight concrete, for slabs/screeds 313
  masonry with, mortar joints 312
  normal concrete 312
  plaster 313
  values 310

wood and wood-based materials 314
insulation materials 24
  cellular glass 323, 352
  cork 323, 352
  EPS 324, 353
  expanded perlite 353
  glass fibre 323, 352
  mineral fibre 323, 353
  XPS 324, 353
integration constants 283
interstitial condensation 8, 183–185
  boundary conditions
    diffusion and air movement 202–203
    diffusion only 201–202
  calculation
    diffusion and moist air movement 204
    diffusion only 203
    example 204–209
inter-zone airflows 276
isobaric 162
isothermal 162
  air diffusivity 137
  drying 233
  water flow in pore
    after water contact 218
    balancing forces 212–214
    horizontal pore, substitute for rain absorption 214–217
    non-isothermal water transfer, after water contact 218–219
    vertical pore, substitute for rising damp 217–218
isotropic open porous materials 139

k

Kirchhoff's law 74

l

Lambert's law 66, 74
laminar forced convection 54
latent heat of transformation 128
life cycle inventory and analysis tools (LCIA) 3

linear thermal transmittance 98
local loss factors 135, 136
long wave emissivity 316
luminosity constant 66

m

mass conservation 176
mass displacement 127
mass flows 55
mass transfer
  air
    airflow in, open-porous materials 135–140
    air permeability and air permeances 133–135
    air pressure differentials 130–133
    dry air 129
    moist air 129
    and moisture, related to durability 127
    and moisture transfer 123–125
    conservation of mass 128
    defined 122–123
    links with, energy transfer 127–128
  moisture 209
    flow, through materials and assemblies 221–225
    simple moisture flow model 225–239
    sources 125–127
    vapour flow in, pore containing water isles 219–221
    water flow, in pore 209–219
  saturation degree scale 123
  water vapour
    in air 156–163
    interstitial condensation 201–209
    in open-porous materials 168–172
    relative humidity on, inside surfaces 165–168
    in spaces 163–165
    surface film coefficients, for diffusion 195–201
    vapour flow by diffusion, and moist air 189–195
    vapour flow by diffusion, in open-porous materials 174–189
    vapour transfer, in air 172–174
measured values, moisture properties
  test results 355–356
    building and finishing materials 342–352
    finishes 353–355
    insulation materials 352–353
    vapour retarders 356
mercury thermometer 18
moist air-related convective vapour flux 174
moisture 121, 209
  content 266
  flow, through materials and assemblies
    mass conservation 223–224
    moisture permeability 223
    starting, boundary and contact conditions 224
    transport equations 221–222
  ratio, at indoor and outdoor conditions 309
  simple moisture flow model 225–226
    capillary suction 227–231
    drying 232–237
    interstitial condensation, limit state 237–239
    wind driven rain 231–232
  transfer equation 221
  vapour flow, in pore containing water isles 219–221
  water flow, in pore
    capillarity 209–211
    isothermal 212–218
    moisture flow, through materials and assemblies 221–225
    non-isothermal water transfer, after water contact 218–219
    Poiseuille's law 211–212
    simple moisture flow model 225–239

moisture properties
    measured values
        diffusion resistance factor, test method   340–342
        insulation materials   352–353
        test results   342–352
    standard values
        building and finishing materials   336–340
        insulation materials   340
multi-layer adsorption   170
multi-layered assemblies   28–31

n

natural convection   58
Navier–Stokes's equations   266
net energy demand for heating   297, 300
Newton's law   53, 212
non-capillary materials
    hygroscopic   125
    non-hygroscopic   125
non-dimensional temperature factor   97
non-hygroscopic materials   189
    non-capillary materials assemblies   272–273
non-isothermal conditions   138
non-steady-state, conduction   38–39
    boundary conditions, flat assemblies
        convolution   48–50
        temperature change, semi-infinitely thick layer   50–51
    periodic boundary conditions, flat assemblies
        methodology   39–41
        multi-layered   47–48
        single-layered   41–46
Nusselt number   24, 57

o

one dimension, flat assemblies   27
    electrical analogy   31
    multi-layered assemblies   28–31
    single-layered walls   27–28
    special cases   31–33
open-porous materials   121

Organization Petroleum Exporting Countries (OPEC)   6
oriented strand board (OSB)   127

p

Pécle, J.-C.   5
Péclet number   149
Planck's law   67, 68
Poensgen method   317
Poiseuille's law   134, 270
polyurethane foam, polyisocyanurate foam (PUR/PIR)   324, 353
Prandl number   57, 196
predicted mean vote (PMV)   10
predicted percentage of dissatisfied (PPD)   10
pressure equilibrium   259
pressure factor   130
    data   131–132

r

radiant flux   64
radiant heat flux   77
radiant heat transfer, variables   64
radiant surfaces
    black surfaces   66–72
    coloured surfaces   74
    grey surfaces   72–74
    simple formulae   74–76
    types   66
radiant temperature   75
related heat flux   33
related hygric capacitances   280
relative humidity (RH)   2
    in moisture content   271
reshuffled conservation equation   269
Reynolds number   57, 58, 136, 211
RH *see* relative humidity (RH)
Rietschel, H.   5, 11
rising damp   217

s

second law of thermodynamics   66
second-order differential equation   213
semi-infinite vertical surface   54

sensible heat   17–18
Sherwood number   195, 196
shortwave absorptivity   316
sick building syndrome (SBS)   10
single-layered walls   27–28
six grey surfaces combination   115
sol-air temperature   107
solar heat flux   88
solar short-wave absorptivity   74
sorption active surfaces and hygric inertia   279–280
  harmonic analysis   284
  sorption-active thickness   280–282
  zone with one sorption-active surface   282–283
  zone with several sorption-active surfaces   283
sorption/desorption isotherm   168–169
  consequences   172
  physics   170–172
  salts, impact of   172
specific gas constant   20
specific heat capacity   20
stack effect   130–133
stack pressures   293
steady-state equivalent diffusion   250
steady state heat flux ambient   80
structural mechanics   4
suction, in moisture content   271
surface condensation   165, 166, 201, 224, 240
  and vapour balance, in spaces   200–201
surface film coefficients and reference temperatures
  indoors   76–78
  methodology   76
  outdoors   78–80
surface temperature   50

t

tangent method   183–184
Technical University Eindhoen (TU/e)   14
temperature damping   45

temperature equilibrium   66
thermal bridges   94–98
thermal comfort   2, 9
thermal conductivity   6, 22, 305
  of fluid   57
  heat transfer modes, property fixing   22–26
thermal diffusivity   53
thermal inertia   101
thermal properties, measured values test results
  building materials   318–322
  insulation materials   323–325
thermal conductivity, test methods   317
thermal resistance   28
thermal stack   133, 291, 296
thermal transmittance   93
thermal water permeability, of pore   219
thermodynamic equilibrium   266
Thompson's law   170, 171
total diffusion resistance   181
total thermal resistance   29
transparent envelope   278
transport equations   221–222
turbulent dissipation   55
turbulent kinetic energy   55

u

U-value   111

v

vapour diffusion, in cold storage walls   8
vapour flow by diffusion, in open-porous materials
  <equivalent> diffusion concept, applicability of   177
  flow equation   174–175
  mass conservation   176
  non-steady state
    equation   187
    hygroscopic materials   187–189
    non-hygroscopic materials   189
  steady state, flat assemblies
    equations   177

isothermal, multi-layered 178–179
isothermal, single-layered 177–178
non-isothermal, multi-layer 181–186
non-isothermal, single-layered 180–181
vapour resistance factor 175–176
vapour permeability 270
vapour pressure
  damping 189
  outdoors 240
vapour resistance factor (μ) 175–176, 235
vapour saturation pressure 158–160, 191, 276, 295
ventilating cavities, veneer drying 198–200
Verein Deutsche Ingenieure (VDI) 11
Vermeir, G. 13
visual comfort 3
volumetric expansion coefficient 58
Vrije Universiteit Brussel (Free University Brussels) (VUB) 14

W

wall's harmonic properties 113
wall surface-related heat balances 116
water flow, in pore
  capillarity 209–211
  isothermal 212–218
  non-isothermal water transfer, after water contact 218–219
  Poiseuille's law 211–212
  vapour flow, in pore containing water isles
    isothermal 219–220
    non-isothermal 220–221
waterproofing 222
water vapour
  in air 156
  changes of state, in humid air 161–162
  enthalpy of, humid air 162
  measuring air humidity 162–163
  quantities 156–157
  relative humidity 157–161
  vapour saturation pressure 157–160
in, open-porous materials
  vs. air 168
  sorption/desorption isotherm 168–172
interstitial condensation, evaluation
  boundary conditions 201–203
  calculation 203–204
  examples 204–209
relative humidity, on inside surfaces 165–168
surface film coefficients, for diffusion
  applications 198–201
  derivation 195–197
vapour balance, in spaces 163–165
vapour flow by diffusion, and moist air 189–190
  isothermal, single- and multi-layered assemblies 190–191
  non-isothermal, single- and multi-layered assemblies 191–195
vapour flow by diffusion, in open-porous materials
  applicability of, equivalent diffusion concept 177
  flow equation 174–175
  mass conservation 176
  steady state, flat assemblies 177–186
  vapour resistance factor 175–176
vapour transfer, in air 172–174
wet interface/wet zone 236
wind 130